The Metaphysics of Technology

What is technology? Why does it have such power in our lives? Why does it seemingly progress of its own accord, and without regard to social or environmental well-being? The quest for the essence of technology is an old one, with roots in the pre-Socratic philosophy of ancient Greece. It was then that certain thinkers first joined the ideas of technê and logos into a single worldview. The Greeks saw it as a kind of world-force, present both in the works of men and in nature itself. It was the very creative power of the cosmos. In the 20th century, German thinkers like Dessauer, Juenger, and Heidegger sought the metaphysical basis of technology, with varying success. French theologian Jacques Ellul argued persuasively that technology was an autonomous force of nature that determined all aspects of human existence, but he neglected the metaphysical underpinnings. Recent writers in the philosophy of technology have generally eschewed metaphysics altogether, preferring to concentrate on constructivist models or pragmatic analyses.

In the present work, Skrbina returns to a classic metaphysical approach, seeking not so much an essence of technology but rather a deep and penetrating analysis of the entire technological phenomenon. Drawing on the Greeks, he argues for a teleological metaphysics in which increasing order in the universe is itself defined as a technological process. On this reading, all of reality constitutes a technical sphere, a 'pantechnikon,' of universal scope. This work—the first-ever book-length treatment of the topic—breaks new ground by providing an in-depth and critical study of the metaphysics of technology, as well as drawing out the practical consequences. Technology poses significant risks to humanity and the planet, risks that can be mitigated through a detailed philosophical analysis.

David Skrbina, Ph.D., is a senior lecturer in philosophy at the University of Michigan at Dearborn.

The Metaphysics of Technology

David Skrbina

Routledge
Taylor & Francis Group
NEW YORK AND LONDON

First published 2015
by Routledge
711 Third Avenue, New York, NY 10017

and by Routledge
2 Park Square, Milton Park, Abingdon, Oxon OX14 4RN

Routledge is an imprint of the Taylor & Francis Group, an informa business

Library of Congress Cataloging-in-Publication Data

Skrbina, David.
The metaphysics of technology / David Skrbina. — 1st Edition.
pages cm. —(Routledge studies in social and political thought ; 94)
Includes bibliographical references and index.
1. Technology—Philosophy. 2. Metaphysics. I. Title.
T14.S57 2014
601—dc23
2014012406

ISBN: 978-0-415-71662-8 (hbk)
ISBN: 978-1-315-87958-1 (ebk)

Typeset in Sabon
by Apex CoVantage, LLC

This book is dedicated to our collective future.

Contents

Acknowledgments

As with any such work, many people contributed directly or indirectly to making it a reality. A few, however, deserve special mention. Adam Briggle offered early support and commentary and helped get this project off the ground. Ken Brady has been a long-time colleague and provided various insightful comments and feedback. Special thanks to my wife Lynn, who gave assistance and advice throughout. And finally, Henryk Skolimowski, who first prompted me to think critically about technology.

Introduction

> An ambitious philosophy of technology has to be ontological, i.e., it has to trace technology in the very fabric of reality.
>
> —Albert Borgmann[1]

My mission in this book is simply stated: to undertake a philosophical investigation of the metaphysics of technology. This is assuredly an "ambitious" task. To this end, I begin with an astonishingly clean slate. Essays and articles on the subject are extremely rare. Scholarly, book-length works are virtually nonexistent. It is not far from the mark to state that this is the first monograph ever written on the metaphysics of technology.[2]

And this is the first of many problems of technology.

The absence of such work is perhaps initially surprising—though upon reflection, less so. At least since the 1920s, professional philosophers have by and large eschewed metaphysical speculations. Analytic philosophy, based on formalisms, logic, and linguistic analysis, found no need for classic metaphysics; it viewed it as largely detrimental to clear and rigorous thought. Analytic philosophy was, above all, *scientific*, and the 20th century was assuredly a scientific century, at least for those of us in the industrialized nations. Science was seen as virtually synonymous with progress; in the same way, scientific, analytic philosophy was the way of progress in its own sphere.

Metaphysics, for its part, did not entirely vanish. It survived, but in thoroughly analytical form. Today we commonly speak of it in terms of 'formal ontology'; it is reduced to dealing with highly abstract topics related to universals and particulars, modalities, individuation, persistence, and time. Intensive focus is placed on the meanings of various designations: relation, property, event, proposition, state. As a consequence, the larger 'classic' metaphysical questions have fallen away. Metaphysics today has little relation to the real world, the *lived* world.

The 20th century was also self-evidently a technological century. Technological advances mirrored or exceeded those of science. In a pragmatic sense, it was technology that had the greater impact upon humanity and the planet.

When the past century began, we were still unable to create a functional flying machine; we rode in horse-drawn carriages and read by candlelight. The century ended with microcomputers, cell phones, nanotechnology, atomic weapons, space flight, and the Internet. One can hardly imagine a greater set of social, psychological, or environmental changes occurring in what was, in principle, the lifespan of a single human being.

Technology thus rightly came under philosophic scrutiny, but it was largely a philosophy that was analytically and scientifically oriented. This had at least two effects. One was that philosophers were—and still are—predisposed to be uncritical of the phenomenon, or at least highly restrained in their critiques. Analytical philosophy is not normative; it does not pass judgment, and it does not proscribe courses of action.

Second, metaphysics as formal ontology had—and still has—virtually nothing to say about technology; the subject was nearly avoided altogether. Consequently, there is a fair body of research on the *philosophy* of technology but virtually nothing on the *metaphysics* of technology. Even a cursory examination bears this out. A search of the main library catalog at my home institution, the University of Michigan, for English-language books on 'philosophy of technology' published since 1900 gives a result of 675 entries. The same search on 'metaphysics of technology' yields four.[3]

Now, this is not to imply that there has been no work done on the topic at hand. There has—much of it in the form of essays or short articles by continental philosophers. The standout example, of course, is Martin Heidegger. Many others have touched on metaphysical themes but have not explored them in depth, certainly not in a book-length treatment.[4] But the technical phenomenon has imposed itself upon us now to a sufficient degree that we can hardly avoid dealing with it on a metaphysical plane. It is too pervasive, too important, and potentially too dangerous to neglect any further.

On the other hand, a survey of the literature on the philosophy of technology finds many writers who, oddly, are not philosophers at all. They have little or no training in the field, and yet have come to dominate the discourse. A representative sampling, reaching back to the 1970s, would include the following:

- Political scientists: L. Thiele (*Indra's Net and the Midas Touch,* 2011), R. Sclove (*Democracy and Technology,* 1995), and L. Winner (*The Whale and the Reactor,* 1986; *Autonomous Technology,* 1977).
- Economists: B. Arthur (*The Nature of Technology,* 2009), J. Mander (*Four Arguments for the Elimination of Television,* 1978), and E. F. Schumacher (*Small Is Beautiful,* 1973).
- Historians: D. Nye (*Technology Matters,* 2006), R. Williams (*Retooling,* 2002), A. Pacey (*Meaning in Technology,* 1999), Smith and Marx (*Does Technology Drive History?,* 1994), and T. Roszak (*Where the Wasteland Ends,* 1972).

- Sociologists: M. Sturken et al. (*Technological Visions*, 2004), and D. Cecchetto (*Humanesis*, 2013).
- Journalists: R. Rhodes (*Visions of Technology*, 1999), K. Kelly (*What Technology Wants*, 2010), and D. Rushkoff (*Present Shock*, 2013).
- Scientists: R. Kurzweil (*The Age of Spiritual Machines*, 1999; *The Singularity is Near*, 2005), and J. Kizza (*Ethical and Social Issues in the Information Age*, 2013).

Perhaps the most striking recent example is Kevin Kelly's *What Technology Wants* (2010). Kelly is a college dropout with no formal training in any academic field, and yet is somehow viewed as among the more prominent spokesmen in the philosophy of technology.[5]

I do not mean to dismiss such interdisciplinary efforts; in fact, I strongly encourage them. All these books make valid insights and important contributions, but all are necessarily deficient in philosophical analysis. The authors lack the requisite philosophical history and context, and are only minimally aware of the larger metaphysical issues at hand. As one measure of this, the books mentioned contain almost no discussion of either Heidegger or Jacques Ellul, the two standard-bearers.[6] None address metaphysics—that is, the underlying nature or essence of the technological phenomenon—in any meaningful way.

I would also add that, on the whole, philosophers have never felt comfortable addressing technology. As with the common man, the subject is largely inscrutable to most of them. They have neither the technical nor the practical training required to understand it at more than the level of the average user. Rare is the philosopher today who comes to his discipline via a path of scientific or mathematical training, let alone engineering or computer science.[7] As a result, technology is, for the contemporary philosopher, by and large, little more than a set of tools with which to research, live, and interact. Like the typical consumer, philosophers primarily see its benefits—faster and better ways to research, write, and exchange ideas. It aids them in their profession, and they profit thereby. Technology is essential to their universities and research centers. Their students are immersed in it from earliest childhood. And despite the periodic hassles and expense, it is generally convenient, reliable, and fun.

When they do think about technology, then, modern philosophers typically see it in a more or less neutral light. As a tool or a set of tools, technology has all the potential of any tool for use, misuse, and abuse. It provides ready communication but allows invasion of privacy. It helps to diagnose our illnesses but also contributes to them. It provides new and cleaner modes of transportation and energy production even as technological society as a whole alters the global climate. It is entertaining but addictive. Philosophers, when they do engage with technology, often focus on a specific problem area and its ethical dilemmas or its impact on larger social values. Not really understanding the technical phenomenon in depth, and being predisposed to see it as a necessary concomitant to modern life, they overlook or undervalue the serious metaphysical issues at hand.

In my own case, I feel relatively well suited to undertake such an inquiry. I came to philosophy via an advanced degree in mathematics, and my diverse undergraduate studies included physics, science, and computer technology, among the requisite courses in humanities and philosophy. I understand, to a reasonable degree, how a computer works. I understand something of information theory, dynamical systems, and field theory. As I said, such a background is rare for contemporary philosophers and goes a long way toward explaining the absence of serious treatment of the subject. Thus, the analysis to follow and my generally critical treatment of technology arise not out of a lack of understanding or some vague apprehension—or least of all, out of 'fear.' In fact, it is precisely because I am comfortable with the subject and feel that I have a grasp of the matter directly that I find it so compelling, engaging—and problematic.

* * * * *

The plan of this book is straightforward. Part I examines a variety of issues in metaphysical theory, including an outline of my own approach that is based on certain ideas originating in ancient Greece. Part II looks at the implications of modern technology for society and nature. The two-part structure is an attempt to lay out the contrast between theoretical and applied metaphysics of technology and to examine the complementary nature of the two. As I will emphasize, metaphysics has consequences, and the necessary linkage needs to be clearly articulated.

Throughout this book, I draw repeatedly from historical ideas. This may seem strange for as 'modern' a topic as technology, but there are at least three good reasons to do so. First, technology itself is old—as old as mankind. Simple stone hand tools go back more than two million years, to the earliest days of the genus *Homo*. To the best of our knowledge, humans have never existed without tools. The benefits and hazards have been with us for ages, and perceptive human observers have therefore always been in a position to make useful reflections upon our various technologies.

Second, the great thinkers of the past were undoubtedly as insightful and penetrating as we moderns. Their abilities to look into the phenomena of the world and to draw valuable conclusions certainly equaled, and in some cases exceeded, that of our best minds today. In fact, living in an era of simpler technology arguably allowed them to see more deeply into the phenomenon; and their capacity for creative metaphysical solutions was undiminished by the complexities of contemporary life.

Third, my own thesis compels me to consider historical views. I argue that there are timeless metaphysical principles at work in technology—and thus they were available for observation and commentary by past philosophers. From my perspective, timeless wisdom is eternally valuable. It would be foolish to overlook the greatest insights from 2,000 years of philosophy.

The metaphysics of technology is a profoundly complex subject; we need all the assistance we can get.

I suspect that many of my peers will hold a contrary view. Most contemporary philosophers of technology find little need for either metaphysics or history. For many of them, old philosophy of technology is little better than old technology itself—quaint, mildly entertaining, but ultimately of no value to our modern world. Old philosophy of technology is only useful, in their view, to show how wrong we were in the past and, by implication, how much progress we have made in recent years. At best, they say, past philosophy is simply irrelevant to the subject at hand. Even Ellul fell victim to this tendency; of the problem of technological determinism and the concomitant loss of freedom, he wrote that "the men of classical antiquity could not have found a solution to our present determinisms, and it is useless to look into the works of [for example] Plato or Aristotle for an answer" (1964: xxix). I disagree. I believe that the men of antiquity, along with those of more recent vintage, had valid and useful insights into technology, including the problem of determinism and freedom. We can learn much from them.

If this is indeed the first book on the metaphysics of technology, it is surely not the last. I cannot hope for a comprehensive analysis in this single work; rather, I offer here a beginning. With luck, this inquiry begins a dialogue with philosophers and others who are willing to engage in the most urgent problem of our time—the problem of technology.

Disliking as I do lengthy introductions, and having already exceeded my own self-imposed bounds, I proceed immediately to the discussion at hand.

NOTES

1. Cited in Olsen and Selinger (2007, p. 8).
2. One could make a case that Dessauer's *Philosophie der Technik* (1927) was the first such book, though this has never appeared in full English translation. Others might point to such works as Ihde's *Technology and the Lifeworld* (1990) or Borgmann's *Technology and the Character of Contemporary Life* (1984), though for various reasons I argue that these books are not truly metaphysical.
3. And even these are misrepresented; none are philosophical studies of the subject.
4. I discount studies *on* Heidegger, for example. Such works offer valuable commentary and critique but do not press the field forward in a significant way.
5. His sole claim to expertise seems to be his past editorial role at *Wired* magazine.
6. With one exception: Winner's *Autonomous Technology*, which leans so heavily on Ellul that it qualifies as a mere recapitulation of his ideas.
7. Historically, of course, there have been many such examples: Pythagoras, Plato, Descartes, Leibniz, Whitehead, Russell, and Husserl, among others.

Part I

Metaphysis

1 Situating Technology

> In spite of all the men of good will, all the optimists, all the doers of history, the civilizations of the world are being ringed about with a band of steel.
>
> —Jacques Ellul[1]

> [T]he essence of technology is nothing technological.
>
> —Martin Heidegger[2]

Technology is mindful creation. It is the realization and reification of thought. Every act of creation is a technological act. Ontologically and linguistically, technology is technê-logos; there is a *technê* of all things, and there is a *logos* of all things. Therefore reality itself, in the broadest sense, is something technological. Thus understood, the metaphysics of technology has profound consequences for all spheres of human thought and action.

In his 1954 essay "The Question Concerning Technology," Heidegger famously said this: "The essence of technology is nothing technological (*das Wesen der Technik nichts Technisches ist*)." Like any good paradox, this statement is both true and false: false because the essence of something must be intrinsic to the thing itself, and thus must, in the end, be something "technological"; and true because this essence is significantly beyond what we ordinarily think of as technology and in this sense is truly "nothing" technological.

In a similar vein, a proper metaphysical analysis of technology cannot be analytical. In one sense, of course, what follows is assuredly analytical; I will certainly be resolving and 'undoing' or 'loosening up' (*ana-lúein*) the various components of the technological phenomenon and striving to uncover its metaphysical roots. On the other, to think analytically today is to think *technologically*—and thus fails to escape to a sufficient altitude for surveying the landscape. To think about technology technologically is to miss its essential nature. And yet this is precisely how we commonly approach the subject today. I submit that all current thinking about technology occurs from within a technological mindset and thus necessarily arrives at certain predetermined results. We are too close to the phenomenon to adequately grasp its nature.

But the reply comes: How else shall we think about it? Conventional analysis is, allegedly, our best tool for thinking carefully and deeply about something—about anything. Apart from mysticism or irrationalism, we seem to have no good alternative. Furthermore, for many spheres of human life, this approach has proven successful—or at least efficacious. Why not stick with what works?

When we come to technology itself, however, a problem arises. There is a cognitive and epistemological conflict of interest. Thinking analytically—technologically—about technology is self-serving and intrinsically biased. It will drive us to forgone conclusions: that technology is a set of tools at our disposal, constructed by mankind, for the betterment of mankind; that it is largely a benign force in the world; that it is morally neutral, in that it may be put to good or bad use depending on the intentions of the user; that it is a humanizing power, a bringer of prosperity and cultural advancement; and perhaps most importantly, that it is devoid of metaphysics. Such are the common views held by philosophers and laymen alike. I find them to be substantively and profoundly mistaken.

We are faced with a dilemma. We are compelled today to think about technology technologically, but this does not and cannot lead us to the truth about the phenomenon. To think clearly about technology, to see into its depths, we must attempt to transcend the very process of thinking that a technological society imposes upon us as the only acceptable means. We are in a bind: We need to think in a language that is both acceptable for a technological society and yet transcends that very language. This seems to be what Heidegger meant when, at the conclusion of his essay, he said,

> Essential reflection upon technology and decisive confrontation with it must happen in a realm that is, on the one hand, akin to the essence of technology and, on the other, fundamentally different from it. (1977: 35)

For Heidegger, this involved the creation of a significantly new terminology and the substantial redefinition of many common words and ideas—with mixed success. Later I will attempt to untangle, to 'analyze,' the main points of his argument in order to find that which is of value and where he has gone astray.

In principle, Heidegger's approach is correct. Proper thinking about technology must be truly metaphysical, in the classic sense. We need not abandon analysis, but it must be incorporated and subordinated within a broader, visionary, even speculative framework. Technology is a profound and penetrating force in the world, and therefore we must think about it profoundly and penetratingly. We need to marshal all our creative and visionary powers and focus them on this maker of reality in order to see into its depths. In fact, our future depends on it.

To distinguish the classical metaphysical approach from the modern one of formal ontology, I will borrow a bit of Kantian terminology. In contrast to the present, formal, analytical metaphysics, I propose something rather different: *synthetic metaphysics*. Where the analytic approach is concerned with definitions and formalisms, synthetic metaphysics seeks to illuminate the essential nature of a thing. Analysis deals with abstract concepts, synthesis with the real world. Where analysis strives to dissect and atomize, synthesis seeks to integrate and capture wholes. This is not just a rehash of the old debate over reductionism versus holism, but rather a move to—or better, a return to—creative, organic, and visionary approaches to metaphysical issues.

Because the synthetic approach is grounded in the real world, it is, unlike analysis, deeply normative. Synthetic metaphysics is concerned with how things affect people and nature and how it aids or hinders their quality of existence. It is ethically as well as descriptively oriented. It passes judgments. It proscribes courses of action. It makes value decisions and defends that which is worthy of defense.

In sum, the synthetic approach that I advocate takes a phenomenon—in this case, technology—and addresses (a) what it is, (b) what it does, and (c) how we ought to respond. This attempts to capture the whole phenomenon and its relationship with the world. It avoids foregone conclusions about objectivity, neutrality, and verifiability—all built into the structure of conventional analysis.

I attempt here to examine, in a philosophically rigorous and careful manner, the nature of technology. But I do so with a minimum of terminological construction and without falling prey to the conventional mode of technological thinking about technology that drives one toward certain predetermined outcomes. In short, I seek a *nonanalytic* analysis of the metaphysics of technology. This is a necessary, but certainly not sufficient, condition for success.

CAN THERE BE A METAPHYSICS OF TECHNOLOGY?

Some will argue that accomplishing this goal is an impossible task. First, analytical thinking is our best and only option, they will say, and any attempt to move away from this—particularly with regard to technology itself—can only lead to obfuscation and pointless speculation. To this I have two replies. If we consider the outstanding thinkers and philosophers of the past, we immediately realize that many of the greatest minds were also the greatest metaphysical speculators. Among the Greeks we have the likes of Pythagoras, Heraclitus, Empedocles, Parmenides, Plato, and the Stoics; and even Aristotle, as analytic a thinker as he was, was in his own way a metaphysical visionary. In later centuries we have Plotinus, Bruno, Campanella, Spinoza, and Leibniz. And again: Hegel, Schopenhauer, Peirce, Nietzsche, Bergson, and Whitehead—bold speculators all. We owe many of our greatest theories and most profound insights to the metaphysical innovators of the past. We neglect that approach at our own hazard.

On the other hand, consider the state of thinking about technology by conventional analytical thinkers. As I noted in the introduction, the field is presently dominated not by philosophers but by others: political scientists, historians, sociologists, economists, and journalists. *Philosophical* book-length analyses of technology are scarce—fewer than two dozen titles in the past 20 years. These rarely, if ever, address metaphysical issues. When they do, it is most often as commentary on past thinkers—particularly Heidegger. The subdiscipline 'philosophy of technology' is barely recognized by academia, and few universities offer courses in it.[3] Classical metaphysics generally has been largely abandoned, or it has been relegated to the slightly disparaging category of 'history of philosophy.' And contemporary metaphysics is now, as I said, formal ontology—entirely abstract, entirely technological.

Apart from the occasional comment on Heidegger, metaphysics is typically shunned by professional philosophers of technology. This occurs both because of the general bias against metaphysics and because the study of technology is, they believe, inherently nonmetaphysical. Usually this stance is unspoken, and betrayed only through neglect of the subject. Occasionally it is explicit. In one recent anthology, Fellows (1995: 1) remarks that "the essays collected here do not constitute a philosophy of technology" in the sense of illuminating "the features of the phenomenon of technology itself" (quoting Don Ihde)—in other words, they openly ignore the metaphysics. "The contributors [to the volume] do not concern themselves with the essentialist enterprise of defining technology." Rather, technology is taken in a naïvely realist sense; it is as it appears. Verbeek (2001: 122), for another, claims that the "new generation" of technology philosophers "no longer thinks in terms of Technology [with a capital 'T'] per se, and finds it problematic to try to understand phenomena in terms of essences." Rather than focus on underlying patterns and characteristics, this "new generation" prefers to concentrate on "concrete technologies and the roles they play in their specific contexts." And not only the newcomers, but the old guard as well: Ihde (2010: 119), for example, declares that "there is no essence of technology"; he opts instead for a "deeper phenomenological analysis."

And then consider the following remarks from another recent commentator. He believes that the whole topic is "misguided." Contemporary philosophers of technology are unconcerned with metaphysics, he says, "because they no longer find the reasons compelling for understanding technology." They prefer to examine "actual technologies, actual devices/machines/systems." They find metaphysical approaches "vague, unhelpful, sweeping, sloppy, undifferentiated, and unconvincing." Such thinkers have "moved beyond Heidegger" and his fuzzy generalizations. Ellul's thesis of technological autonomy—which I examine at length later on—is likewise considered "no longer a viable, credible approach. It is too overarching, too remote, too undifferentiated." On the whole, "the movement away from metaphysical issues is the virtue of contemporary philosophy of technology. It is the result of deliberate choice and reasoned argument."[4] Here we have one reading of the conventional view, in very explicit terms. Speculative metaphysics of technology—and by inference, metaphysics generally—is passé, unhelpful, and largely pointless. We no longer seek fundamental principles but rather only detailed analyses of specific applications. The real world is too complex and too diverse to be encompassed by any one set of metaphysical concepts. Metaphysics, in short, is a waste of time.

What can we say about the conventional view? Useful, perhaps, in its own way. Very pragmatic, very analytic, and very narrow. Technology will likely be viewed either as a "social construct," driven by and contextually dependent upon the cultural setting, or in instrumental terms, possibly intrinsically neutral. In either case, technology is unlikely to be a danger or even a threat. After all, most individual technologies can only pose limited and restricted risks—risks that can be easily assessed and mitigated. Only the most powerful of weapon systems or the most nefarious of terrorist devices presents a real danger—and this fact says nothing about technology at large, on the orthodox view. A focus on particular, actual technologies necessarily precludes metaphysics; only a Heidegger dares to suggest that there is a metaphysics of, say, a hammer. And he is now outmoded.

One can also turn the picture around. This very deemphasis on metaphysics suggests that technological thinking has dominated our approach to technology. Technology 'prefers' that we focus on specifics, on details, on structure and function. This is indeed how it advances. Technology in the physical world is about machines, devices, invention, amusement, communication, information, profit—and power. Philosophy is almost meaningless in this context; metaphysics is wholly so. Technology bids us to pay no attention to the man behind the curtain. There is nothing 'behind' technology, nothing there to see.

But we must ask: Is this so? Is it not true that all phenomena possess an inner or intrinsic nature, subject to some kind of guiding principles? I believe it to be self-evident that there are such principles of nature, principles that account for the order and complexity of the cosmos. Without them, there would be chaos—and consequently no philosophers around to speculate about their absence. And if such principles exist, by what criterion do we deny their metaphysical status?

What, after all, is metaphysics? Here we have a question fraught with difficulty. We can readily agree with Loux's (2006: 2) admission: "It is not easy to say what metaphysics is." Is it a body of principles? Essences? Higher realities? Or is it abstract concepts and definitions, as contemporary philosophy would have it? There seems to be little consensus on the matter.

Even the word itself has an obscure origin. Long designated as the title of Aristotle's work of the same name, he himself never used the term. Nicholas of Damascus, circa 25 BC, was evidently the first to coin the phrase *metà tà physicá* in reference to Aristotle.[5] In his own usage, Aristotle preferred "first philosophy" (*prôtê philosophía;* e.g., 1026a30). Its subject matter is "being qua being" (*ontos hê on;* e.g., 1003a23), that is, being in itself. On other occasions, he referred to it as the study of "original" or "first causes" (*arches aitíon; prôtên aitían;* e.g., 983a24–25). As such, he means the original cause or source of cosmic order. In all cases, he intends to examine eternal and universal principles that account for motion and change in the physical world—change occurring within *physis* or nature, and the principles existing *meta-physis,* beyond nature.

In a more modern sense, some have defined metaphysics as "the nature of ultimate reality"—though this is not a particularly helpful definition. Others equate it with ontology, which is perhaps more to the point. At a minimum, and rather obviously, it would seem to refer to something other than 'the physical.' If by physical we mean 'composed of matter or energy,' then many things would conceivably count as metaphysical: mind, consciousness, emotions, ideas, principles, even physical laws. Leibniz, for example, was adamant that the laws of physics were themselves above physics and thus something metaphysical; in 1702 he wrote that "although we say that everything in nature is to be explained mechanically, we must exempt the explanation of the laws of motion themselves, . . . which should be derived . . . from a metaphysical source" (1989a: 255). Later in life he reiterated the point even more forcefully: "[I] have at last shown that *everything happens mechanically in nature, but that the principles of mechanism are metaphysical*" (1989c: 319). Kant wrote, "It is evident that the very first sources of nature's operations definitely have to be a subject of metaphysics" (2012: 57). Schopenhauer explicitly agreed.[6]

In a similar vein, the "first philosopher of technology," Friedrich Dessauer, wrote that "the laws of nature . . . have a reality of a very high order. Of course, they are not themselves perceptible; they are not objectively concrete" ([1927] 1972: 318). Lest we think this a strictly archaic view, consider the recent and related observation by Ray Kurzweil, who remarked that "most scientific laws are not physical laws"; rather, they "result from the emergent properties of a large number of events at a lower level" (2012: 267). If scientific laws are "not physical" and yet address physical reality, they are perforce metaphysical.

Even in a straightforwardly materialist universe, then, metaphysical principles exist; they describe the essential nature and overarching rules of material reality. Evidence suggests, however, and in more than one sense, that we do *not* live in such a universe. Mind and consciousness remain huge problems for conventional physicalism—specifically, their emergence and their qualitative nature. The multiplicity of fundamental particles and forces resists satisfactory explanation—the Higgs boson notwithstanding. Gravity remains to be integrated into quantum theory. And all this is not to mention bizarre substances like dark matter and dark energy, of which we know virtually nothing, and yet they are jointly said to comprise some 95% of the universe. These entities alone may point to a kind of metaphysical reality to the cosmos, one that dwarfs the nonmetaphysical component.

Generally speaking, we must accept that any real, concrete phenomenon must have some intrinsic nature. Existence is never per se; it is always of a kind, as a manifestation of something. The fact that we may not know this nature should not deter us from accepting that such a nature exists. Intrinsic natures are not empirical; we can have no direct sensory or objective data on them. Rather, they are subject to rational analysis, broadly conceived, and to metaphysical speculation. The movement away from metaphysics over the past century, and typified by the commentators cited, does not negate the reality of intrinsic natures. On the contrary; it is merely a sign that analytic philosophy has abstained from engaging in the subject matter, having little to say about it, and thus consigning it to irrelevance. Carnap (1935) famously declared his "rejection of metaphysics," and Popper (1962) argued that metaphysical theories, being nonfalsifiable, are of no value. Philosophers ever since have largely followed their lead.

In truth, however, it is not possible to avoid metaphysics. Even to declare that there is no intrinsic nature or metaphysical reality behind phenomenal appearances is itself a metaphysical claim. To state that reality is wholly and mechanistically physical is to adopt a metaphysical position. To be an eliminativist is to hold a metaphysical stance. Metaphysics is here to stay; our only choice is to address it or to set it aside.

Some philosophers of technology recognize and accept this fact. They give serious consideration to Heidegger, or Dessauer, or Ellul, and to their quest for a technological essence. They embrace the notion of technology as a real and concrete phenomenon, but one that must have an inner nature or underlying principles. One of the early contemporary philosophers of technology, Henryk Skolimowski, stated as much already in 1973. Pulling no punches, he wrote:

> This process [of the development of modern technology] contained a number of preconditions which had to be fulfilled before technology could acquire its full potential. . . . Technology is thus not so much a collection of tools as a state of Western mind. Heidegger's declaration that technology is the last stage of metaphysics becomes meaningful in this context. Our inability to come to terms with technology is no doubt related to our infantile naiveté about its nature. It is still assumed that technology is 'indifferent,' that it contains no metaphysics, and that its relation to traditional metaphysics is none. All these assumptions are an expression of an infantile mind. (1973: 53)

"Technology," he continues, "is a historical phenomenon born of a certain idea of nature, of a certain idea of progress . . . and also related to specific social ideals and specific ends of human life." By these facts alone, "it is laden with the elements of traditional metaphysics."

A metaphysics of technology, then, must refer to eternal and universal principles that account for the coming-to-be and operation of the technical phenomenon. By inquiring into its fundamental principles, we will gain insight into its nature and purpose. And then we will know how best to respond to its many challenges.

ON NATURALISTIC MONISM

At a fundamental level, I take it that the cosmos is a unity. The ultimate metaphysical description of the universe must refer, in the end, to a single substance, principle, or entity. We simply have no rational basis for assuming an ontological duality or plurality. Problems of interaction, generation, and explanation seem insurmountable for any nonmonistic schema. Monism is parsimonious, and it is elegant. Only a radical change in our modes of thinking could compel us otherwise. As such, reality is wholly natural. There is no supernatural realm. My entire approach here is naturalistic and monistic.

Everything that exists in the physical realm is, in the deepest sense, natural—that is, of nature. All technology is natural. Of course, we commonly think of it as artificial, as *artificium*, *ars facere*, something made by human art or craft. But to state the obvious, humans are natural beings, and thus our doings and makings are also natural, that is, nature acting on nature. All human contrivance, all man-made technology, is nature-made-nature, a second-order product of nature itself.

Thus understood, technology is the appearance of order from within the realm of nature. It is nature manifesting itself in a directed manner. Here was one of Heidegger's central insights: the essence of technology is intimately bound to the essence of reality, to being itself. Consequently, technology cannot be merely that which is man-made. It must be something more.

As long as we think of technology as a human contrivance, we overlook its essence. On even the simplest levels of analysis, we can see that technology exceeds human scope. Animals have been tool users and constructors since long before the existence of the human race. Wild chimpanzees in West Africa use modified twigs to hunt for termites and have been observed "fashioning deadly spears from sticks, and using the hand-crafted tools to hunt small mammals."[7] Captive rooks can make and use hook-shaped tools to get food.[8] Bottle-nosed dolphins use sponges to aid in hunting.[9] Octopuses carry and use coconut shells as shelter.[10] Humpback whales use "bubble nets" to catch krill and herring.[11] Even such ordinary structures as a bird's nest, a beaver's dam, or a spider's web qualify as technological. In fact, any animal that creates a home utilizes technology. If technology is fundamentally related to the process of construction and creation, as I claim, then every form of animal life possesses and deploys technological capabilities.

And not only animals: Plants also make use of objects around them and thus put things into their service. Carnivorous plants attract, trap, and digest insects for food. Vines and other tendriled plants seek out and grasp onto objects around them for use as support. Parasitic plants (hemiparasitic, holoparasitic) utilize others plants as nourishment. Plants can even produce and emit chemical signals as a means of communication.

And yet the common view remains that technology is something human, *essentially* human, the trademark of the human animal. Carl Mitcham offers to define technology as "the human making and using of material artifacts in all forms and aspects" (1978: 232). Ferre (1988: 131) says this: "From the viewpoint of this book . . . technologies themselves [are] defined as practical implementations of intelligence—*human* intelligence." Pitt (2000: 11) defines technology simply as "humanity at work." And Johnson (2006: 196) states: "The very idea of technology is the idea of things that are human-made." More specifically, she offers the following general definition: "Technology is a combination of artifacts, social practices, social relationships, and systems of knowledge" (197)—a decidedly anthropocentric conception.

So conceived, we open ontological divides between humans and other life forms and between 'artificial' and 'natural.' I don't deny the reality of such distinctions; they are obvious and have been a focus of study for at least two thousand years. I do deny, however, that they are of metaphysical significance. Metaphysics seeks general principles that cut across categorical distinctions—in this case, it seeks the common ontological foundation of all constructed things. When we introduce distinctions of human/animal or artificial/natural, we have to justify those distinctions in a non–question-begging way. This is a difficult task at best. Even Heidegger tripped on this divide and, as I will argue, it introduced an irreconcilable difficulty into his philosophical system.

From a metaphysical standpoint, therefore, we ought to be radically non-anthropocentric. At the same time, however, we are unavoidably bound to our human nature. We see the world as only humans can. Four billion years of evolution have left us indelibly human centered, animal centered, and life centered. This is not a complaint but simply a trivial observation. This fact poses a substantial challenge to our metaphysics. Almost unconsciously, we fix our attention on ourselves, our status in the world, and the objects of our direct construction. For most fields of human study, this works very well. Even scientists and physicists have little problem seeking universal principles because they effectively treat the human being as an embodied system of mass/energy, as a physical object, and thus adopt a physical standpoint from which to examine the world. When we see ourselves as physical objects in a physical universe, we then have a common basis of operation.

But metaphysics operates beyond the objective realm—hence the difficulty. We typically see ourselves as the only enminded beings, as the only intellect in the cosmos, and thus our intellectual comprehension of non- or meta-physical reality is profoundly biased. If we could view ourselves as more fully integrated into the cosmic order, as more a part of the natural evolution of the universe, we would then have less difficulty examining general metaphysical principles. But here is the dilemma: it requires an understanding of such principles in order to see ourselves thusly integrated.

Perhaps this lies at the root of the difficulty of metaphysics: a human nature that seeks to understand, from a human perspective, the essence of the more-than-human.

Nonetheless, philosophers have in the past, and will again in the future, make precisely such attempts at a general metaphysical understanding of reality. Humans and their constructions are part of a monistic universe, one governed by universal metaphysical principles. They can only be properly understood in such a context, I claim. And to do so, we must examine the origins of order in the cosmos—origins that relate, to a surprising degree, to the Greek concepts of technê and logos.

NOTES

1. *The Technological Society* ([1954] 1964: 127).
2. "The Question Concerning Technology" ([1954] 1977: 35).
3. When they do, it is usually under the rubric of 'science and technology studies' (STS) or, more generally, philosophy of science.
4. Anonymous reviewer of an early proposal for the present work (April 2012).
5. See Chroust (1960: 602).
6. See Schopenhauer (1992: 21).
7. "For first time, chimps seen making weapons for hunting," *The Washington Post* (February 23, 2007; p. A1).
8. Bird and Emery (2009).
9. Mann et al. (2008).
10. Finn et al. (2009).
11. Wiley et al. (2011).

2 Technê, Logos, Theos

"In the beginning was the Word, and the Word was with God, and the Word was God" (John 1:1). There is deep wisdom hidden behind this wholly inadequate translation, and it is nothing Christian or biblical. These words, written by an anonymous author in the 90s AD, were likely drawn from the Jewish philosopher Philo, who in turn borrowed from the more ancient wisdom of the Stoics, Plato, and Heraclitus. This passage is an entry point on the path to a comprehensive vision—a fundamentally *Greek* vision—of the metaphysics of technology.

The passage, the first line of the Gospel of John, was originally written in Greek. The original text is as follows: Ἐν ἀρχῇ ἦν ὁ λόγος, καὶ ὁ λόγος ἦν πρὸς τὸν θεόν, καὶ θεὸς ἦν ὁ λόγος. In Latin transliteration, we read:

> *En archê ên ho logos, kai ho logos ên pros ton theon, kai theos ên ho logos.*

'Arche' means not merely the 'beginning,' as in the beginning of time that is commonly understood here, but rather something more: an *original ruling principle*. 'Theon' and 'theos' are rightly understood as God—not the Christian God, not a personal or moral God, but God as the Divine Source, the creator and sustainer of the universe. But the key term here is 'logos'—a complex Greek concept that is exceptionally difficult to translate. It can mean 'word,' but it has various other connotations, including 'speech,' 'reason,' and 'logic.' It suggests a kind of inner meaning, a deeper truth to something that is not readily apparent. In context, it is a kind of hidden, underlying order to the cosmos. Bearing this in mind, it is perhaps best to leave the word untranslated and allow the contextual meaning to emerge.

As a further point, we see that the usual translation reverses the original ordering of the concluding phrase; it is not that "the Word was God," but rather, "God was Logos." The author is defining not Logos as God, but God as Logos.[1] Thus we can construct a more nuanced, if less poetic, interpretation of the passage:

> "The original ruling principle of the cosmos was Logos, and Logos was with God, and God was Logos."

There is a tension here: How can Logos be "with God" and, at the same time, *be* God? This is the Christian dilemma; they draw the problem of reconciling a personal God with an impersonal cosmogenic ordering principle.[2] But it was no problem for the Greeks. From the time of Heraclitus through the early Stoics, Logos was a divine ordering principle, present in the cosmos at large and, writ small, as the logos of men. With Plato, the logos concept first became connected with techne (τεχνη). With Aristotle, the terms were merged into one, and the wider meaning of technology began to appear. The Stoics then completed the picture, applying the concept of technê to all processes of becoming—a world-sphere conceived, in my terminology, as a 'Pantechnikon.' Retracing this development allows us to sketch the outline of a metaphysics of technology, one that has important implications through the present day.

EARLY PREMONITIONS—HERACLITUS

The first clear intimation of Logos as a divine ordering principle comes from Heraclitus, ca. 450 BC. He opens his only known book with the phrase, "The Logos is as here explained" (fr. 1). He continues with the observation that "all things come into being in accordance with the Logos" (*ginomenon gar pánton kata ton Logos*). This is key: Logos is the generating source of all things, the orderer of creation—not merely at some universal beginning but throughout all time. Furthermore, the Logos brings to the cosmos a metaphysical unity: "It is wise to listen, not to me, but to the Logos, and know that all things are one" (*en panta einai;* fr. 50).

In spite of its omnipresence, the Logos is somehow hidden away. It is "separated from us" (*pánton kechorisménon,* fr. 108; *tà hóla dioikoûnti,* fr. 72). As embodied in nature, it is removed from our sight: "Nature likes to hide" (*physis krúptesthai phileî;* fr. 123). This "unseen harmony" (*harmonin aphanes;* fr. 54) is stronger than any visible harmony or structure in the cosmos, which is only an imperfect manifestation of the divine Logos.

Heraclitus gives his highest of gods different names: the Wise One (*to Sophon;* frs. 41, 108), the Only Wise One (*to Sophon Mounon;* fr. 32), Zeus (*Zenos;* fr. 32), the Eternal One (*Aion;* fr. 52), the Thunderbolt (*Keraunós;* fr. 64). Indeed, "The Thunderbolt steers the universe" (*tà dè pánta oiakízei Keraunós*). But the most important alternate name relates to *pyr*—a term commonly translated as 'fire' and widely seen as the monistic substance of which all things are made. This, however, is an incomplete view, on two counts. As Cleve (1969: 39–40) explains, Heraclitus' *pyr* is sometimes referred to as 'air' (*aer*), sometimes as 'the ether-like body' (*to aitherion soma*). His contemporary, Anaxagoras, viewed elemental fire as a kind of ether; Aristotle writes, "To him, fire and *aither* are the same thing" (*De caelo,* 302b4). It is thus plausible to conclude that *pyr* is more than fire; it is ether, a kind of shining cosmic fire, of which earthly fire is one manifestation. For Heraclitus, *pyr* is the Logos become tangible.

Secondly, *pyr* is not simply some inert elemental substance; it is alive:

> This cosmos, which is the same for all, was created neither by [lesser] gods nor men; but it was and will always be an ever-living *pyr* (*pyr aeizoon*), growing in measures and fading in measures. (fr. 30)

Logos in physical form is thus a *pyr aeizoon,* an ever-living ether-like fire. This is the material substance of the physical universe. Things come from this living ether and ultimately return to it: "All things are exchanged for *pyr* [*aeizoon*], as well as *pyr* [*aeizoon*] for all things, just as merchandise is exchanged for gold, and gold for merchandise" (fr. 90). *Pyr aeizoon* is the universal currency by which material objects purchase their existence.

Implicit in the cited passages is the fact that the Heraclitean God was endowed with two fundamental aspects. On the one hand, it is luminous ether, forming the material substance of the world. On the other, it is the Wise One, a living and intelligent entity.[3] The Logos grants physical beings their material existence from out of itself, even as it grants them their *psychic* being. Just as our physical bodies reveal the depths of the *pyr aeizoon,* our psyches reveal the depths of the mind of the Logos.

> You would not find the limits of psyche, even wandering down every path—such a deep Logos does it have
> (*psychês peírata iòn ouk àn exeúroio pâsan epiporeuómenos hodón: hoúto bathùn Logos éxhei;* fr. 45)

> The Logos of the soul increases itself
> (*psychês esti Lógos heautòn aúxon;* fr. 115)

The wisdom of the Logos acts through each individual thing as it steers the cosmos: "The Wise One knows the plan by which everything is directed through everything" (. . . *ekybérnatai pánta dià pánton;* fr. 41). As a result of this intelligent Logos action, "the thinking faculty is common to all" (*xynón esti pâsi tò phronéein;* fr. 113). Everything, living and nonliving, embodies both material and psychic dimensions. Cleve (1969) explicitly acknowledges this panpsychic aspect of Heraclitus: "[He] still panzoistically takes consciousness and body as two sides of the same . . . [C]onsciousness is bound to everything as a function" (41, 58).[4]

If the Logos steers all things *through* all things, as a kind of thunderbolt, this raises the question of free will. Can any living creature, humans included, truly act autonomously if guided and directed by a cosmic power? Heraclitus offers us a rather sophisticated conception here. The Logos, the One, does indeed direct all our actions, and we are compelled to follow: "And it is law also, to obey the will of the One" (*nómos kaì boulê peíthesthai Henós;* fr. 33). The common man does not understand this truth, and thus believes in his own absolute freedom: "we must follow the Common [i.e. Logos], yet the many live as if they had a wisdom of their own" (*dió deî epesthai to {Xyno, toytésti to} koino Xynòs gar ó koinós;* fr. 2). However, we, like all things, are merely a sliver of the cosmic Logos itself, and it clearly cannot be bound, as there is nothing greater. To the extent that we embody Logos, we do in fact have a component of freedom. But insofar as we are individual beings, we are directed and ordered by the Logos. In one aspect we are bound by the Law of Logos, but in another, as beings *of* Logos, we possess a measure of freedom. Thus, determinism holds throughout the universe, but one that is compatible with an element of freedom—a freedom that is manifest, perhaps, only when we understand our true nature.

Finally, there is even a glimmer of evolutionary thinking in Heraclitus. The four Empedoclean elements—fire, air, earth, water—seem to undergo a series of transmutations. On the one hand these can be cyclical, as in his fragment 76 that indicates a movement from earth to water to air to fire, and then back to earth. By contrast, fragment 36 suggests something different, a *progression:* "From earth comes water, and from water, soul" (*ek gês dè húdor gínetai, ex húdatos dè psyché*). This is not surprising, as we find similar but more explicit thoughts in Empedocles himself—like Anaxagoras, a contemporary.[5] And we have one more hint. Evidently Heraclitus believed that the cosmos has gone through innumerable cycles of generation and destruction; Aristotle states that "Empedocles and Heraclitus . . . believe that it alternates, being sometimes as it is now and sometimes different and in a process of destruction, and that this continues without end" (*De caelo,* 279b17). During any given lifetime, the universe is progressing, evolving, from some initial state to an end state. But unfortunately we lack the details of this particular story of cosmic evolution.

In Heraclitus, then, we find many of the basic pieces of the larger picture that I hope to spell out: the divine Logos as original ordering principle of the cosmos and as a mind/matter composite, generating and directing all things; a monistic physical universe as embodiment of the *pyr aeizoon,* the ever-living ether; a clear determinism, but a compatibilist one that allows for a measure of freedom; and a suggestion of evolutionary progression. We humans, and all things on Earth, are part of the Logos revealing itself to itself. Ultimately there is only the self-revealing Logos, and when we understand this and our role in it, we then gain a small degree of free thought and action. The Logos moves in us and through us, and ultimately it will move beyond us—or so Heraclitus suggests. We would do well to contemplate this situation.

THE INSIGHT OF PLATO AND ARISTOTLE

For all that he accomplished, Heraclitus neglected one thing: the specific means, the actual process, by which the Logos-creation comes about. The later Greek philosophers identified this process; they called it *technê.* Plato was evidently the first to give it serious consideration, and he was also the first to connect it, explicitly, with the logos (human, not divine). Technê for Plato, as for Aristotle, was a kind of art or skill. On Plato's view, this includes the manufacturing crafts (carpentry, house building, pottery, painting, cookery) as well as skilled activities and performance arts (music, medicine, rhetoric, mathematics, animal husbandry). For the most part, his discussion of the various *technai* is situated within the context of an analysis of *episteme,* or knowledge—but that lies beyond the present study.

There is one instance, however, in which Plato makes an explicit connection between technê and logos. At the beginning of *Gorgias*, Socrates is questioning the lead character, a sophist, about the name of his particular technê. The answer comes: it is rhetoric. The dialogue continues (449e):

SOCRATES With what is rhetoric concerned?

GORGIAS With *logos*.

SOC What sort, Gorgias? Such *logos* as would teach the sick how they might get well?—No.

SOC Then rhetoric does not treat of all kinds of *logoi*?—Certainly not.

SOC And yet rhetoric makes men able to speak?—Yes.

SOC And to understand that about which they speak?—Of course.

SOC But does not the art of medicine, which we were just now mentioning, *also* make men able to understand and speak about the sick?—Certainly.

SOC Then medicine is also concerned with *logos*?—Yes.

SOC Concerning diseases?—Just so.

SOC And isn't physical training also concerned with *logos*, regarding good and bad condition of the body?—Very true.

SOC And the same, Gorgias, is true of all the other *technai*: All of them are concerned with a particular *logos*, namely, the one that corresponds to each particular *technê.*—Clearly.

SOC Then why don't you call all the other *technai* 'rhetoric'? They are concerned with *logos* too!

GOR Because, Socrates, the knowledge of the other *technai* has only to do with some sort of external action, such as with the hands; but there is no such manual action in my rhetoric, which works and takes effect only through the *logos*. And therefore I am justified in saying that rhetoric is the particular *technê* that treats of *logos*.

Every technê has a corresponding logos, which is the underlying pattern or order to the skill at hand. Technê is the process by which the specific logos is manifest and made tangible. This is almost incidental on the human plane, but on a larger scale it has important consequences.

Near the end of his life, we find evidence in two dialogues that Plato did indeed find some deeper meaning to this connection. At the conclusion of *Sophist*, the Visitor remarks that "production (*poiêtikês*) has two parts"

(265b), namely the man-made and the divine—*poiësis* here taken as synonymous with technê. Production is then defined as "any capacity (*dynamin*) that causes things to come to be that previously were not." Nature creates not haphazardly but "by a cause that works by reason (*logou*) and divine knowledge (*epistemes theias*) derived from a god." A few lines later Plato makes an explicit shift in terminology: "I'll assume that divine expertise (*theia technê*) produces the things that come about by so called nature" (265e). Hence the divine Logos produces the things of nature via a process of technê, entirely comparable and analogous to the human.

A similar account occurs in Book X of *Laws*, where Plato provides an extended argument for the existence and nature of the gods. Early on (888e) he observes that all things come into being in one of three ways: nature (*physis*), chance (*tyche*), or technê ("art"). The common man believes that technê arose last, but this is incorrect. Likewise, he erroneously believes that soul (psyche) came into being after physical matter, when in fact the reverse is true. Psyche was a primal being, and brought all things into existence via two processes: *technê* ("art") and *nous* ("reason"):

> It's the soul, my good friend, that nearly everybody seems to have misunderstood, not realizing its nature and power. . . . It is one of the first creations, born long before all physical things, and is the chief cause of all the alterations and transformations. . . . In particular, the grand and primary works and creations, precisely because they come in the category 'primary,' will be attributable to technê. Natural things, and nature herself . . . will be secondary products from technê and nous. (*Laws* 892a–b)

Plato here makes the slight alteration from logos to nous, emphasizing the active, rational power of the aboriginal psyche rather than simply the notion of an underlying order. But the intent is clearly the same: technê and logos/nous are the cause of all existence.

This alteration and primacy of soul are consistent with the fact that Plato's creator-god was not Logos but rather the Demiurge (*dêmiourgos,* lit. 'public doer').[6] Upon constructing the cosmos, the Demiurge endowed it with a secondary soul—the world-soul, *anima mundi.*[7] The cosmos was thus a complete and perfect living being. And not just the cosmos: the Demiurge also saw fit to ensoul the stars (*Timaeus* 41e), the sun (*Laws* 898d), the Earth (*Timaeus* 40c), plants (*Timaeus* 77b), and even the Form of Being (*Sophist* 249a)—as well as humans and animals, of course. Widespread ensoulment by the creator is another point of commonality with Heraclitus.

As we saw, rhetoric seems to have had a special standing for Plato; it is the one technê that directly addresses the logos, whereas all other *technai* are merely associated with a logos. This point evidently impressed Aristotle, and not only because he wrote a book on the topic. In *Rhetoric,* Aristotle chooses to literally merge the concepts of technê and logos, coining a new term in the process. The book opens with some reflections on the "treatises on rhetoric" (*technas tôn logon;* 1354a12)—that is, the technê of the logos. A few lines later, the phrase becomes, for the first time, a single word: *technologousin* (τεχνολογοῦσιν; 1354b19). Notably, the word is never translated as 'technology'; instead we find such phrases as "rules of art" or "theorizing." Three more times Aristotle presents this concept, each a slight variation. At 1354b26 we read of "treatises" or "rules of art" (*technologein*). Later in the opening section we find *technologousi* (1355a19). Finally in Book 2, Aristotle writes of "treatises on rhetoric" (*technologountôn;* 1356a11). We can perhaps understand why this word never appears as 'technology'; it is, after all, simply the "technê that treats of logos"—that is, rhetoric—and nothing more.

In the late work *Nicomachean Ethics*, Aristotle returns to the topic of technê, offering up a more precise definition. Book 6 examines five ways of knowing, of which the first named is technê (1139b16)—misleadingly translated as 'art.' Section four (1140a1–23) then elaborates: There is a distinction, he says, between things *made* (*poiêsis*) and actions *done* (*praxis*); only the former is truly technê. Both are alike, however, in the sense that they each involve a "rational quality" or "a reasoned state of capacity" (*meta logou hexis*)—that is, a condition or process by which an underlying logos is transformed into something tangible, something visible. Technê, accordingly, is "a rational quality concerned with making" (*meta logou hexis poiêikês*); this would include, e.g., architecture or construction. Here, then, he parts with Plato, who believed that all human activities constituted technê. But it is notable that, in the first formal definition of the word, Aristotle *defines technê in terms of logos*; the two concepts are intimately related.[8]

In fullest form, just a few sentences later, the definition includes an addition: the word *alêthous* (ἀληθοῦς), literally meaning 'unconcealed,' that is, 'that which is revealed' or, more broadly, 'truth.' The full definition then is: *hexis meta logou alêthous poiêtikê* (1140a10). Technê reveals into being that which was hidden within the logos of the creator. Technê is the process of coming-to-be, guided by the logos.

Also notable in this passage is Aristotle's subsequent definition of atechnê ('non-technê'). Atechnê for him is the process by which nature, not man, brings things into being. Technê is necessarily a human activity, on his view, and so the corresponding natural process must be an opposite, in some sense. It is a 'false-making.' It is therefore appropriate for Aristotle to exchange *alêthous* with an opposing term, *pseudous*; atechnê thus becomes *meta logou pseudous poiêtikê hexis.*[9]

We can see, then, that Aristotle even more tightly circumscribes the concept than Plato. Technê is a human (only) process, concerned strictly with making, not doing. And we are not surprised that no broader conclusions are drawn. Logos plays no cosmic role; Aristotle's god is the Unmoved Mover, pure mind, not formally a creator. Also worth noting is the fact that the social milieu of the time was not dominated by technology. Ellul

(1964: 29) identifies that era as one in which the leading figures deliberately held technology in check—the last such era, in fact. Techné was a low-class activity of manual craftsmen, unworthy of philosophical examination. Furthermore, says Ellul, the Greeks somehow knew that tools and machines had the potential to disrupt society and thus were intentionally shunned.[10] These reasons, then, suggest an explanation why neither Plato nor Aristotle examined more deeply the connection between techné and logos.

THE SYNTHESIS OF THE STOICS

The final step toward a complete metaphysical system came with the early Stoics, who combined the insights of their predecessors into a comprehensive worldview—one that would strongly influence the next four centuries of intellectual thought. The essential principles were laid down by Zeno of Citium (334–264 BC), the founder of the movement. Logos was restored as a central concept. It was the grand unifying principle, connecting the rational order of the cosmos with human reason: "Cosmic events and human events are . . . not happenings of two quite different orders: in the last analysis they are both alike consequences of one thing—logos" (Long 1974: 108). For the Stoics, Logos is God.[11] It is the principle of self-motion and change, and the vivifying intelligence in matter. Seneca called it *logos spermatikos:* "seminal reason."[12]

The Logos-principle cannot exist alone; it requires a material substrate. Drawing from Heraclitus, Zeno identified *pyr* as the sole basis of physical reality. But Zeno's *pyr* was not simply ether-like and alive; it was also a builder, a creative force, the dynamic and energetic substance of the world. The earliest surviving reference to this entity, Cicero's *De natura deorum* (Book II, 57), calls it *ignem artificiosum*—a "creative fire." Elsewhere it is referred to as "designing fire," "creative reason," and "fire that is an artificer."[13] Zeno himself called it *pyr technikon*—the artistic fire. The definitive phrase comes to us from Diogenes Laertius (7.156): φύσιν εἶναι πῦρ τεχνικόν, ὁδῷ βαδίζον εἰς γένεσιν (transliteration: *physin einai pyr technikón, hódo badízon eís genesin*). In Long's translation:

> "Nature is an artistic fire, going on its way to create."

Everything in the physical universe is a mixture of divine Logos and *pyr technikon*; neither exists without the other. Logos is the universal rational principle, "the indwelling cause of all things,"[14] and the creative fire its physical embodiment. Invoking the important term 'pneuma,' meaning spirit or hot air, Diogenes adds that nature, on the Stoic view, is "pneuma fiery (*pyroeides*) and craftsmanly (*technoeides*)."

For present purposes, the central point is that Zeno identifies his fire as a *technikon:* a creative force, a process of techné, but one that realizes the intelligent Logos. (A striking insight: fire is not a destructive force, as commonly understood, but rather a constructive and creative one.) This ontological connection between technê and logos, first established in the human sphere by Plato and Aristotle, is now expanded to a universal principle. Energetic fire is the material reality, Logos the guiding mind, and technê the process by which all things are formed. Technê and Logos are inextricably combined in everything, and indeed they have no existence independent of the other. The world-sphere can therefore rightly be called a *Pantechnikon*: an all-technê, a material unfolding of universal reason.[15]

As with Heraclitus, the directing Logos implies a form of determinism, which the Stoics embraced. Chrysippus—a leading Stoic thinker, along with Zeno and Cleanthes—put forth the first known causation-based argument: "Prior events are causes of those following them, and in this manner all things are bound together with one another. . . . From everything that happens, something else follows depending on it by necessity as a cause."[16] Fate is a dominating factor in the unfolding of the universe, to the degree that it was virtually equated with the Logos itself: "God and mind and fate and Zeus are one thing, but called by many different names" (*"En t' einai theòn kaí noún kaí eímarménhen kaí Día*).[17]

We are confronted again with the dual-aspect nature of the will. This is something shared by most any form of metaphysical monism, but the problem is especially vivid with a strong form of pantheism, such as the Stoics put forth. Heraclitus' pantheism was ambiguous, and so was his stance on free will. Despite the material monism—the *pyr aeizoon*—of the physical universe, the divine Logos was somehow "separated from us," hidden away. It directs the cosmos from afar. For the Stoics, however, there is no separation. Nature is God.[18] In such a case, there is no distant puppet master pulling the cosmic strings. The possibility of some sort of free action remains.

Such in brief is the Stoic worldview, a culmination of thinking from Heraclitus, through Plato and Aristotle, to Zeno, Cleanthes, and Chrysippus. Each step along the way articulates a more refined view of the relation between technê and logos. The Stoic system is a comprehensive metaphysic, addressing cosmic origins, ontology, theology, the human person, ethics, and teleology. It was not, of course, a metaphysical theory of technology per se, for the reasons stated previously. Technology was not a compelling issue for the Greeks and posed no serious metaphysical challenges. But today, this is no longer true. Technology does pose serious and urgent problems for humanity and the Earth, and by understanding its metaphysical basis, we may better respond to the situation.

I suggest that this Heraclitean/Stoic worldview—the cosmos as a Pantechnikon—can serve as the basis for a relevant metaphysics of technology. I plan to take up their view, place it in a contemporary context, and lay out the implications. In present terminology, it is a substantivist and (soft) determinist position. Such a stance has few defenders today and several critics. I will offer supporting arguments and defend it against competing theories.

As a metaphysical view, it compares well with Heidegger's, which I will examine in detail. In particular, it conflicts with him on several important points, but positively, in the sense that it is more coherent and avoids certain unresolved problems. From a pragmatic standpoint, a pantechnical metaphysics involves specific consequences for society and nature—consistent with those implied by Ellul. But Ellul's work, brilliant as it was for the 1950s and 1960s, is outdated today. And his view has furthermore been criticized, rightly, for lacking any metaphysical basis. The thesis of the Pantechnikon supplies such a foundation.

It is a trivial observation to note that we are surrounded today by advanced technology. It is another matter altogether to cite technology as the metaphysical basis for the world. Rightly understood, it can point to new conception of ourselves and all existence. Wrongly understood, it may mean the end of humanity, and even life on this Earth. It is a matter of greatest importance.

NOTES

1. All of the standard English translations of the Bible have this reverse order. In German, though, the Luther translation retains the original: "... *das Wort war bei Gott, und Gott war das Wort.*"
2. On some Christian interpretations, Logos is read as the preembodied spirit of Jesus. A similar problem occurs with attempts to accommodate the important Greek concept of the pneuma. In the New Testament, pneuma becomes 'Holy Spirit'—which, like Logos, has an awkward and ambiguous status.
3. This is perhaps the earliest known instance of a true dual-aspect monism in Greek philosophy.
4. For more on Heraclitus' panpsychism vis-à-vis other pre-Socratics, see my *Panpsychism in the West* (2005).
5. Empedocles' fragments 57 and 61 describe bizarre creatures and parts of organisms that existed in some early period of the earth. Fragment 62 mentions the "undifferentiated shapes" present at that time, which Fire (*pyr*) "sent up" into higher forms; see Chapter 3 for a further elaboration.
6. See *Timaeus* 28a6, and the discussion in Chapter 3.
7. In addition to *Timaeus,* see *Philebus* 30a.
8. See Mitcham (1994: 117–129) for a good analysis of the meaning of technê and logos in Plato and Aristotle. However, Mitcham does not situate the terms historically; he has no discussion of Heraclitus or, subsequently, of the Stoics. As a result, he misreads Aristotle's definition as "a habit . . . with a true logos" (120), and finds his usage of the combined technê-logos terminology in *Rhetoric* perplexing (128–129).
9. The usual translation is awkward at best: "a rational quality concerned with making, that reasons falsely."
10. See Chapter 7 for an elaboration.
11. Diogenes Laertius (7.134) wrote: λόγον τὸν θέον ("Logos is God").
12. *Ep.* 65, 12. Similarly, Diogenes referred to the Stoic God as the "spermatic Logos of the cosmos" (*spermatikòn lógon ŏnta ton kósmou*); 7.136.
13. Respectively: Aetius 1.7.33, in Long and Sedley (1987: 274); Long (1974: 112); Sandbach (1975: 73).

14. Long (1974: 165).
15. British readers may call to mind the term 'pantechnicon,' which was the name of a London warehouse in the early 1800s that contained a variety mechanical devices, contrivances, and furniture. It later came to mean a moving van or truck associated with such a warehouse. I trust it is clear that my term has no connection to either of these meanings.
16. Cited in Long (1974: 164). Cf. SVF ii.945.
17. DL 7.135.
18. "The terms refer to the same thing in Stoicism" (Long 1974: 108).

3 Cause, Teleology, and Transcendence

The Greeks established the foundational concepts of a philosophy of technology and cast them in terms appropriate for their age. They lived in an exceptional society, at an exceptional point in time. They had the intelligence, sophistication, creativity, and will to push philosophy to its highest peaks. They created a sufficiently refined civilization to allow sustained contemplation and yet were close enough to nature to still feel an intimate connection to it. The human condition and the grandeur of the cosmos were there before their eyes. Collectively, they were able to lay down many of the essential philosophical principles and concepts. Whitehead's (1978: 39) well-known remark that all of philosophy is only footnotes to Plato rings true.

Regarding technology as tools and devices, the Greeks lived at a time of technological progress but not revolution; there were no major advances during the height of Athenian society. Technology was, relatively speaking, well under human control, and the technicians of the day were generally low-status craftsmen—in noted contrast to the present time, where high-tech king makers, corporate executives, and billionaires carry so much influence. Greek society was sufficiently transparent and the objects of their world sufficiently mundane that they could see into the essence of things, and express these essences in philosophical terms. It is our task and our challenge to extract the wisdom of that age, and apply it to the complex and evolving nature of modern technological society.

Their core insights remain valid. Technology, qua technê, is a process of creation. It is a coming-to-be. Mankind creates, and nature creates, but it is creation all the same. Durable structures of matter and energy come into being, persist for a time in a more or less stable condition, and then pass away. From our biased vantage point, we see our own creations as of a different order, but they are not. All coming-to-be is creation, and all creation is technê.

Heidegger grasped this very point. He wrote:

> The Greek for 'to bring forth or to produce' is *tiktô*. The word technê, technique, belongs to the verb's root, *tec*. To the Greeks technê means neither art nor handicraft but, rather, to make something appear. . . . The Greeks conceive of technê, producing, in terms of letting appear. (1993: 361)

With perhaps the exception of Aristotle, who defined it in strictly human terms, technê was simply the process of making, of creating. For what is creation? It means to produce (*creâre*), and more generally, to grow (*crêscere*)—that is, to gradually bring into being. We create, animals create, plants create, forces of nature create, and the universe creates. And creation, for the Greeks, was technê.

For its part, logos is reason, a manifestation of the Logos or cosmic reason that guides all coming-to-be in the universe. As explained in the previous chapter, Logos was seen as a God-like being, endowed with the divine power of creation. In its more ordinary usage, logos is the bringing into appearance of something hidden, the revelation of some obscured underlying meaning. The term itself derives from *legein*—to collect, to gather up. Thus, *legein* or logos is to bring together, in a coherent fashion, that which is hidden or obscure. It is then made apparent. In collecting and gathering, the cosmic Logos brings together the unformed substance of the universe into an ordered, organized, coherent form. Logos, then, is an organizing and coordinating power—and hence rational and intelligent. Indeed, it is rationality *as such*.

Why, on the Greek view, does Logos do this? Why does it bother with gathering up from the primordial chaos, from unformed matter, and produce things that become visible? One answer is that this is simply its intrinsic nature, that it cannot help but bring into being the things that are. Evidence for this is found in our own existence, the existence of the myriad structures around us, and the cosmos itself. A somewhat more complete answer, however, would include something like intentionality: that the Logos creates due to some *desiring* or *striving* on its part. In effect, it *wants* to create. As the primordial force in the universe, Logos surely 'does what it wants.' There is nothing to block it, no greater power or force to obstruct its desires. There is resistance—creative friction, as it were—and there are natural laws working at counterpurposes, but by clear appearances the Logos is indeed willing and able to construct objects that endure. We ourselves are proof of this, as are the stars, planets, and all beings that we see around us on Earth. In fact, apart from our own ability to experience the world, nothing is more self-evident than the fact that structure exists in the cosmos.

Resistance to creation is the strongest argument in support of cosmic striving. We know that there is inertial resistance to movement, and that energy is required to bring objects into proximity and to fuse them into enduring structures.[1] We also know that laws of entropy and thermodynamics insist that isolated systems, removed from a unifying power, will decay toward randomness. Were there no continuous, unifying, coherent forces in the universe, durable structures would never have appeared. The fact that there are such structures and that there are also evolving systems of increasing complexity suggest that there is in reality some power of coherence, some force toward order and structure. A frictionless universe could be excused for denying a cosmic Logos, but in our world, inertia and entropy must be overcome, both initially and continually, for ordered structures to exist. That this happens is evidence of the fact that some organizing power—call it Logos—is indeed striving for order, working upstream as it were, against the forces of disintegration and decay. And it is succeeding.

This has two further implications: that creation is *satisfying* and that creation is for some *purpose or reason.* If Logos strives to create, in opposition to the resistive forces of the universe, and succeeds, it has accomplished something in the face of adversity. On the human level, this would of necessity be described as satisfaction. Under the assumption that the human logos, or the rational human psyche, is a manifestation of the cosmic Logos, we may infer that parallel actions yield parallel consequences. Perhaps it is only an analogue of that which we call satisfaction. The point here, after all, is to grasp the ways in which humans are fully natural beings and are fully integrated into the natural order. This is a subset of understanding how all structures, all technology, are natural and how they are all to be understood in a similar manner—namely, as integrated and persistent beings in an evolving cosmos.

A further elaboration of the Pantechnikon requires an exploration of how, rationally, the universe can function in a self-directed manner. On my thesis, all of reality is a process of Technê-Logos; this accounts for natural and man-made structures alike. My focus in this book is, of course, human technology, but this cannot be fully understood without situating it within the larger universal context; hence the need for this somewhat lengthy discussion on the goal-oriented nature of the cosmos. This chapter makes the case for a teleological universe, and the chapter to follow connects this to the workings of modern technology.

TECHNOLOGICAL CREATION IN A TELEOLOGICAL UNIVERSE

If the Logos strives to create and succeeds in doing so against the various forces of resistance, one may postulate that it does so *for a reason.* Mindless creation or pointless creation would be entirely arbitrary and capricious. Creation, it seems, must be heading somewhere—if only toward ever-greater order, structure, and coherence. This can only happen by a process of universal movement from less-complex to more-complex states. Each new state incorporates and surpasses the preceding one; it is a process of forward movement. In other words, the cosmos is a self-transcending thing.

There are various ways to view this directed cosmic process. Some have speculated that the divine Mind is seeking to recreate itself, others that we are realizing God or some Absolute Spirit. Perhaps so. But we need not be concerned with these speculations at the moment. My point here is to argue that *some* purpose of cosmic creation, whatever it may be, is more plausible than no purpose at all.

This of course is a teleological argument. On my view, a pantechnical metaphysics is indeed teleological: the universe is working toward some end, a process that can be most plausibly and conservatively described as a *striving toward order.* In doing so, the cosmos continually supersedes and transcends prior states. Science, evolution, and philosophy all provide support for such a view.

Aristotle well understood these issues. He was greatly concerned with explaining how things came into existence and how they contributed to the overall mission of the cosmos. In his terminology, he sought the *aition,* or cause, of being. As a reason or explanation for the coming-to-be of some object, an *aition* must account for its essential characteristics. We recall his theory of the four causes of all things, first articulated in *Physics* (II.3): material, formal, efficient, and final. Material cause is the matter that comprises the object, for example, the marble of a statue, the wood of a table. Shape or structure is the formal (*eidos*) cause. Efficient cause is the agent or active mover, that which "starts the material on its way to the product, changing it from what it was to what is to be" (194b30). Final cause is the end (*telos*) served, the reason the thing was created, its purpose. These notions differ, of course, from the modern scientific meaning of 'cause,' which has been largely reduced to the efficient. And yet matter, form, and purpose clearly all play an essential role in the coming-to-be of a given thing. We are simply no longer accustomed to referring to them as causes.

Consider an Aristotelian reading of a simple technical act—baking a loaf of bread. The process begins in a condition of want: *I am hungry.* From this emerges an intentional state of will and desire: *I need to eat, I need food.* The condition of want contains within itself the end goal. From this emerges a more specific and articulated end goal: an edible and nutritious loaf of bread. To this end I deploy my logos-reason to gather up the necessary components: the material causes (flour, yeast, salt, and water), and the formal (the bread pan). Through efficient causation I combine the ingredients and work them into a dough. I step through a rational process, always dictated by my end goal. As each stage is successfully attained, I gain a degree of satisfaction. The bread progresses from an unformed and dense ball of dough, to a risen ball, to an uncooked but loaf-shaped dough. As it moves through the various stages, it grows ever closer to the desired end state; it thus progressively *increases in value.* The evolution of the bread toward the end goal is a story of increasing value. Value, which exists in the ingredients only potentially, becomes actualized over time. Efficient causation, in the service of a rational end goal, is in fact the production of value.

Finally it goes into the oven and comes out as a completed object, in conformance with my end goal. If my rational process was correct and no other factors intervene, I succeed. There is a sense of satisfaction. The value of the material and effort is realized; it is a completed entity—an entelechy, as Aristotle would have it.

The four causes give one account, but another, more modern one exists at the level of energetics. On the scientific view, I am a dynamic, living, metabolic system of matter and energy. To persist, I need regular inputs of low-entropy matter (food, drink). A portion of my condition for existence must be dedicated to maintaining myself through the acquisition of food. This requires the expenditure of effort, of energy. I must *expend energy* in order to *obtain energy*—either directly (by producing or preparing the food) or indirectly (earning money to pay for food). Crucially, the energy input I receive from the food must exceed the effort expended in obtaining it. Were this not so, it would be a net loss—an untenable condition over the long run.

There may be many reasons for me to bake a loaf of bread: to hone a skill, to make money, to demonstrate self-sufficiency, for an aesthetic sense of pleasure, as an offering to the gods, and so on. But at a fundamental physical level, bread baking exists to supply the energetic needs of humanity. The Earth–sun system produces a superabundance of low-entropy plant matter that serves to sustain all animal life. The light of the sun and the soil of the Earth combine to yield the surplus energy that we all need to survive. Our status as ordered, structured, 'far from equilibrium' beings is entirely dependent upon this surplus energy flow. Each of us is a constructed being, a construction of energy.

Unsurprisingly, both views are teleological. This much has long been acknowledged: that living organisms seek specific ends that serve their well-being or that of their species. But when viewed against a background of energetics and a broader conception of life, we move toward a comprehensive vision of teleology, one that incorporates the inanimate world. This was clearly Aristotle's intention.

The theory of causes, driven by a telos, is evidently valid for human action. Without a purpose or end, we would do nothing. Only when the telos is established can the other causes come into play. But what about nature? It, too, constructs things. Hence a question of importance: Does nature also construct on the basis of ends? And if so, how can we best conceive of this process?

We know the answer, at least from Aristotle's perspective. *All* construction, all coming-to-be, whether man-made or natural, is teleological. Nature always acts with a purpose or goal, one that may be described as toward an increase in value; in Aristotle's words, "the better." Thus we arrive at the conception of the self-transcending cosmos, striving for order and the production of value.

Unsurprisingly, this idea, so important in Aristotle's thinking, found prior expression in Plato and Socrates. *Phaedo* describes Socrates' interest in Anaxagoras and his theory of Mind (*Nous*) as the goal-directed cause of all things.[2] "I [Socrates] thought that if this were so, the directing Mind would direct everything and arrange each thing in the way that was best" (97c4). He continues:

> If then one wished to know the cause of each thing, why it comes to be or perishes or exists, one had to find what was the best way for it to be,

> or to be acted upon, or to act. On these premises then it befitted a man to investigate only . . . what is best (*beltistos*). . . . Once [Anaxagoras] had given the best for each as the cause for each and the general cause of all, I thought he would go on to explain the common good for all (*koinon pasin agathon*). (97c6–98b2)

Alas, Socrates is disappointed that this does not happen. But there is the clear expectation by Plato that this condition of striving toward the best is a universal condition. Only a few lines later he remarks that this is the "real cause" of things, and refers to all things' "capacity (*dynamin*) of being in the best place they could possibly be put" (99b2, c2). Plato chides those who "do not believe that the truly good and 'binding'" are what organizes the material universe.

These initial thoughts on teleology are further developed in the creation story of *Timaeus.* The creator, Demiurge, is the wise and good craftsman of the cosmos, the cause of order. Beginning with an end in view—the best and most beautiful cosmos possible—the Demiurge exercises his "divine foresight" (*theou pronoian;* 30c) in bringing order from chaos. "He was good," declares Plato, "and he wanted everything to become as much like himself as was possible" (30a). Thus the Demiurge "brought [the cosmos] from a state of disorder to one of order, because he believed that order was in every way better than disorder." It is notable that the divine craftsman, unlike the Judeo-Christian God, did not literally bring the cosmos into being from nothing; rather, he brought it from an amorphous and incoherent state to a better, more beautiful, more ordered one. The cosmic pronoia is an organizer and an orderer. In doing so, he "uncovered" (*alêtheia*), made real and true, the present beauty of the world.

Plato is insistent that the Demiurge strives for order because it is "the best." It is not possible, he says, "that one who is supremely good should do anything but what is best" (30b). The creator "wanted nothing more than to make the world like the best of the intelligible things" (30d). And what did Plato have in mind as "the best"? That which was most like the Demiurge himself—that is, living, intelligent, and ensouled.

> Accordingly, the god reasoned and concluded that in the realm of things naturally visible no unintelligent thing could as a whole be better than anything which does possess intelligence as a whole, and he further concluded that it is impossible for anything to come into possession of intelligence apart from soul. Guided by this reasoning, he put intelligence in soul, and soul in body, and so he constructed the universe. He wanted to produce a piece of work that would be as excellent and supreme as its nature would allow. This, then, in keeping with our likely account, is how we must say divine foresight (*pronoian*) brought our world into being as a truly living being, endowed with soul and intelligence. (30b1–c2)

Thus was the cosmos constructed by a god, *as* a god (34b1, 34b9, 55d6). The Demiurge then proceeded to create many lesser gods (41), who in turn crafted all living beings, including humans.

The point here, of course, is that Plato adopts an explicitly teleological view of the origins of the universe. It was fashioned with a view to "the best," to the greatest order and harmony possible, and to be most like the divine creator himself. A parallel process extends to the celestial objects—sun, stars, Earth—and to the variety of living and ensouled creatures. In giving form to the formless, the Demiurge acted out of desire and will and with an end in mind. Upon completion, "he was well pleased" (37c9).

In Aristotle's worldview, this teleological principle also assumes prime importance, but with one crucial difference: there is no Demiurge. Aristotle's world requires no craftsman who works upon unformed matter to create order, because the source of order is intrinsic to nature itself. This was one of his great metaphysical innovations, and is a persistent and recurrent theme throughout his corpus. Nature, of its own accord, acts teleologically. This idea is embodied in his famous phrase, "Nature does nothing in vain." It always acts with purpose, with a goal, toward some established end. This telos is the most fundamental of the four causes and drives the others.

As for the specific character of the end itself, Aristotle finds much agreement with Plato: Nature strives toward form, toward ensoulment, and toward "the better." The canonical description of Aristotle's view is found in Book II of *Physics*, but the basic idea exists throughout his works. A brief sampling of the relevant passages, in roughly chronological order,[3] would include the following:

- *Progression of Animals:* "nature creates nothing without a purpose, but always the best possible in each kind of living creature" (704b15).
- *On the Heavens:* "nature always follows the best course possible, [and is] ordered in the best way possible" (288a2–9).
- *On Generation and Corruption:* "For in all things, nature always strives after the better" (336b27).
- *Physics:* "there is something divine, good, and desirable [that nature] desire[s] and yearn[s] for. . . . [W]hat desires the form is matter" (192a16–22).
- *On Sleep:* "we assert that nature operates for the sake of an end, and that this end is a good" (455b17).
- *On Youth and Old Age:* "nature does nothing in vain" (476a14).
- *Eudemian Ethics:* "all existing things . . . seek [their] own special good" (1218a30).
- *Politics:* "nature makes nothing incomplete, and nothing in vain" (1256b20).
- *Parts of Animals:* "that cause is first which we call 'that for the sake of which' [i.e. the telos]. For this is the account of the thing . . . alike in the works of technê and in works of nature. . . . Now in the works of

nature, the good and 'that for the sake of which' is still more dominant than in works of technê" (639b14–20).
- *On the Soul:* "for nature, like thought, always does whatever it does for the sake of something, which something is its end" (415b15).
- *Generation of Animals:* "all the operations of nature . . . are for a final cause and for the sake of what is best in each case" (789b3).

This shows something of the breadth and significance of the concept in the Aristotelian mindset. Despite what some believe, teleological thinking is not limited to his early or middle period, nor to *Physics*. It covers several subject matters, and persists throughout his lifetime.[4] And it is vitally important, because it is a key to understanding how all things—"alike in the works of technê and in works of nature"—come to be.

Nature, then, constructs nature. It is fundamentally a self-building, self-revealing, and self-transcending process. There is no external force, no Demiurge who works tirelessly to organize the world. The agent is hidden within matter itself. But nature is nonetheless goal oriented, striving toward increases in order and, consequently, value. It is, says Aristotle, like "a doctor doctoring himself" (199b30), one who works upon himself to make himself better; "nature is like that."

This is not to say, of course, that Aristotle's universe was a godless one. His conception of the Unmoved Mover, a kind of cosmic Mind at the center of the cosmos, provided the vital function of motion to all existence. And it furthermore seems to be intimately connected to "the good" toward which all things strive. As articulated in Book 12 of *Metaphysics*, the prime Mover is indeed a god (*theon*), and the ongoing cause of cosmic motion. Himself unmoving, he "initiates motion by being an object of love (*erômenon*)" (1072b4)—that is, by a manifestation of desire. Love or desire reflects a value standing; things are desired for their worth, because they are "the best."

For its part, the prime Mover exists in a state of complete self-realization and self-actualization. This condition—*energeia*—is an important term for Aristotle and is related to our 'energy.' *Energeia* derives from *en-ergon,* meaning roughly 'at work'—not necessarily in the sense of actively doing something, but rather as being fully realized, present, and efficacious; 'energized' might be an appropriate translation. It is contrasted with *dynamis* or potentiality, normally a state on the way to realization. *Dynamis,* incidentally, is further relevant to the present discussion because it is explicitly identified with technê. As Aristotle states in *Metaphysics*, "all the 'arts' (*technai*), that is, all productive forms of knowledge, are potentialities (*dynameis*)" (*pasai hai technai kai hai poiêtikai epistêmai dynameis eisin;* 1046b1). This is so because technê, like *dynamis*, is a "principle of change," a process by which something transitions into a new state of being. They are modes of transcendence.

Given that he is continually able to self-actualize through cosmic motion, the divine Mover is in a constant state of satisfaction or pleasure, which is part of his actuality: *hêdonê hê energeia toutou* (1072b17). In his essence, god is mind (*nous*). His activity, such as it is, is pure thought.[5] As an actively thinking being, the god must also be counted as alive: "And life also belongs to God; for the actuality of thought is life, and God is that actuality" (*kai zôê de ge hyparkhei: hê gar nou energeia zôê, ekeinos de hê energeia;* 1072b26). God is thus "a living being, eternal, and most good." But he is a mover, not formally a creator. He provides the impulse that activates the self-building tendency in nature.

ON EVOLUTION AND UPWARD STRIVING

If nature always strives for the better, and if it is evidently able to achieve this, is this not at least the outline of an evolutionary theory? Indeed so—and the concept arose neither with Darwin nor even with the Athenian Titans but rather from the earliest days of Greek philosophy. Anaximander conceived of the idea that humans came from fish or fish-like creatures that in turn came into being in water ("the moist"): "Humans originally resembled another type of animal, namely fish."[6] And according to other reports, "Anaximander (held) that the first animals took origin in the moist, wrapped in bristly shells, that, however, with advancing maturity they went upon the more dry and that . . . they shifted their way of life (accordingly)."[7]

These early ideas were elaborated by Empedocles, who declared that living creatures changed over time, proceeding from simple but complete living beings to the more complex, through a process of union and natural selection. The first elemental and self-contained organisms he called *oulophyeis*, "whole-growns".[8] Composed in part of fire (*pyr*), they were driven "upward" by heat energy to unify with other whole-growns, forming crude but isolated "limbs" and "organs" (frag. 57). Over time the limbs and organs joined together (frags. 59 and 61)—both as normally constituted organisms and as deformities and monstrosities: "ox-headed men," "man-headed oxen," "two-faced" creatures, hermaphrodites. The fittest survived, the unfit perished. It was a striking anticipation of modern theory.

Plato offers only a single relevant passage on this subject. At the end of *Timaeus*, he discusses the origin of different classes of animals. Birds, he says, are "the products of a transformation" (91d8). They have "descended from innocent but simpleminded men." Land animals with four legs "came from men with no tincture of philosophy." And fish came from "men who were without question the most stupid and ignorant of all" (92b3). Thus it was a kind of karmic-based theory of devolution—intriguing in its own right but less visionary than Empedocles.

Aristotle took the topic more seriously, concerned that some such theory of 'random' natural selection might contradict his teleological view. In Book II of *Physics* he asks, Why should nature not operate "of necessity," that is, by chance and coincidence? On this view,

> Whenever all the parts came about just what they would have been, if they had come to be for an end, such things survived, being organized spontaneously in a fitting way; whereas those which grew otherwise perished and continue to perish, as Empedocles says his "man-faced ox-progeny" did. (198b29–32)

Again, a straightforward recognition of the process of natural selection. Unfortunately, he says, it is impossible for this view to hold. As we know, body parts "invariably, or for the most part, come about in a given way," that is, with regularity. The regular patterns of nature—the parts of animals, meteorological and astronomical cycles, and so on—are prescribed for the sake of an end, and it is only the exceptions that may be attributable to chance. Chance exists, but it represents deviation from the teleological rule. As with technê, natural creation is prone to error; "mistakes are possible in the operations of nature also" (199b1). Being imperfect, nature is occasionally subject to "the corruption of some principle." This functions as an "impediment" to the normal, goal-based activity of nature. When it happens, deformities result. All this may be granted, says Aristotle, but it does not undermine the general teleological function of nature.

The outcome of such reflections led to something like a consensus among many of the ancients. Nature is indeed teleological and acts toward an increase in value, toward that which is best. Other factors do come into play—chance, necessity, spontaneity—but these are mere perturbations on the dominant goal-directed tendency. Nature's process of creation is either technê itself or technê-like (Aristotle), but in any case always strives for greater perfection—conceived as beauty, intelligence, life, and ensoulment. Finally, the teleological process requires a *guiding intelligence,* either in the form of an external craftsman-god (Plato) or as an innate proclivity responding to a cosmic mind (Aristotle). As I explained in the previous chapter, the Stoics adopted and perfected many of these ideas. Their influence persisted for centuries. As late as circa 60 AD, Seneca could pose the question, What is god?, and answer, "The mind of the universe" (*mens universi*).[9]

THE QUESTION OF ENSOULMENT

In the previous chapter I touched briefly on the tendency toward widespread attribution of soul by certain of the Greeks. In fact this 'widespread' applicability was rather more like universality, and the adherents included nearly all the major thinkers—arguably even Plato and Aristotle. Something like panpsychism seems to have been essential to the Greek worldview, and it bears directly on the nature of technê and logos, and on the matter of teleology. A brief overview is in order.[10]

Panpsychism was present in Greek philosophy at the outset. Thales is known for his *arche* of water but also for his observation that psyche is a motive force and thus that the causes of motion are conceivably of a psychic nature. Explaining his view, Aristotle wrote, "the lodestone has a soul because it moves iron" (*De anima*, 405a19). On the one hand, a rock with the power to induce motion must have been a miraculous thing; on the other it was, after all, just a rock—suggesting that the power to move may have existed in lesser degrees in all rocks, and thus that they might all be ensouled. Today, interestingly, we accept this idea of a motive force present in rocks; we call it gravity. For the Greeks, however, the power to move was a psychic force. If plain rocks are ensouled, it is plausible that psyche pervades the universe and consequently resides in everything. That this was Thales' conclusion is confirmed by another important fragment recorded by Aristotle: "Certain thinkers say that psyche is intermingled in the whole universe, and it is perhaps for that reason that Thales came to the opinion that all things are full of gods (*theon*)" (411a7)—evidently equating cosmic psyche with divinity.

We find a number of related comments in several of the pre-Socratics. For Anaximenes, it was the pneuma that was the cosmic *arche*. Pneuma is another multidimensional concept, conceived variously as air, breath, spirit, or soul. As the central organizing principle of the cosmos, it follows that pneuma is present in all things, and even that it serves to integrate or coordinate everything. This notion appears in a fragment documented by Aetius: "As our soul . . . being air, hold us together and controls us, so does [breath] and air enclose the whole world."[11] Of Pythagoras we have mostly indirect knowledge, but Cicero remarks that "[he] held mind was present and active throughout the whole universe, and that our own minds were a part of it."[12] The cosmic mind was "diffused and imprisoned in the world," presumably meaning all parts of the universe, thus implying a form of panpsychism. Parmenides is obscure on this matter, as on most topics, but one fragment is of interest: "For it is the same thing to think and to be" (*tó gàr autó noeîn èstín te kai eînai*).[13] If being and thinking are "the same," then thought, which is a psychic attribute, is omnipresent, and panpsychism obtains.

Heraclitus was examined in some detail in the previous chapter, where I noted his view that "the thinking faculty is common to all." Further to the point, Diogenes Laertius remarks on his belief that "all things are full of souls and spirits" (*pánta psychôn eînai kai daimónôn plére*).[14] And of course his notion of the "ever-living fire" as *arche* implies a life force inhering everywhere. For Anaxagoras, mind (*Nous*) is the guiding entity in the universe: "whatever things were to be, and whatever things were, as many as are now, and whatever things shall be, all these mind arranged in order" (frag. 12). As expressed by Aristotle, he believed that "just as in animals, so in nature, mind is present and responsible for the world" (*Met.*, 984b15). On Cleve's (1969: 207) reading of Anaxagoras, "every molecule is surrounded by Nous on all sides"—a clearly panpsychic view.

Empedocles, though, was perhaps the most explicit of all the pre-Socratics. His duality of ruling principles, Love (*Philotes*) and Strife (*Neikos*), immediately suggests the ubiquitous presence of psychic properties. Fragment 103 states, very directly, that "all things have the power of thought." This fact seems to be related to his theory of the four elements—fire, air, water, and earth—which were themselves ensouled, as attested by Aristotle: "Empedocles declares that [psyche] is formed out of all his elements, each of them also being psyche" (*De anima,* 404b11). And his lovely and poetic fragment 110 concludes thusly: "for know that all things have wisdom and a portion of thought."[15] Guthrie sums up the situation: "it was in fact fundamental to Empedocles' whole system that there is no distinction between animate and inanimate, and everything has some degree of awareness and power of discrimination."[16]

As previously noted, Plato explicitly calls out a number of ensouled objects, including the sun, stars, the Earth, plants, and the Form of Being, along with humans, animals, and the cosmos as a whole. He does not offer a systematic categorization of ensoulment, but it is difficult to conceive of any coherent theory that would grant psyche to this impressively broad range of objects and *not* include all things. And in fact his last work, *Laws*, seems to resolve the question for us. Upon acknowledging that, as with Anaxagoras, "psyche manages the universe," Plato adds this:

> Now consider all the stars and the moon and the years and the months and all the seasons. . . . A soul or souls (*psyche ê psychai*) . . . have been shown to be the cause of all these phenomena, and whether it is by their living presence in matter . . . or by some other means, we shall insist that these souls are gods (*theous*). Can anybody admit all this and still put up with people who deny that "everything is full of gods"? (899b)

Note his frustration with those who are dismissive of Thales' evidently famous pronouncement.

Technically speaking, Aristotle is no panpsychist because he understands psyche in a restricted sense, as the form of a living being—plant, animal, or human. His writings, however, display numerous and perhaps surprising associations with such a view. These begin with the idea discussed earlier, that matter embodies a kind of upward striving toward "the better." This inclination toward transcendence and change was described by him as a quasi–life force in nature. As he says in *Physics,* this movement is "in fact an immortal never-failing property of things that are, a sort of life, as it were, to all naturally constituted things" (205b12). Elsewhere, in his discussion of the upper realm of the cosmos, he remarks that "the heaven is animate and possesses a principle of movement" (*De caelo,* 285a28)—in evident sympathy with Plato's view. And like his teacher, Aristotle allows that we should think of the stars "as enjoying life and action" (292a21), that is, as ensouled. Indeed, "we must think of the action of stars as similar to that of animals and plants."

But perhaps most intriguing is Aristotle's flirtation with the concept of the pneuma. Appearing prominently in his last three biological works (*Parts of Animals, Motion of Animals, Generation of Animals*), it provides a theoretical basis for something very near to panpsychism. Perhaps drawing inspiration from Anaximenes, he describes pneuma as a "vital heat" (*thermoteta psychiken*), "the faculty of all kinds of soul," and "the principle of soul" (GA, 736b29ff). Like the celestial ether, pneuma functions as a conduit to and conveyer of the psychic power of the unmoved Mover. Just as the ether penetrates and animates the heavens, pneuma penetrates all earthly things, granting them the "sort of life" and upward striving that they so clearly display. Lest we doubt this, Aristotle offers us a strikingly explicit statement:

> Animals and plants come into being in earth and in liquid, because there is water in earth, and pneuma in water, and in all pneuma is *thermoteta psychiken* ["vital heat"], so that in a sense all things are full of soul. (GA, 762a18–20)

All things are not literally ensouled, because they are not biologically alive. But nonetheless they are, "in a sense," full of soul (*panta psyches einai plere*) because they are interpenetrated and informed by the soul-like pneuma. This important aspect of Aristotle's thinking has been generally neglected by recent commentators.[17]

Finally, to briefly reconsider the Stoic view, we recall that Logos resumed a major role in their cosmology and served as a central element, the guiding intelligence, to their generally pantheistic outlook. The substance of the material world was the *pyr technikon*, an artistic fire-energy that formed the substance of all things. Like Aristotle, the Stoics developed a theory of the pneuma, which they saw as a combination of elemental fire and air. And in fact they seem to have used the concepts of pneuma and *pyr technikon* interchangeably. Such things were the substrate that the cosmic Logos worked upon and, via the process of technê, brought all things into being.

Pneuma was furthermore the cohesive force, or *tonos*, that held objects together. *Tonos* seems to have functioned at three distinct levels: of the soul (*psyche*), of living beings (*physis*), and of inanimate objects (*hexis*). As with Aristotle, psyche was a restricted property, but it was representative of a universal quality.

The combined effect of an intelligent Logos working on a fiery pneuma meant that all things possessed an aspect of ensoulment. Diogenes Laertius explains the view: "The cosmos is administered by mind (*nous*) and providence (*pronoia*) . . . since mind penetrates every part of it, just as soul does us" (DL 7.138). The end result is clearly a form of panpsychism. Sandbach (1975: 130) notes that, for the Stoics, "a 'life-force' could be recognized everywhere." In a similar vein, Tony Long (1996: 228) remarks that "All things in the Stoic universe are combinations of god and matter, stones no less than humans."

TELEOLOGY INTO THE MODERN ERA

This, then, was the viewpoint of the ancient world. A Logos or cosmic mind guides the creation and evolution of the material world, imposing a progressive kind of order and harmony. The material substance of things is either in part or wholly fire (*pyr*), something with the capacity to accept the formative power of the Logos. And the process itself by which things come to be is technê, or something analogous. Such was the Greek insight into existence, God, and the process of technê-logos.

What is the legacy of these ideas? In the theological world of the Christian era, God served the combined role of Demiurge and cosmic Logos. The Bible contains a number of intriguing references to Greek philosophical concepts, most notably the pneuma. In both the Greek translation of the Old Testament—the Septuagint—and throughout the New Testament (which was originally written in Greek), virtually every occurrence of the English 'spirit' or 'Spirit' is in reality 'pneuma,' or some variant thereof. Even the contextual usage is similar: pneuma as God, pneuma as fire and air, pneuma as mind, and pneuma as omnipresent.[18] Furthermore, the so-called design proofs for the existence of God were clearly teleological. Both Augustine's *City of God* and Aquinas' *Summa Theologica*[19] employed such arguments, even as they took their cues from the Greeks. On the other hand, anything like panpsychism was antithetical to church teachings, as it threatened the presumed human monopoly on ensoulment.

Into the Renaissance, panpsychist ideas flourished again within the Italian naturalist movement, led by Cardano, Telesio, Bruno, and Campanella. Cardano's references to the pneuma and to his concepts of universal sympathy and antipathy (recalling Empedocles) demonstrate the Greek influence. Through Telesio's dual principles of Heat and Cold, all things possessed some ability to sense and perceive and thus were aware; all were sentient beings. Bruno generalized the Aristotelian conception of form as psyche. For him, all natural objects had a form, and thus all had a psychic aspect: "not only the form of the universe, but also all the forms of natural things are souls" ([1584] 1998: 42). Campanella's metaphysical system centered on his three primalities: power, wisdom, and love. These were universal qualities of all things, from the greatest (God) to the lowliest rock. Knowledge, for him, meant knowledge of one's own existence; all things know themselves. Love was manifest primarily as a self-love, as a defense of one's being, and as a striving to persevere—thus serving a teleological function.

Spinoza ([1677] 1989) embraced and developed several of the Greek themes; he was a panpsychist, a pantheist, and a teleologist, though of a unique variety. The latter point is of particular interest, as it suggests a mildly

Aristotelian orientation that was likely influenced by Campanella. Spinoza's teleology is most explicit in his conatus ('striving') principle, as articulated in Part III of *Ethics*. The view is summarized as "Each thing, as far as it can by its own power, strives to persevere in its being" (Prop. 6) and "The striving by which each thing strives to persevere in its being is nothing but the actual essence of the thing" (Prop. 7). By this he means that all things—humans, animals, plants, rocks, tables, stars, and so on—somehow function to persist, to fend off decay and disintegration, as best they can. This is clear for living things, which obviously seek to stay alive, but ambiguous for nonliving or composite bodies. It is perhaps best understood as a simple structural cohesiveness and integrity: things resist pressure, hold together, and generally make their presence felt in the world.

Such a tendency in matter is a muted form of Aristotle's view; there is no upward movement, no seeking of soul or form, no real striving after the better. Things simply want to survive. They have no higher aspirations. There is no evolutionary imperative here. Spinoza is explicit: "Nature has no end set before it" ([1677] 1989, Part I, Appendix)—by which he means nature as whole is working toward no end—specifically, no human-related end. His primary concern was theology, and in particular the Judeo-Christian theological belief that God is a personal being who promotes human well-being. As we know, for Spinoza, God is Nature, and it has no concern whatever for human welfare. God/Nature simply acts according to its essence.

And yet, things do seek a goal—their own persistence. Every particular thing is enminded ("[all] individuals . . . in different degrees, are nevertheless animate"; IIP13S), and the conatus principle is the primary manifestation of this mind, at least for nonliving objects. Even if there is no Teleology for Spinoza, there is teleology—a teleology of things.

Leibniz too saw striving in nature, in both the material and spiritual realms. His monads exhibit two fundamental psychic qualities, perception and appetite. Appetition, or desire, is a kind of striving for new states of perception, or toward complete perception of the universe.[20] First discussed in 1695, he describes his "formal atoms" (not yet monads) as possessing an efficacious force that necessarily entails both sensation, or perception, and appetite—something "on the model of the notion we have of souls," as he says.[21] This is so because souls act teleologically and are thus driven by final causation. Bodies, on the other hand, act mechanistically and are therefore subject to efficient causation. Hence Leibniz's "two kingdoms," which are in harmony yet never interact.[22]

Additionally, in the realm of the divine, God acts teleologically, guiding the universe toward that which is best—defined as "fitness," or "degree of perfection." God chose "the best possible plan in producing the universe, a plan in which there is the greatest variety together with the greatest order." As a consequence, we have today "the most perfect actual world possible." Furthermore, the very existence of God can be proven by the fact that mechanical (efficient) causes are insufficient to account for the properties of the universe: "by a consideration of *efficient causes* alone . . . we cannot give the reason for the laws of motion. . . . For I have found that we must have recourse to *final causes* for this, and that these laws [depend] upon the *principle of fitness*, that is, upon the choice of wisdom."[23] The cosmos is on a progressively upward path from "nature" to "grace," drawn forward by God's divine guidance.

The two great 17th-century philosophers thus make for an interesting comparison. Both were panpsychists, both were teleologists (of sorts), and both saw striving as intrinsic to material reality. Both found it useful to incorporate God into their metaphysical schemes. Leibniz's active deity recalls more of a traditional Christian viewpoint, but without the heavy moral baggage of heaven and hell. Spinoza, given his Jewish upbringing, had no burden of Christianity to overcome and thus was perhaps freer to recall the Platonic/Stoic conception of pantheism. Neither found it important to examine further the classic concepts of logos or technê, and neither saw any metaphysical significance in technology per se. Leibniz did find the microscope—invented circa 1590—to be most helpful in proving the existence of microorganisms,[24] and both men could surely appreciate the benefits of the printing press, which was already some 200 years old. But otherwise the 17th century was surprisingly bereft of dramatic technological advances, which perhaps accounts for their lack of attention to the matter. The steam engine and coal furnace did appear toward the end of Leibniz's life, but these evidently did not yet merit philosophical discussion.

Let me close this historical recap with a brief look at Hume's religious skepticism, as it relates to universal teleology. His *Dialogues on Natural Religion* ([1779] 2007) lays out a well-known and scathing critique of the design argument for God, seriously undermining this approach to teleology. However, there is still purpose and striving and mind in the universe. In fact, there is still a God in Hume's universe, albeit one stripped of all traditional moral qualities. God, for him, was simply the creative power of the cosmos: "the original cause of this universe (whatever it be) we call God." This originating power obviously constructed a vast and ordered system, very roughly analogous to the creative abilities of mankind. Thus, of God we can say precious little, save that he possesses an intelligence that is "somewhat similar to the mind of man." God as cosmic mind functions rather like the human soul, maintaining order and persistence within the whole. Hume suggests that we can perhaps best view the cosmos as a living organism, as an animal, with God as its soul—just as Plato would have it.

Hume closes the work by acknowledging the truth of Aristotle's dictum, "nature does nothing in vain." Nature works toward a mindful end. God exists, but we can do little more than refer to him as mind or intelligence. Without surrendering his famed skepticism, Hume nonetheless reaches a specific conclusion about the cosmic Mind: "*the cause or causes of order in the universe probably bear some remote analogy to human intelligence*"

([1779] 2007: 93) There is order in the cosmos; this order is not arbitrary but is the hallmark of intelligence. A form of teleology thus reigns.

Since Hume's time, as we know, the growing dominance of the scientific method, the declining intellectual standing of religion, and the increasing efficacy of technology all conspired to push aside talk of teleology. As early as 1620, at the dawn of the scientific era, Bacon could write that "the Final cause is a long way from being useful; in fact it actually distorts the sciences, except in the case of human actions."[25] Aristotle was increasingly viewed as suffering from terminal anthropomorphism. Only humans acted with ends. Nature was simply matter driven by mechanistic natural laws.

Eventually Darwin articulated a theory of natural selection that severely undermined divine teleology. But even here, in evolutionary theory, a kind of teleology remains. In selecting certain organisms over others, nature increases the preponderance of one while decreasing the other—yielding a better ecological fit. Nature increasingly fine-tunes ecosystems so as to make optimal use of the available matter and energy. In fact, this 'efficiency' is arguably the goal or purpose of natural selection. Ayala (1970: 5) gives a concise explanation:

> Natural selection has been compared to a sieve which retains the rarely arising useful and lets go the more frequently arising harmful mutants. Natural selection acts in that way, but it is much more than a purely negative process, for it is able to generate novelty by increasing the probability of otherwise extremely improbable genetic combinations. Natural selection is creative in a way. It does not "create" the entities upon which it operates, but it produces adaptive genetic combinations which would not have existed otherwise.

Nature's form of creation, he argues, is more a mixing and recombining than a godlike *ex nihilo*. Nature creates by rearranging. In fact, we can observe that all natural creation, including the human, is precisely of this kind. Ultimately, all creation is reconfiguration. All coming-to-be, all revealing of new structure, is a recombination of existing elements of matter and energy. This view of creation is as old as Plato's Demiurge, who was an orderer and organizer; he too created nothing *ex nihilo* but simply reordered the material at hand into a more perfect and more complete pattern.[26]

Evolution is teleological in another, Aristotelian sense: it works toward "the better." It is clear that, over the broad sweep of history, nature has moved progressively from the simple to the complex, from disorder to order, and from the less enminded to the more mindful. It is *not* a monotonic increase; there have been setbacks and regressions at the level of individual organisms, species, and ecosystems. But on the whole there is no doubt that evolution has progressed and is progressing toward greater articulations of physical, biological, and psychology complexity. It is a self-transcendent process.

And yet the question of volition remains. Is it intentional? Is it truly a striving, as Aristotle, Spinoza, and Leibniz claimed? The pronouncement of science is clear: there is no striving in nature. Such a thing would entail the "unacceptable" consequence of panpsychism. And yet panpsychism is now recognized as a viable approach to the mind. It does not undermine science, nor even conflict with it in any direct way. Neuroscientists like Stuart Hameroff, Christof Koch, and Giulio Tononi are openly expressing panpsychist ideas.[27] In the realm of physics, John Conway and Simon Kochen (2006, 2009) have argued that elementary particles must have a modicum of free will—if indeed we ourselves do.[28] And a growing number of philosophers, psychologists, and others are exploring variations on panpsychism.[29] If the barrier of panpsychism is removed, perhaps the way is clear for a more modern form of teleology.

NAGEL ON MIND AND COSMOS

More recently, we have Thomas Nagel's book *Mind and Cosmos* (2012). Apart from technology, Nagel raises many of the issues discussed in this chapter and attempts to outline a kind of structural integration. He has made a career of exposing the inability of the dominant, mechanistic materialism to account for mind and consciousness, and this book is no exception. And here he extends the thesis to include rationality, cognition, value, life, and order generally. Such things, he argues, are vanishingly improbable under current metaphysical assumptions; thus a new scheme is required, one in which mind, reason, and value are fully integrated into nature, and are expected, if not required, outcomes of cosmic evolution:

> A satisfying explanation [of the universe] would reveal mind and reason as basic aspects of a nonmaterialistic natural order. . . . The essential character of such an understanding would be to explain the appearance of life, consciousness, reason, and knowledge neither as accidental side effects of the physical laws of nature nor as the result of intentional [i.e. divine] intervention in nature from without, but as an unsurprising if not inevitable consequence of the order that governs the natural world from within. (32–33)

As a result, "an expanded . . . form of explanation will be needed," one which is likely "to include teleological elements."

Nagel declines to specify a candidate for such an expanded explanation, other than to outline a few parameters. Given the many philosophical deficiencies of theism, he leans strongly toward an atheistic, naturalistic worldview. The propensity toward order, life, mind, and reason must be structurally integrated into our metaphysics, as any brute emergence of such things is highly improbable, if not incoherent. And the actual evolution toward order and complexity must be a fundamental aspect of existence—hence his support for teleology.

On the issue of consciousness and mind, Nagel has a compelling argument. At present we have no reductive materialist account of subjectivity. There is no physical explanation of why our qualitative experiences are the way they are, nor, for that matter, why we have them at all. There seems to be no requirement for consciousness in a physical world. As it stands, mind is simply presumed to appear, as if by magic, at some unspecified level of biological complexity. Even basic questions about it remain unanswered: Physicalist philosophers can't explain when in evolutionary history conscious experience first appeared; they can't tell us when a developing fetus suddenly and miraculously acquires subjectivity; and they can't tell us which organisms on Earth today are conscious. These are not minor quibbles. They are major philosophical deficiencies in the physicalist worldview. Mind, the most self-evident and most real of all aspects of existence, is still largely a mystery in a materialistic universe.

Hence we need to reconceive our assumptions, says Nagel. Perhaps experientiality and the corresponding subjectivity do not emerge, but are intrinsic parts of reality. Perhaps all of reality is experiential, in some sense. "Perhaps," he says, "the basis for this identity [of the mental and the physical] pervades the world" (42). Indeed, a complete explanation of reality "may have to be something more than physical all the way down" (53). This of course is the panpsychist conclusion that I sketched earlier. Though he declines to elaborate, Nagel could have pointed to its lengthy philosophical pedigree and quoted some of its prominent defenders, to help ease any concerns of the unconventionality of such a view. Instead he cautiously sneaks up on the idea, tentatively suggesting that it has promise. "So this reductive [monist] account can also be described as a form of panpsychism: all the elements of the physical world are also mental" (57).[30]

Like any approach to mind, panpsychism is not without difficulties. It is unclear that mental properties of elementary particles, for example, are truly intelligible. What can we meaningfully say of atomic experiences? There is also the part-whole problem: How, for example, do the mental properties of the molecules or cells in my body relate to my mind? Do they compose it somehow? If so, how would this work? Do they relate to it in any way? Or was Nietzsche correct when he said that "our body is but a social structure composed of many souls"?[31]

Furthermore, are there any outward signs of the inner mental life of inanimate structures? Nagel thinks there would have to be. Just as with humans, wherein our mental lives correlate with behavioral action, "something analogous must be true at the micro level" (63), if panpsychism is true. Thus, "the protomental will have behavior implications." In fact we do have some recent evidence of such behavioral implications, namely an attribution of free will to quantum action—as cited earlier in the work of Conway and Kochen. And they are not alone among physicists. In the 1970s, British scientist Freeman Dyson argued that mechanistic laws hold most firmly at the largest scales, but at smaller scales nature behaves progressively more organic and spontaneous. At the quantum level, "the laws [of physics] leave a place for mind in the description of every molecule" (Dyson 1979: 249). In other words, "mind is already inherent in every electron, and the processes of human consciousness differ only in degree but not in kind from the processes of choice between quantum states which we call 'chance' when they are made by electrons."

David Bohm was another prominent physicist who endorsed such a view. He was struck by the "basic similarity between the quantum behavior of a system . . . and the behavior of mind" (1986: 130). Consequently, he came to believe that "the mental and the material are two sides of one overall process." The fact of mental experience will "ultimately reach the level of the wavefunction and the 'dance' of the particles. . . . It is implied that, in some sense, a rudimentary consciousness is present even at the level of particle physics" (131). Refining his view a few years later, he noted that "quantum theory . . . implies that the particles of physics have certain mind-like qualities" (Bohm 1990: 272). Such statements suggest at least some basis for the "behavioral implications" that Nagel calls for.

But it is the teleology that draws the most attention. Sober (2012: 55) calls it the "most radical idea in Nagel's book." On Nagel's view, some teleological principle is a necessary supplement to the ordinary mechanistic, causal laws of nature—an evolutionary principle that drives matter toward greater degrees of complexity. "A teleological account will hold that in addition to the laws [of physics,] there are also principles of self-organization or of the development of complexity over time that are not explained by those elemental [physical] laws" (Nagel 2012: 59). Though a conceptual throwback to Aristotle, such a naturalistic teleology is at least "coherent" and "certainly shouldn't be ruled out a priori" (66). Nagel's version is nonpurposive and nonintentional, in the sense that no God or divine actor is required to direct nature to specific ends. And he is clear that it operates jointly with ordinary physical laws: "natural teleology would mean that the universe is rationally governed in more than one way"—through the conventional laws of efficient causation, and "also through principles which imply that things happen because they are on a path that leads toward certain outcomes" (67). These outcomes include life, consciousness, and the sense of value that they bring to the world. Indeed, there is a "cosmic predisposition" (123) to the creation of these things, a teleological drive that fills in the missing gap in standard evolutionary theory.

As shocking as it may sound to modern ears, all this is old hat, philosophically speaking. Plato, Aristotle, the Stoics, and even the pre-Socratics understood many of these basic issues and would likely have agreed with much of Nagel's broad outline. Leibniz's "two kingdoms" of efficient and final causation mirror exactly Nagel's two-part schema. And Hume's cosmic mind as the cause of order, not to mention the Greek Logos, reflect his call to see "mind and reason as basic aspects of a nonmaterialistic natural order." But since Nagel declines to offer any further specifics regarding his teleological approach, we are left wanting.

Be that as it may, even the mere mention of teleology in a modern context brings a shower of criticism from defenders of the mechanistic worldview. The response is almost visceral, as if teleology spelled the end of rational society and threatened to unravel two thousand years of progress. Consider this comment from a popular 1960s college biology text:

> A series of cause and effect relations such as this [explaining the opening and closing of plant stomata] illustrates the complexity of superficially simple plant responses. . . . Contrast this *cause and effect* explanation with some of the *teleological* explanations of stomatal movement sometimes seen in print: that stomata open in the morning "so that the plant can secure carbon dioxide for photosynthesis" and that stomata close at night "in order to save water." Such teleological explanations, crediting the plant with intelligent and purposeful behavior, are easy to formulate but totally inadequate in explaining plant responses. Teleological explanations get the cart before the horse by converting a possible result into a cause. If botanists were satisfied with teleological explanations for plant behavior, research aimed at discovery of the actual course of events would cease. (Greulach and Adams 1967: 261)

One senses the utter contempt the authors have for such a notion. First they suggest that teleology is antiscientific because it reverses cause and effect—that is, that an end or effect can never serve as a cause. This concern, sometimes expressed as backward or reverse causation, is seen by some as a fatal logical flaw in teleology. Second, they imply that it is irrational because it provides no explanation for action. Third, they look with disdain on the idea that plants, though living beings, might actually possess anything like intelligence; clearly panpsychism was not a viable option for these men. And finally they state baldly that admitting teleology would mean the virtual end of science.

Such concerns are baseless. Teleology is not some anticausal 'just so' story about natural behavior, nor is it a mystical or supernatural phenomenon. As should be obvious, *human* teleology happens all the time. People constantly act on behalf of ends, and with goals or purposes in mind. Arguably *all* conscious behavior is, in fact, goal oriented and thus teleological. And not only humans: animals too—certainly the so-called higher ones—act with purpose and toward specific ends. Teleology is an entirely natural phenomenon and is arguably a hallmark of animal behavior. This in no sense entails the end of scientific analysis. Teleology is logically coherent and does indeed provide an account, and thus contributes to an explanation, of behavior and action.

And yet similarly specious arguments are deployed against Nagel. Leiter and Weisberg (2012) link teleology with "supernatural causality," suggesting that they are one and the same. Modern nonteleological science, they claim, has proven its worth by its ability to *predict* and *control* aspects of nature. Efficient causation suffices for explanation. The study of complexity and self-organization of matter, they say, can be adequately understood via reductive science and nonteleological cause and effect. And yet Nagel's points hold: We have no good physicalist account for the presence of life, consciousness, or rationality. These things seem to be intrinsic to matter (hence panpsychism), and they grow in complexity and subtlety over evolutionary time, even in the face of thermodynamic laws to the contrary (hence teleology).

Sober (2012: 54) accepts the idea that "things tend to change in the direction of certain types of outcome" but declines to label this as teleological. Quoting Nagel, he claims that a truly teleological principle must be a nonphysical law. For example, the thermodynamic tendency toward entropy, or randomness, is not goal directed, he says, by virtue of the fact that it is a physical law. Thus, what Sober and Nagel demand for teleology is a "natural law" that is not a "law of physics"—presumably meaning a higher-order principle that dictates certain parameters or operation of physical laws. In other words, they demand a metaphysical principle.

Fair enough; we can supply such metaphysical laws of teleology. And yet even the physical laws themselves are not unambiguously nonteleological. The second law of thermodynamics says that certain physical systems—namely, physically isolated ones—will inevitably evolve toward states of increasing randomness. Pure randomness is a final goal or end state toward which all such systems tend. Likewise the tendency of an object to accelerate when subject to a force, in the direction of that force, is arguably an end-directed action. Materialists will object that there is no "because of" or "for the sake of," no intentionality or will, involved in such instances. And yet we are in a poor position to make such judgments. Physical systems evolve in particular directions, depending on conditions and forces at work. They move toward logical, coherent, preestablished ends. This is sufficient to qualify as teleological, in the physical realm.

In the end, Sober's objection is not to teleology per se but to the implication that teleological statements "cannot be explained by a purely causal/materialistic science." Sober accepts teleological action by humans, animals, and (surprisingly) plants but claims that all such instances have fully physical causal explanations. In effect, he is arguing for the causal closure of the physical realm: all physical events can be completely described by physical causes. This view is well known and widely held. But given how little we know about much of the physical realm—specifically, things like dark matter, dark energy, quantum gravity, and so on—it is a rather presumptuous claim. But so be it. We can accept physical causal closure. This still leaves us, though, with several mysteries, including (a) the unexplained nature of mind and consciousness, (b) the unexplained tendencies of universal evolution, and (c) the existence and nature of metaphysical laws. Clearly there is more to be said.

Finally, Orr (2013) attacks Nagel's teleology on three grounds. First, he says, it is highly implausible. Without citing any argument, Orr quotes Darwin, who saw little reason to invoke a design-based theory of nature. But of course Darwin was addressing the question of intentional, theological design, and that is not at issue here. The implausibility critique is thus little more than an intuition, one that works both ways on this count.[32] Second, Orr claims that Nagel is preoccupied with such high-level properties as sentience, reason, and value, and that these have little relevance to the vast majority of life forms on Earth. Specifically he names "fungi, bacteria, and flowering plants"—none of which is sentient, all of which are "mindless." One wonders how Orr knows these things. Once again, we see the failure to even contemplate a panpsychist alternative, one that would render this objection irrelevant.

Lastly, Orr dismisses Nagel's references to nature's tendency toward higher and more complex forms by citing apparent exceptions: cave fish evolved eyes and then lost them; parasites lost evolved traits after taking up a parasitic way of life. He might also have mentioned the case of marine mammals or, more recently, that of dust mites.[33] If there is a teleology at work, Orr asks, why does nature "keep changing its mind?" However, cases of biological devolution do not invalidate teleology. It is no more surprising than the fact that, in circulating river eddies, some of the water goes "upstream" or that a rock tumbling down a mountainside occasionally bounces up, opposing gravity. The exceptions simply show that other forces are at work in nature. No single force is all powerful, and all action is the result of numerous simultaneous pressures and impulses. Evolutionary niches open up that may allow certain life forms to devolve into them. But this does not negate the general, long-term trend toward increasing complexity and order.

In a footnote, Orr acknowledges that "organisms are on average more complex now than they were 3 billion years ago," but this is attributable, he says, to the logical necessity of life starting from a maximally simple configuration; it has nowhere to go but up. But that's not true—life could have bounced all around: up, down, flat, extinct, reemerge, extinct again . . . all with no long-term tendency. But in fact it does have a multibillion-year trend toward greater complexity and order, both at the level of the organism and as a global ecosystem. This is not the random walk that physics and chemistry would predict but a clear inclination toward an end: complexity and order.

WHITHER TELEOLOGY?

Where does all this leave the case for teleology? The three most commonly deployed arguments against it—that it requires backward causation, that it entails a God or some intentional divine being, and that it implies a nonviable panpsychism—are all shown to be inadequate. Teleology is a real and naturalistic phenomenon, readily apparent in human and animal life, at a minimum; since it clearly occurs, the logical impossibility suggested by the backward causation claim is irrelevant. In theory God could, of course, be a source of teleological agency, but since so few commentators today entertain this option—and I certainly do not—we can safely set it aside; at issue here is an atheistic, naturalistic teleology. And panpsychism is considered unworkable primarily by those with no knowledge of its history or of its supporting arguments. Hence there are no compelling arguments against naturalistic teleology.

Supporting evidence comes from the long-term trend toward increasing structural complexity. In its roughly 14 billion years of existence, the universe has undergone a steady overall progression from a state of highly concentrated and undifferentiated energy through phases of continuous (but nonlinear) expansion, concomitantly with a steadily growing material complexity. So too the Earth. Over its 4.5-billion-year history, our planet evolved inorganically to the point where it could support life after roughly 1 billion years of existence, and then organically via the biosphere to a present-day complexity that is unequaled in the known universe. Such progressions occur in the face of the seemingly inviolable second law of thermodynamics, which demands an increase in entropy over time. But, as we know, this applies only to theoretical, isolated systems—*of which there are none in the universe.* There are no closed systems in the real world. Even the universe as a whole is an undecided proposition, to the point where we have no coherent conception of either "total universal entropy" or whether it is increasing or decreasing. Open physical systems—that is, every definable physical system—may undergo either a decrease or an increase in order over time, depending on the net flow of energy into or out of the system in question. In the case of the Earth, the surplus of solar energy combined with radiant heat from within provides the planetary surface with an (evidently) ideal condition for an increase in order. Likely every planet undergoes some form of progression, albeit at dramatically varying rates.

The most 'useful' aspect of the second law is that every energy transaction exacts a fee. Energy is irretrievably lost, primarily in the form of heat, in all interactions. Thus any given dynamical system will inevitably run down and ultimately stop without an ongoing injection of surplus energy—sufficient to offset the transaction costs. And for the system to grow in complexity, there must be an even higher energy injection—sufficient to build new forms of order. But a balancing act must occur. Too much energy, of course, would destroy a given system. Too little, and the system grinds to a halt. Order grows only within certain broad limits of energy consumption.

Why, ultimately, is there so much opposition to teleology? At one level there is the evident and historically well-motivated scientific suspicion that teleology would invite the return of an oppressive, antiscientific, theological mindset. Leiter and Weisberg close their critique of Nagel by stating that his book seems targeted at "evolution deniers, intelligent-design acolytes, religious fanatics," and other nonscientific types. The book will serve mainly as "an instrument of mischief" (2012: 31), they say. But Nagel clearly and repeatedly disavows any endorsement of intentional teleology.

At a second level, though, there is a vague intimation that teleology is somehow magical, or mysterious, or spooky, or simply a mythological anachronism. On the one hand, this is an odd concern in a universe of quantum nonlocality, black holes, dark energy, wave–particle duality, and any number of other bizarre entities. But more generally, a naturalistic teleology could be broadly conceived, as I have been suggesting, in a relatively mundane fashion: as a straightforward process of evolution toward complexity and order. Both in the evolution of the universe and in life on Earth, we see consistent long-term trends toward greater structural complexity, more intricate hierarchy, and greater sensitivity to surrounding conditions. The cosmos is biased toward order.

An image I have in mind is that of a child's wind-up toy that, once set, runs around the floor in a random, spinning motion. Such a toy has no directional bias but simply traces out a random path. Now set one on a tabletop and tilt the table very slightly. The toy will spin randomly but with a bias downward; it will gradually and inevitably progress down the table. Take a hundred such toys and place them on a large, slanted table: the spinning mob will gradually diffuse, but their center will inevitably shift downward over time.

This movement, of course, has a simple explanation. A physical force—gravity—that was not in play on the level tabletop[34] comes to be 'activated' by tilting the surface. Slanting the table introduces a teleology into the system, which is enacted by a preexisting gravitational attraction. The toy system moves toward a directed end—in this case, merely the edge of the table. But it provides us with a useful image.

The universe is evidently biased in an analogous way. The cosmic table is slanted toward complexity. Physical systems evolve 'downhill' toward order. This happens not everywhere and not at all times; frequently physical systems degrade, run down, disintegrate, die, and generally act in opposition to the overall teleology—much as some of our random spinning toys will move uphill in spite of the downward bias. In the universe there is movement in both directions, but clearly the progression toward order wins out overall—at least so far. And we have no reason to suspect it will change in the foreseeable future.

This notion of a bias toward order is not new. It was implicit in Plato's and Aristotle's "striving toward the better" and in Hume's discussion of the appearance of order in the universe.[35] Darwin gave a boost to such ideas. The late 1800s saw the publication of such works as Peirce's article "The Architecture of Theories" (1891), in which he articulates the view of matter as a progressive solidification of mind via "habits," or recurrent patterns of thought. In the same year, Ferdinand Schiller produced *Riddles of the Sphinx*, arguing for a universal progression from chaos to order, and from matter to spirit. Bergson's *Matter and Memory* (1896) and *Creative Evolution* (1907) promoted similar views, as did Samuel Alexander in *Space, Time, and Deity* (1920). A culmination of this line of thinking occurred in Teilhard de Chardin's *Phenomenon of Man* ([1955] 1959); his notion of complexity-consciousness and of the "interiorization" or "organic doubling-back" of the universe brought a grand teleological sweep to cosmic evolution. In his short piece "Reflection of Energy" ([1952] 1970), he proposes a fundamental law, "reflective energy," that drives evolution forward and increases order despite the second law of thermodynamics.

More recently we find scientifically formalized treatments by such individuals as Prigogine (1977, 1980), Prigogine and Stengers (1984), Jantsch (1980), Kauffman (1993, 1995), Goodwin (1994), and Bak (1996). Kauffman's "order for free" argues that the universe is a kind of deterministic chaotic system that locks into certain dynamical patterns called strange attractors; these impose order on an otherwise unruly nature. All such works seek to elaborate naturalistic, nonmystical, nontheological solutions to the issue of growing order and complexity. In this sense they articulate a naturalistic teleology.

Three questions remain. First, what is a possible mechanism or means by which the cosmos is biased toward order? What is 'tilting the table'? Second, what are the implications for human society and the planetary ecosystem? And third—to close this lengthy tangent—what does it tell us about the nature of technology? I think we can make progress on all three issues, as I explain in the following chapter.

NOTES

1. Energy may be produced upon joining—as in nuclear fusion. But there is still significant resistance and numerous physical hurdles to overcome before completing the union.
2. We also find some evidence of Socrates' belief that nature works toward "the better" in the early dialogue *Apology*. Near the end of the piece, Socrates comments on his just-announced death sentence: "Things were bound to turn out this way, I suppose, and I imagine it is for the best" (39b; Gallop translation). Thus he has in mind the idea of a cosmic destiny that works toward the good—at least, in his case.
3. Any chronology for Aristotle is problematic, but here I follow Rist (1979).
4. A complete list of similar references would also include: 271a32, 290a31, 432b22, 434a30, 471b25, 691b4, 695b19, 708a9, 711a16, 717a15, 738b1, 744a36, 788b20, and 1253a9.
5. At 1074b34, Aristotle famously describes the god's cognitive action as a kind of self-contemplation or self-reflection on the very process of thought: "its thinking is a thinking on thinking" (*estin hê noêsis noêseôs noêsis*).
6. Barnes (1987: 72).
7. Cited in Cleve (1969: 145).
8. Frag. 62, following Cleve (1969).
9. *Natural Questions*, I.13.
10. See my *Panpsychism in the West* (2005) for a complete account, or my 2003 and 2009a articles for more concise discussions.
11. In Kirk et al. (1983: 158–159).

12. *De nat deor,* I: 26–28.
13. Fragment 3, Freeman (1948) translation.
14. DL 9.7.
15. Guthrie (1962–1981, vol. 2: 230).
16. 1962–1981, vol. 2: 233.
17. The main exception in recent years is the work of Abraham Bos; see for example his 2003 and 2008 works.
18. See Skrbina (2005: 58–63) for an elaboration.
19. The best-known instance for Aquinas is his "fifth proof" for God (Article 3, Question 2).
20. This is only vaguely articulated by Leibniz. See "Principles of Nature and Grace" (sec. 2), and "Monadology" (sec. 15).
21. "New System of Nature," Leibniz (1989b: 139).
22. Summarized in "Monadology" (sec. 79): "Souls act according to the laws of final causes, through appetitions, ends, and means. Bodies act according to the laws of efficient causes or of motions."
23. "Principles of Nature and Grace" (secs. 10, 11).
24. Such "animalcules" proved, in his mind, that tiny forms of life were ubiquitous, at all scales.
25. *New Organon,* II.2.
26. In doing so, he necessarily ensouled it. There is a parallel process here: the higher orders of organization correspond to the higher orders of soul. For the Demiurge, there *had* to be a world-soul; it was not possible otherwise. The world-soul is the source of the ongoing Logos that maintains order and harmony in the cosmos. Without it there would be disorder and chaos. And it would have been "impossible," as Plato said, for the Demiurge to construct a universe such as this.
27. See for example Hameroff (1998, 2009), Koch (2009, 2012), and Tononi (2008).
28. As they say, their "Free Will Theorem" presents us with a world "in which fundamental particles are continually making their own decisions" (2006: 1468). Harking back to Epicurus, they declare it likely that atomic will "is the ultimate explanation of our own [free will]" (2009: 230).
29. See Skrbina (2009c) and Blamauer (2011) for two recent collections.
30. Unfortunately, Nagel vacillates on this subject. "Without something unimaginably more systematic in the way of a reduction, panpsychism does not provide a new, more basic resting place in the search for intelligibility. . . . It offers only the form of an explanation without any content, and therefore doesn't seem to be much of an advance on the emergent alternative" (62). But he immediately adds, "yet the proposal is not empty" because it is amenable to naturalistic reductive options that have proven successful in the past. "On the other hand," he states a few lines later, "the idea of reducing the mind to elementary mental events or particles seems unnatural in a way that physical atomism doesn't." One can be excused for not knowing where Nagel comes down on this issue.
31. "*Unser Leib ist ja nur ein Gesellschaftsbau vieler Seelen.*" *Beyond Good and Evil,* sec. 19 (Nietzsche [1886] 1973: 49).
32. The creationist, of course, can ask this: How likely is it that complex, conscious organisms such as ourselves came into being by mere physical laws . . . ?
33. See Klimov and O'Connor (2013).
34. Except, of course, to hold the toys down in the first place.
35. "for ought we can know *a priori*, matter may contain the source or spring of order originally within itself . . ."

4 Technology and Cosmic Evolution

Life evolves, and the universe evolves. The former is uncontroversial, but the latter is still disputed by many scientists. Evolution, some say, is the process by which life forms change and adapt over time; since the universe is not alive, it cannot evolve. Furthermore, they can argue that individual entities cannot evolve, only classes of living things, such as species, can; we don't say that an individual child evolves into an adult, but we do say that early primates evolved into humans. The universe changes, develops, and perhaps even progresses, but it does not evolve—some might say.

To hold such a view is to adopt a very restricted definition of evolution. Common usage is more generous: "a process of change in a certain direction."[1] Only secondarily and relatively recently has it become "the historical development of a biological group (as a race or species)." Evolution is literally an unfolding (*ex-volvere*), and its original meaning relates to the unrolling of a papyrus roll. Hence a curious fact: evolution originally meant to reveal into sight *that which is already written.* Here there is no structural change over time, only a change in that which is made visible; the reality itself is fixed. On this literal view, evolution is epistemological, not ontological.

In the previous chapter, I examined some background concepts on causation, order, and naturalistic teleology. I argued that teleology is a viable concept and is an appropriate description of our universe as it evolves on a trajectory of increasing complexity and order. In fact, the universe is biased toward order, and it achieves this even in light of opposing thermodynamic laws. For this universe, it is 'downhill to order.'

Technology and technological artifacts are obviously part of the naturalistic cosmos, but their precise role is an open issue. As I argued previously, the Greek conceptions of technê and logos provide a useful context. Logos is the hidden order of the cosmos, the mind or intelligence of the universe, that reveals itself in the unfolding of events. Technê is the process by which all things come into being. It is the revealing of the logos-order, as manifest in the material world. And the substance of the world itself is a fiery, ethereal, living energy; conscious and alive, in some sense, it expresses and embodies the will of the logos-mind. The world is thus a Pantechnikon, a sphere of creation, order, and complexity that reflects a hidden, underlying order to things. Ordinary technology, as we commonly think of it, is one phase—the latest phase—in this universal process of becoming.

If this is an appropriate picture of reality, we should find supporting evidence in a number of areas of inquiry. Previously, I cited several examples from past and present philosophy that endorsed aspects of a pantechnical worldview. I also cited the work of scientists and physicists, several of whom supported either a broadly teleological outlook or a panpsychic conception of reality.

And yet we should expect more. Physics should be able to describe cosmic evolution toward order in a unified manner and find consistent patterns throughout all physical structures—if indeed reality is a monistic substance of some kind. This worldview should be rooted in fundamental laws that have operated throughout the history of the universe. And ideally, it should find a place for the evident power of technology to act on and reshape the world. In fact, all these points converge in the work of senior Harvard astrophysicist Eric Chaisson.

ENERGY DENSITY AND THE EVOLUTION OF COMPLEXITY

Beginning in 1997, Chaisson published a series of pieces articulating a "grand scenario of cosmic evolution," incorporating the entire progression of order in the universe from the Big Bang to the present day. His presumption was there must be, at root, a simple explanation for the growth of order and complexity, and therefore science should be able to establish a quantitative basis for it. Working strictly from the laws of physics and thermodynamics, Chaisson formulated the basic outline of his theory, one that would be elaborated in his book *Cosmic Evolution* (2001) and other subsequent works.[2] All complex, dynamic structures in the universe, he says, are related by their need to process energy. Every physical system—from a star to an amoeba—takes in energy from some source, metabolizes it, and excretes waste energy. In doing so, the system maintains its complex structure in the face of thermodynamic tendencies that would otherwise lead to its demise. Under appropriate conditions, a system may even grow in complexity as it feeds off a continual stream of available ("free") energy. Such a thesis allows for a unified and comprehensive picture of the growth of order throughout all of history, and it has intriguing implications for our technological society.

Like most of his colleagues, Chaisson's central physical quantity is energy. This, he says, is the universal currency of physics, the one quantity that is well defined, measurable, and objective. In a dynamic universe, though, what matters is not energy per se but energy *flow*—the flux of energy as it moves through physical systems. But even this is not sufficient as a universal metric, because of the huge disparity in mass between physical objects. To take an extreme comparison, the mass of the sun is roughly 10^{45} times that of a single bacterium. To compare such disparate objects with a single unit of measure, we must look at energy flow per unit mass, which results in the concept of energy *density*. Chaisson's single metric is thus the "free energy rate density" (ϕ_m). It is a measure of the energy flow, per second and per unit mass, through any given object or system.[3] As such, it can be calculated as a single number and applied to everything from the Milky Way galaxy to a grain of sand.

As an example, consider a typical star: the sun. Based on its mass and luminosity, Chaisson calculates a figure of 2 erg/s/g.[4] This, as it turns out, is rather low in the grand scheme of the universe. The sun obviously puts out a tremendous amount of energy, but it is also extremely massive (mostly hydrogen), and thus its energy per unit mass is quite small. A similar number obtains, incidentally, for any collection of stars, such as a galaxy.

Importantly, low numbers indicate not only low energy flux densities but, correspondingly, low orders of complexity. Simple objects are physically unable to process high energy densities. Despite their huge mass and size, stars and galaxies are relatively simple objects in terms of structural complexity. Inversely, complex dynamic objects have high energy densities and thus high values of ϕ_m.

Consider next a planet such as the Earth, and in particular its highly dynamic climasphere—that is, the atmosphere and upper oceans. This system absorbs as heat energy about 70% of the sunlight hitting the Earth, which in turn drives the weather cycles and ocean currents. By a similar calculation, Chaisson derives a figure of about 75 erg/s/g; the climasphere is both significantly more complex and significantly more energy dense than the sun is.

Narrowing down to the biosphere and the mass of plant life, he finds a higher number yet: around 900 erg/s/g. This higher energy density is directly attributable to that decisive development known as photosynthesis.[5] Among animal life, the ectotherms—reptiles, fish, amphibians—live at a typical rate of 4,000 erg/s/g, whereas the warm-blooded mammals are much higher, averaging 40,000. Humans, interestingly, are below average among mammals, at around 20,000 erg/s/g while resting. During strenuous physical exertion, however, we rise to an impressive 400,000. Birds, on the other hand, have remarkably high energy rates; they range from 100,000 to 500,000 and during energetic flight maneuvers can reach 1,000,000 erg/s/g.[6]

Chaisson was also able to look at energy flux rates for individual organs, such as the brain. This is important because, for all animals, the brain consumes a disproportionate share of the organism's energy budget. Insect and lizard brains process at a rate of 50,000, which is 5 to 10 times their bodily rates. Mice brains come in at 100,000. The much-exalted human brain is high, but not dramatically so; it averages about 200,000 erg/s/g—much above our bodily rest rates but significantly lower than ourselves, other mammals, and even birds when under physical exertion. Thus, neither our bodies nor our brains stand out as particularly 'special' in the larger picture described by energy flux densities.[7]

Most relevant for the present study, Chaisson examines nonbiological structures, including human society and technological devices. Regarding the former, early hunter-gatherers created simple social structures that consumed, collectively, little more energy than did each individual person. Therefore their primitive societies had ϕ_m values of about 40,000 erg/s/g, virtually the same as that of the human body. Agricultural societies deployed both fire and animal power, raising the average social level to perhaps 100,000. People of the early industrial age accessed yet more energy, primarily via fossil fuels and hydropower; this brought a further increase, to roughly 500,000. Finally, those of us living today in technologically advanced industrial societies are able to increase this figure by a factor of three or more. Specifically for the United States, Chaisson calculates an energy flux rate of about 2,000,000 erg/s/g.[8]

Technology, though, is the most interesting aspect of his study. Technological devices are amenable to the same energy flux analysis as all other objects in the universe and thus fit into the natural evolutionary progression of complexity. In their most complex forms, they equal or exceed all other known structures. A few examples are instructive. A 1970s–era automobile comes in at about 600,000 erg/s/g. Present-day autos, being lighter but equally powerful, rate a higher value of about 1,000,000. And, just as flying animals (birds) rate highly, so do flying machines; a modern 747 airliner metabolizes energy at a rate of 27,000,000 erg/s/g, and the future generation of military jet fighters rates an even more impressive 82,000,000.[9]

Unsurprisingly, the technological peak is represented by the computer. Early personal computers, such as a typical 1990 desktop model, rate a respectable 200,000 erg/s/g. A present-day laptop consumes less energy but in a much smaller, lighter space, thus achieving a value of about 280,000. But the really impressive numbers come from examining not the box itself but the brains of the computer, the microprocessor. Even the earliest chip designs were capable of processing relatively large amounts of energy in extremely small areas. The Intel 8080 processor, released in 1974, consumed about 1 watt of power with a chip mass of only 1 milligram; this yields an astonishing flux density value of 10 billion (10^{10}). A present-day Pentium II chip has about 1,000 times as many transistors in the same space, and consumes 10 times as much power; correspondingly, its ϕ_m value is roughly 100 billion.[10] This is by far the highest flux rate of any known structure in the universe.[11]

The overall picture that Chaisson paints for us is one of steadily increasing energy flux densities throughout the history of the universe, using his metric ϕ_m as a measure of evolutionary progress. The Big Bang produced a vast number of elementary particles and atoms, primarily hydrogen, which quickly coalesced into galactic masses of very low complexity. Gravity brought masses of such particles together and initiated stellar fusion—but even stars are of relatively simple construction and relatively low energy densities. Over time, planets emerged, atmospheres formed, and complexity—and energy flows—increased again. Then came plants, animals, humans, human society, and technology, each in turn generating increased energy density and increased complexity. When this is plotted over time, we get an illuminating graphical depiction of universal evolution; see Figure 4.1.

It is an impressive picture, but not without difficulties. The upward trend, as Chaisson emphasizes, is visible only at the largest scale; the story has wide variation within types, numerous outliers and apparent exceptions, and many curiosities. Stars, for example, have ϕ_m values ranging from 0.001 (white dwarves) to 2,000 or more (for the brightest red giants). Plant values range

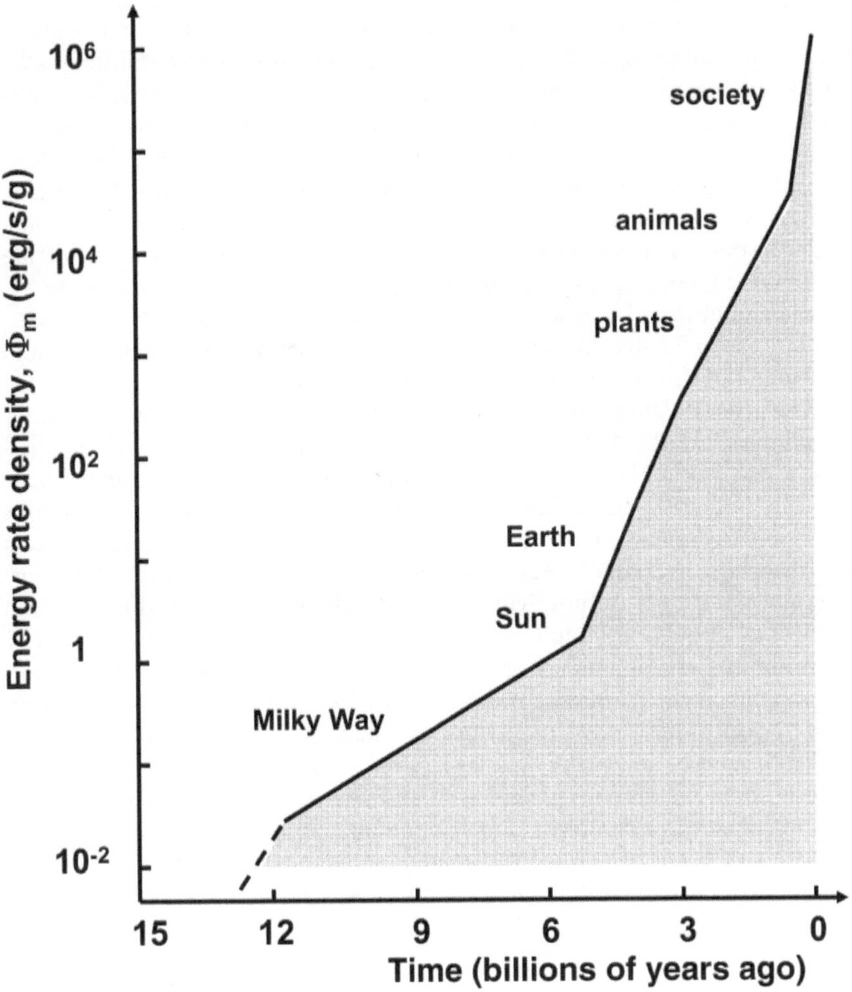

Figure 4.1 Progression of energy density and complexity over time. The source was the author and creator, Eric Chaisson. It describes the steady increase over evolutionary time of the ability of physical structures to process energy more densely; and it places technological society in line with broader universal trends.

from a few hundred to more than 20,000. The most energy-dense plants, such as corn and sugarcane, have higher flux densities than do mammals at rest. Such facts cast some doubt on the usefulness and relevance of a single metric.

And there are other issues. A system or object must be actively processing energy in order to possess a flux rate, but many complex objects are energy static and thus do not count at all. A crystal would seem to be more complex and ordered than a typical amorphous rock, but both have ϕ_m values that are zero. By contrast, an apparently simple process, like a candle flame, produces substantial energy with very little mass; correspondingly it rates very highly, on the order of 1 million—comparable to a modern automobile.[12]

One intriguing aspect of this analysis is the fact that life loses some of its cosmic uniqueness. But in a positive sense: certain key characteristics of life are found to be nearly universal in extent. In his discussion of dissipative structures, Chaisson observes that "all ordered objects—living and non-living—more or less maintain their being by means of a regular flow of available energy from their outsides to their insides" (2001: 52). The energy differential, or influx minus outflow, goes toward building and sustaining order in the system or object. In the most general sense, we can define life as "an open, coherent, spacetime structure maintained far from thermodynamic equilibrium by a flow of energy through it" (121). Only in the particular case of Earth is life also a carbon-based system operating in a medium of water, but this is our local peculiarity—as far as we can tell.

The general definition, though, fits a multitude of physical systems other than biological life. As Chaisson emphasizes, it applies equally well to a galaxy, a star,[13] or a planet.[14] "Life likely differs from the rest of clumped matter only in degree, not in kind" (122). Both living and nonliving structures "are able to circumvent locally and temporarily the normal entropy process by absorbing available (free) energy from their surroundings" (180). Furthermore, not only do nonliving systems undergo evolution, they are even subject to a kind of nonbiological natural selection: "systems are selected by their ability to command energy resources. . . . Broadly considered, selection does occur in the inanimate world" (150).[15] One recalls Aristotle's dictum that "a sort of life" exists in all natural objects.

Furthermore, Chaisson's account is, as expected, a strictly physicalist story of evolution. He disavows teleology[16] and does not address the thorny subject of human consciousness. But for present purposes, these are not major concerns. What is important is that he has given us a quantitative formulation of evolution that integrates technology within the larger cosmic scheme.

COSMIC EXPANSION

And in fact he has gone further than this: He has offered something of an *explanation* for the progressive increase in energy density. Recall the basics of cosmic history. For the first few seconds of its existence, the universe was essentially pure radiant energy; the cosmos consisted of nothing but high-energy photons. Very quickly, though, it expanded and cooled, allowing subatomic particles to come into being. In rapid succession came quarks, leptons (electrons), protons, neutrons, and then hydrogen and helium nuclei—all within the first 10 minutes or so. Thus we had a universe of both matter and radiation, as we do today.

For a long time thereafter—about 100,000 years—the energy density of radiation exceeded that of matter, even as the average temperatures of the two were roughly equivalent. This first phase of the cosmos Chaisson calls the "Radiation Era." But as the universe continued to expand, the temperatures of matter and radiation began to diverge.[17] Also around this time, the energy density of matter began to exceed that of radiation—resulting in a new cosmic phase, the "Matter Era." It was during this period that hydrogen and helium atoms came into abundance, setting the stage for the formation of stars, a process that would begin some 500 million to 1 billion years later.

Most importantly, the Matter Era witnessed an increasing divergence between the temperatures of matter and radiation. Whenever there is a thermal gradient between two things, energy (heat) flows between them. And it was this energy flux, this free energy, that accounted for the rapid growth of material structure and complexity, on Chaisson's view. In his words, "the Universe self-generates a thermal gradient, and increasingly so with time, suggestive of an ever-powerful heat engine" (2001: 128)—an engine that produces order in matter.

But the question then arises: Where does this critical thermal gradient come from? Ultimately, its source is the expanding universe itself. The expansion creates differential cooling rates and differential energy densities. The growing universe is the efficient cause of the thermal divergence and thus of the production of order. Hence Chaisson's remarkable conclusion: "*cosmic expansion itself is the prime mover for the construction of a hierarchy of complex entities throughout the Universe*" (126).

And what causes universal expansion? We are tempted to say "the Big Bang," but that was an *event*, not a motive force. Physicists are divided on this question, but it seems that expansion, at least at present and perhaps at all times, is driven by that mysterious entity called dark energy. This substance, known to physics only since the late 1990s, has just one salient physical property: a "negative pressure" that acts to accelerate the expansion of the universe.[18] For decades we have known that the universe is expanding, but it was presumed to do so either at a steady rate or in a decelerating fashion—leading perhaps to a point of maximum size followed by an ultimate collapse. Recently it was determined that, contrary to expectation, the expansion is actually speeding up. This astounding discovery had no known cause, and thus physicists were compelled to speculate about the existence of an entirely new substance—dark energy—that accounted for the accelerating expansion but that did not interact with the other basic forces.[19]

Though sparse and elusive, dark energy must be ubiquitous and thus constitute a large portion of the cosmos. In fact, recent estimates suggest that it comprises fully 68% of the total energy content of the universe; dark matter accounts for another 27%, and ordinary matter and radiation only 5%.[20] In retrospect this is unsurprising: that the ultimate source or cause of order in the universe is the dominant aspect of reality.

All these ideas have counterparts in my thesis of the Pantechnikon, as derived from the Greeks. There is an unseen energetic force at work in the cosmos (dark energy) that is the cause of order (Logos). This energy brings a "sort of life" to all things, driving them upward toward order, toward "the better." Nature is indeed "an artistic fire going on its way to create," as the Stoics believed. The accelerating cosmic expansion demonstrates that the universe is in fact biased toward order; the cosmic system is running 'downhill,' at an ever-faster rate, toward complexity, order, and increasing energy densities. Thus we arrive at a truly naturalistic teleology, a directional cosmic progression, without the need for a god. And to the main point, technology is seen to be a pinnacle of this evolutionary progression and a likely portent of its future.

Regarding the specific place and role of technology, Chaisson offers some suggestive comments. The present Matter Era, which began early in cosmic history,[21] is one in which the energy density of matter exceeds that of radiant energy. This condition has held for the vast majority of the life of the universe. Now, however, with the coming of modern technological society, we are gaining power over both matter and radiation. This is truly a milestone event, Chaisson says.

> The emergence of technologically intelligent life heralds a whole new era: a Life Era. Why? Because technology, despite all its pitfalls, enables life to begin to control matter, much as matter evolved to control radiation more than 10 billion years ago. (2001: 122)

The present generation, he adds, is "participating in an astronomically significant transformation, a second magisterial change in the history of the cosmos" (147). Technology—automobiles, computers, jet aircraft, and the social order that puts them to use—constitutes the cutting edge of evolution. "Our technology society is currently and undeniably positioned at the most complex part of the cosmic-evolutionary scenario" (197). Technology is the profoundest instantiation of the fundamental cosmic process of complexification and concentration of energy flow. At least in our region of space, it represents the pinnacle of a fundamental universal process.

This has an immediate and striking implication: that technology is, in an ultimate sense, beyond human control and thus may well be unstoppable. This bears directly on the question of technological determinism, which I will examine in detail later on. For Chaisson's part, he is undecided whether this is a good thing. Technological society, he says, "will likely continue rocketing upward, either successfully to achieve new heights of complex artificial things, or self-destructively to ends unknown" (202). I take it as obvious that we have an interest in which of those outcomes materializes; hence the need for the present inquiry.

THE FUTURE OF TECHNOLOGICAL CIVILIZATION

In his 2006 book *Epic of Evolution*, Chaisson provides some additional sobering thoughts about our technological future. The present generation of humanity "stand[s] on the verge of slowly becoming a meaningful factor in the future evolution of the Universe" (436). If, he says, we collectively develop a sense of our common planetary society and a concomitant global ethic, we may survive to fulfill our destiny. But this is not predetermined. It may be that biological human beings are merely "way stations on the path to systems of greater complexity" (439). It may be that some kind of man–machine symbiosis is in our future, a cyborg humanity. This alternative relates both to the so-called transhumanist movement and to the notion of the singularity, items to be discussed in further detail.

But there is the other possibility: Chaisson asks, "Do complex, technical systems like ourselves naturally self-destruct?" The power of modern technology is, of course, also the power of collective suicide. He cites overpopulation, nuclear warfare, genetic degeneration, and environmental destruction as among the more pressing problems facing us, all of which have been brought to bear through modern technology. It is theoretically possible to solve all of these, but "it's unclear if we humans have the wisdom to achieve and sustain that lofty plateau in the cosmic hierarchy" (440) that we hope to attain.

In fact the evidence is not promising. On the one hand, we seem unable to rationally resolve any major problem facing us. War, pollution,[22] global climate change,[23] nuclear weapons, political corruption, deforestation—all proceed apace, despite decades of effort. Technology promotes and enables all these problems, even as it is lauded as a means to solving them. Technology functions as a metaproblem, complexly intertwining itself at all levels. If we cannot handle the relatively straightforward issues of the day, we have little hope of addressing the much more subtle and pervading problem of technology itself.

On the other hand, we have the interesting problem of cosmic loneliness. A basic understanding of the universe tells us that there ought to be thousands, if not millions, of other technologically advanced societies. And yet we have no evidence that they exist, despite decades of searching the skies. Conceivably our planet has been visited by advanced alien civilizations in the past—but again, no definitive evidence is forthcoming. Contemplating this issue in 1950, Enrico Fermi allegedly asked the obvious question, "Where is everybody?" Hence the issue that came to be known as Fermi's Paradox: Why the striking and complete absence of advanced civilizations?

If in fact "the longevity of technological civilizations everywhere is inherently small" (2001: 125), as Chaisson speculates, then this is a troubling state of affairs indeed. The nascent Life Era, defined earlier as the dominance of technological society over matter and radiant energy, ought to be the glory of humanity—unless it never arrives. "Regrettably, perhaps no one, anywhere in the Universe, makes it into the Life Era" (2006: 440). If no one, anywhere, survives technological society, then technology is clearly a force of unstoppable destructive power.

There are a number of conceivable replies to the Fermi Paradox. One category includes what I would call extreme solutions: the borderline science fiction (they are among us, we are them, etc.) and the only slightly less far-fetched (they all transcended to another dimension, another universe, or some virtual reality). One problem with such theoretical alternatives is that they allow for no exceptions; a single slip-up or deviation from the rule would invalidate them. In other words, if they are among us, it would be nearly impossible that not a single such being was ever exposed. And if even 99% of civilizations transcended to another dimension, that would still leave thousands of nontranscendent ones to contact us. Given that we have no evidence of contact, despite billions of years of Earth history and the fact that millions of such civilizations should exist, these solutions are highly unlikely.

More reasonable possibilities exist. Perhaps the vastness of interstellar space makes travel or communication impracticable. Perhaps we simply reside in some far-flung corner of the galaxy, and it is our luck—good or bad—to have no nearby neighbors. Perhaps they do exist and are communicating, but we are too primitive to comprehend them. Perhaps they are deliberately avoiding contact, in order to study us, or keep us 'pristine.' Perhaps we have made contact, but the information is being suppressed. These are all theoretical options but for various reasons are unlikely as well.

So the problem persists, and the paradox ought not be lightly dismissed. Most scientists acknowledge that many such civilizations should exist at this time and that it is likely that at least one would have made contact in a recognizable fashion. The fact that this has not happened bodes ill for technological society generally and specifically for those researchers who advocate rapid technological development. Ultra-technophiles like Ray Kurzweil promote technology as the destiny of humanity, and they predict our rapid expansion to galaxywide or near-universal cyber-beings. Of course, they must hold that any intelligent beings on other planets would also have followed a similar track, and therefore it is highly problematic that we have no evidence of such a thing; the paradox looms large and threatens to seriously undermine their grandiloquent vision.

Kurzweil seems to understand the severity of the problem and thus is driven to an extreme conclusion: we are alone in the universe. He writes, "it is likely (although not certain) that there are no other civilizations. In other words, we are in the lead. That's right, our humble civilization . . . is in the lead in terms of the creation of complexity and order in the universe" (2005: 357). This position, hardly more credible than the theological equivalent, is Kurzweil's only option. Technology is such an expansive and uplifting power that any organisms discovering it would quickly span the universe, and since this has evidently not happened, we must be the first—as improbable and far-fetched as that might seem.

But we need not accept either the science fiction options or the technophile "we must be in the lead" scenario. More likely are two other alternatives: Either technological societies literally destroy themselves, or, they survive . . . *as nontechnological societies.* The threat of self-destruction has been evident for decades, certainly at least since the advent of nuclear weapons in the 1940s. Recent developments—most notably genetic engineering, nanotechnology, and robotics (GNR)—have made the dangers exponentially worse.[24] There is no reason to think that any advanced civilization would not also come upon such technologies. Mastering their use and avoiding all other risks to become long-term stable would likely lead to detection—but this has not happened. Therefore, barring future evidence, we must conclude that there is no path forward to a sustainable technological society. Advanced technology most likely means the death of civilization.

To survive, it would seem, a society must of necessity become 'nontechnological'—meaning, of course, not zero tool usage, which is impossible, but rather a *minimally technological society*, one that exists prior to the emergence of civilization-destroying capabilities. This would mean prior to GNR technologies, of course, and prior to nuclear weapons. Practically speaking, a civilization would have to back off even further, so as to not attempt to survive on the brink of such destructive technologies; too close to the threshold, and a rogue nation or individual might cross it and wreak havoc. It would also have to exist in a condition not dependent upon unrenewable fossil fuels if it sought long-term existence. Its population would ultimately reach a zero-growth level, one that would be sustainably in balance with the remainder of the planetary ecosystem. And its economic, political, and social systems would all conform to these realities. I emphasize: These are not arbitrary policy positions, nor are they planks in some conceptual political platform; they are *biophysical necessities for long-term survival*, based on our best current knowledge of the universe.

Such practical issues will be discussed further in Part II of the present work. For now I simply cite them as consequences of a pantechnical metaphysics of technology, one that situates the technical phenomenon firmly within the cosmic evolutionary order. These conclusions obtain from both philosophical and physical analysis and from plausible speculation about the likelihood of various future outcomes. They are not arbitrary, and they are not unfounded. But before addressing the pragmatic issues, I will examine in some detail the more strictly metaphysical analysis of technology, with a focus on three important German thinkers: Dessauer, Juenger, and Heidegger.

NOTES

1. Merriam-Webster online (def. 2).
2. Chaisson's initial presentation was his (1997). See also his later works (2006, 2011a, 2011b, 2012, 2013).
3. The unit is: erg s^{-1} g^{-1} (or equivalently, erg/s/g). An 'erg' is the unit of measurement of energy in the CGS system. Energy rate is expressed as ergs per second, and the mass density is calculated per gram. Thus the rate density unit is "ergs per second per gram."
4. Chaisson (2001: 136). See also his (2011a: 29).
5. Individual species, however, can range to many times this figure. Deciduous trees metabolize energy at a rate of 7,000 erg/s/g, and hybridized crops like corn and sugar cane can exceed 20,000 erg/s/g.
6. The implication here is that birds are at least as biologically complex as mammals, if not more so.
7. Chaisson writes, "Despite our manifest egos, our bodily selves do not have the highest energy rate density among animals, nor are we likely demonstrably more complex than many other mammalian species" (2011b: 52).
8. Chaisson (2011a: 36). Earlier (2001: 204) he had estimated an even higher figure of 3.2 million.
9. (2011b: 56).
10. These calculations come from Chaisson (2001: 201). I must note that he came to have second thoughts about the appropriateness of calculating flux rates for component parts of a unified functioning system like a computer. In his 2011b (57), he calls it "probably futile" to calculate such rates, because "chips do nothing by themselves." But this is inconsistent with his other analysis of, for example, organs like the brain; clearly the brain "does nothing by itself," but only within a functioning body. And yet if the brain value is relevant, so too must be the microprocessor.
11. With one qualification: of any *stable* and *persistent* structure. Unstable energy flows, as in explosions, do in fact yield higher densities but only for very short times. For example, a 4,000-kg atomic bomb releases roughly 10^{24} ergs in a fraction of a second, resulting in a ϕ_m of 10^{23} or higher. But an explosion is not a structure; it is rather an antistructure, a destructuring force. Hence the calculation is not comparable.
12. Chaisson (2001: 144).
13. Elsewhere Chaisson adds, "as regards ubiquitous energy flow, external environmental interaction, and internal structural modification while experiencing change, stars have much in common with life. . . . [C]lose parallels are apparent, including stellar populations, variations, adaptation, selection, and perhaps even a kind of crude replication" (2011a: 37).
14. The planet as alive is essentially the Gaia hypothesis, though Chaisson supplies an independent basis for that theory.
15. See 2001 (161) for an elaboration.
16. From his brief comments, it is clear that he means the intentional or theological variant of teleology. Of his theory, he states that "there need not be any designer agents or mystical acts" (2001: 142); we do not see "any evidence for obvious design or overt purpose" (143). Later he remarks that "there is no known specific goal or ultimate stage that the Universe aspires to achieve" (219). And yet it is clear that the process he describes is oriented toward order and complexity, as a part of the intrinsic nature of reality: "To be sure, the cosmos is expanding in a directed fashion; in short, it is evolving" (83). This can be read as a form of naturalistic teleology.

17. There is disagreement among astrophysicists as to whether the temperature of matter exceeded that of radiation or vice versa. See Chaisson (2001: 123–126).
18. Actually, the concept of opposing universal forces of attraction and repulsion was envisioned by Empedocles circa 450 BC. He called them "Love" (*Philotes*) and "Strife" (*Neikos*) and attributed to them the source of cosmic order—a striking anticipation.
19. Apart from gravity, which it counteracts. Another contender, incidentally, for the source of expansion is the Higgs field.
20. "First Plank results" (*Arstechnica,* March 21, 2013).
21. Circa 100,000 years after the Big Bang, as discussed previously; note that this represents just 0.001% of the present age of the universe (about 14 billion years).
22. Recent studies suggest that air pollution alone, in the form of fine particulates and ozone, is killing between 2.6 million and 4.4 million globally per year. Most of this is in Asia and India, but even in the industrialized West we have significant fatality rates; Europe experiences at least 190,000 deaths and North America 77,000 deaths annually. See Silva and colleagues (2013).
23. A 2009 UN report, *The Anatomy of a Silent Crisis* (UN Global Humanitarian Forum), estimated 300,000 deaths annually due to climate change, most attributable to "gradual environmental degradation."
24. See Joy (2000a) for one elaboration—and the discussion in Chapter 8.

5 German Metaphysical Insights
Dessauer, Juenger, Heidegger

Surely one is unsurprised to learn that the early stirrings of a technological metaphysic sprang forth from German soil. With its twin legacies of philosophical inquiry and technical leadership, in both invention and industry, German culture was perhaps fated to combine these two fields of endeavor. Of the philosophical legacy, one need only mention the obvious names—Leibniz, Kant, Hegel, Schopenhauer, Nietzsche, Heidegger. On the technical side we find the likes of Gutenberg, Siemens, Daimler, Zeppelin, Benz, Diesel, Bosch, Geiger, and von Braun. Thus it is to be expected that something approaching a metaphysics of technology should take root there.

As I stated at the outset of this work, however, one finds not a fully developed metaphysical view but rather a series of observations and insights, individually incomplete but collectively valuable. This is perhaps understandable, given that technology became a topic of serious inquiry only from the late 1800s, just as the German philosophical tradition was fading from its lofty peak of the previous two centuries. A number of German thinkers took up preliminary aspects of the issue, ranging from Marx, Hegel, and Nietzsche to lesser-knowns like Ernst Kapp, Fred Bon, and Eberhard Zschimmer.[1] None of these men, though, pressed deeply on the metaphysical aspects of technology. It was not until the early 20th century that the topic became the subject of extended inquiry, for at least a few minds. Three of these are particularly relevant: Friedrich Dessauer, Friedrich Juenger,[2] and Martin Heidegger.

My intention here is not to provide a full recounting of their views; this has been done elsewhere, particularly for Heidegger. Nor do I want to offer a decisive critique. Rather, my objective is to trace the development of metaphysical ideas that pertain to the thesis of the Pantechnikon. Each of the three men offers ideas consistent with that outlook, even as each makes significant deviations. I will argue that a pantechnical metaphysic provides a better context for understanding their views and, in particular, shows where they have gone astray.

DESSAUER: THEOLOGIAN, PHYSICIST, METAPHYSICIAN

Born in 1881 in Aschaffenburg, Germany, Dessauer came to maturity as a businessman and devout Catholic. His interest in the newly emerging field of

X-ray technology prompted him to study physics at the university in nearby Frankfurt, earning a doctorate in 1917. He was a prolific writer in several fields, addressing the philosophy of technology in some half dozen books between the early 1900s and the mid-1950s. Like many of his radiology colleagues of the day, he died of cancer, in 1963.

Of particular interest is an early work, *Philosophie der Technik* (1927).[3] It was at this time, in the 1910s and 1920s, that the unique power of modern technology and its union with modern science was becoming apparent to the more perceptive individuals—Dessauer, of course, but also the likes of Max Scheler and Alfred Whitehead, whom I will discuss in Part II. Dessauer, though, was among the very first to inquire into the metaphysical essence of the technical phenomenon. He opens the second part of his book with the following words:

> Now is the time to come closer to [technology's] essence, and to discover how technology is possible. This advance toward the origin of possibility and power . . . leads beyond sense perception. It signifies a step toward critical metaphysics. ([1927] 1972: 317)

We must not be misled by his use of the word 'critical.' He speaks here not in terms of a negative assessment—in fact, precisely the opposite. Dessauer is a boundless optimist regarding technology and perhaps its greatest philosophical proponent of the 20th century. His is a critical metaphysics in the sense of an incisive and penetrating analysis into something intrinsically nonphysical, one that bears directly on human nature itself.[4]

Dessauer's readiness to transcend the physical realm is striking. Perhaps it was his theological background, or maybe his extensive work with those ethereal and mysterious X-rays, that inclined him to think so naturally in transphysical or metaphysical terms about something as concrete as technology. No one before had so clearly articulated a view of technology as something fundamentally beyond the physical world.

Only by viewing technology from a "central standpoint," he says, can we grasp the entire phenomenon and thus begin to understand its essential nature. Doing so "means to approach the essence of technology"; and we thereby grasp "the polymorphic character of an apparently meaningless process [which] can be seen from the vantage point of its essence" (317). Ordinary language is not up to the task, and thus the meanings of key terms must be altered or expanded. "Existence" (*Existenz*) has different meanings in the physical and metaphysical worlds. "Realm" (*Reich*) means not a geographical region but, now, an aspect or dimension of reality. Concepts such as "being present" (*Dasein*), "being-such" (*Sosein*), "dependence-relation" (*Abhängigkeit*), "ordering" (*Ordnung*), "fulfillment" (*Erfüllung*), and "working out" (*Bearbeitung*) gain new importance.

Technology clearly operates within the laws of nature, but it also aids us in transcending these laws, and this fact is a clue to its metaphysical standing. Analogies are useful here. Airplanes allow us to fly, thus opposing gravity; "gravitation is overcome but not negated" (320). Engines magnify human power many times over. Myriad physical limitations are surpassed. Technology accomplishes this, not by copying or imitating nature but by an altogether different process, something supranatural. Planes fly, but not like birds. Automobiles transport us, but not by walking. Ships traverse the seas, but they do not swim. Generally speaking, technology achieves its ends via nonnatural or even antinatural means. "*The ordering of the means*," says Dessauer with emphasis, "*is alien to nature*" (321).

The essence of technology is bound up with the process of invention, and we cannot understand this essence without thinking carefully about the act of inventing. This act is not as men normally conceive it. Inventors do not literally create their inventions *ex nihilo*; rather, they draw forth something that is, in a very real sense, already existent. Inventors themselves understand this. They "see" or "grasp" or "envision" the new device, as if it were presented to them through some mysterious means. The inventor looks within himself, and by a process of "inner working out" (*innere Bearbeitung*), he intellectually seeks the solution to his problem—which is already there, waiting for him, as a kind of transcendent ideal. Because man is imperfect and limited, he cannot fully grasp the ideal but is only able to approach it to a greater or lesser degree. Thus we see a "progression" of technological artifacts, each striving in turn to more closely approach a preexisting ideal form.

As an example, Dessauer mentions the bicycle. He recalls from his youth, circa 1890s, a wide variety of early designs: frames, hubs, wheels, rims, all of varying dimension. These were so many imperfect visions of an ideal. Over time, designs progressed toward this ideal, as it became clearer and more transparent to the inventors. Eventually they approached "the ideal solution, of which there is only one" (322). A similar process holds for all technical inventions, he says; they all proceed along a curve, approaching asymptotically the theoretical ideal.

Thus we have the rough outline of Dessauer's metaphysical schema: Technology exists, originally and independently of mankind, in a nonphysical dimension—a "fourth realm" (*vierte Reich*), as he calls it, to distinguish from the three Kantian realms of nature, morality, and aesthetics. This realm is a vitally important aspect of the universe and indeed of existence itself. Thus, "the essence of technology (*das Wesen der Technik*) emerges as something special; it opens a glimpse into a locked-up depth of Being (*Sein*)" (324). And even more than this: The act of technological invention allows us to *grasp the thing-in-itself*, something that has long eluded us in the other realms.

As we recognize today, a central aspect of modern technology is force or, more precisely, power. The same holds in Dessauer's system: "The great characteristic of the structures of the fourth realm is abundance of power (*die Fülle der Macht*)." When the inventor reaches into this realm to transfer the technological forms back to the realm of nature, "he thereby opens gates through which power flows." A link is established, not to be broken; this power "continues to operate with the inexorability of a law of nature" (324).

Furthermore, the whole process operates *teleologically*, as if planned or fated. Inventors, especially in the early stages of technical development, can be likened to "children who play with the monstrous control panel of fate" (325). The decisive historical role of technology is not of our doing; its significance "does not stem from human planning." We do not control the progression of technology. It is essentially dictated to us by God or fate. We are thus powerless to oppose it. Dessauer presciently remarks that "no one opposes this power because that is already hopeless" (328)—though he means it in a positive sense, as in, Who would want to oppose it? Regardless, it is a clear statement of technological determinism, attributable to the innate power of technology as a literal force of nature.

But clearly we do play a central role in the overall process. We are the creatures fated to enact technical discovery. This is *our* essence, and thus we are intimately connected to the technological essence. Every day, inventors are reaching into the fourth realm and opening up new lines of power, altering the world. Since we are obviously part of the world, "we are transformed along with it." As the world changes, so we change.

Dessauer summarizes his main points with admirable force and clarity:

> [T]here enters in our time, on an overwhelming scale, a metacosmos resonant with power (*ein dynamikdurchpulster Metakosmos*)—the technological world. . . . The power of technology over men is . . . constituted like some power of nature (*eine Macht der Natur*), such as a mountain range or the Gulf Stream. Such things transform humanity from without. People *must* react. . . . [T]hus technical inventions have always altered human existence. The power of a new technical structure is as autonomous as if creation were to be increased by a [new] continent. . . . [T]he power of a newly created form of technology has basically the same autonomy as the generation of a mountain range, a river, an ice age, or a planet. . . . It is a colossal fate . . . continuing to operate with inconceivable autonomous power (*unvorstellbarer autonomer Gewalt*). It is *the greatest earthly experience of mortals.* (330–332)

One is struck by the almost religious sense of awe in the face of this world-changing phenomenon.

Because technology is unique in its ability to allow us direct contact with things in themselves, the fourth realm should rightly take priority in our metaphysical scheme over the other three, which are decidedly inferior on this count. It alone allows us to glimpse "something of the secret of Being." It alone teaches us how to directly access the eternal forms. In doing so, technology will literally pave the way for us to other, perennially more-elusive forms, like beauty, justice, and truth: "technology is the school whereby mankind learns by illustration how a reality of a different sort [operates]" (334).

As such, it is the key to grasping reality. Technology as "a realization of ideas" is the means by which we can "investigate *realization in general*"—that is, of existence in itself. It should thus serve as the very foundation of our philosophical worldview.

Needless to say, this is an exceedingly positive view. "The keynote of such a philosophy," says Dessauer, "is *heroic and optimistic*" (333). It achieves the long-sought unification of the insights of Plato and Aristotle, valuing both the forms themselves and their beneficial objectification and seeing them as deeply interconnected. On the pragmatic side, technology is obviously very useful. It not only provides comfort and convenience but also promises to cure diseases and other causes of human suffering and to generally draw "unimagined fruitfulness" from the Earth. The significance of this fourth realm can scarcely be overestimated.

Readers familiar with Heidegger's outlook will see a number of direct precursors and anticipations, beginning with the very concept of a metaphysics of technology. Apparently no one prior to Dessauer had conceived of such a thing, but only with Heidegger did it become famous. And not only a metaphysic, but something much more: Both men saw technology as leading the way to the very core of Being (*Sein*), and both focused on the "essence" (*Wesen*) of technology as a means of achieving this. In analyzing the situation, both found it necessary to alter the common meaning of many words. Certain terms, such as *Dasein* (being-there), had a long history of philosophical use, dating back at least to Hegel, but both men deployed the term extensively.[5] Both emphasized the concept of "presencing," Dessauer with the phrase *ins Dasein trittt* and Heidegger via *Anwesen*. "Ordering" is present in both writings, as *Ordnung* (Dessauer) and *bestellen* (Heidegger). And finally, the idea that human essence is inextricably connected to the technical phenomenon plays a central role for both men. The parallels are striking but seem to be generally overlooked by Heidegger scholars. Most recent books on Heidegger contain no reference to Dessauer at all, even those that emphasize the question of technology. Academic articles elaborating on this connection are almost nonexistent.

From the perspective of the Pantechnikon, we find many points of comparison in Dessauer. Technology is indeed something profoundly metaphysical, and it is much more fundamental than merely human contrivance. We obviously have a central role to play, but it is more as reagent or catalyst than creator and controller. There are no Platonic forms in the pantechnical worldview, but there is an operative intelligence, a Logos, that produces order in the cosmos; as Dessauer said, the Greeks do offer many important insights. Likewise, technê is a key to the nature of Being and thus of decisive philosophical importance.

And there are other important areas of agreement. Technology, on both views, is indeed about power and the transformation of energy. It is indeed a deterministic entity, operating with virtual autonomy in the world; it is nearly "hopeless" to attempt to oppose it. Therefore it proceeds teleologically, as a kind of cosmic destiny or fate. And on both views, technology does say something vital about our own human essence—and, more, threatens to change it profoundly.

Conversely, Dessauer's "four realms" bespeak a metaphysical pluralism that is at odds with the naturalistic monism that I am advocating. The concept of multiple realms opens up a variety of interaction problems and arguments from parsimony that are avoided under a monistic scheme. Though not explicit, one senses of kind of Hand of God lurking behind Dessauer's metaphysics—again, something completely eliminated in a naturalistic pantechnical universe. And perhaps most notably, we have Dessauer's boundless optimism about technology contrasted with the pessimistic realism of my view. For Dessauer, technology works almost unconditionally for the good of humanity. It has no unanticipated side effects and no negative impact on nature. The only adverse consequence is that it imposes itself on humanity, willingly or unwillingly; but this is a minor inconvenience—more than offset by the myriad gifts it bestows upon us.

In the roughly 90 years since Dessauer wrote his book, we can clearly see the consequences of advancing technology. The evidence does not justify his unconditional optimism. Since that time, humanity has suffered through a(nother) horrendous world war and more than two dozen other major wars and genocide events, resulting in perhaps 100 million deaths—all enabled or caused by modern technology. During the Cold War, and even today, the world sits under the threat of an even more catastrophic nuclear conflict. Chemical and biological weapons, along with other so-called weapons of mass destruction, are directly attributable to advanced technology. Technology has allowed the human population to grow from around 2 billion in Dessauer's day to more than 7.1 billion today, with the virtual certainty of exceeding 9 billion by 2050. The direct and indirect impact on human suffering of this unprecedented growth is difficult to estimate, but it is unquestionably profound. And simply our everyday modern technological lifestyles cause a wide range of physical and psychological ailments, many of which we are only beginning to understand.[6]

Additionally, we have the issue of environmental impact, which, unsurprisingly, Dessauer completely neglects. By nearly any measure, humans are dominating the global ecology and adversely impacting it in multiple ways via our advanced technology. Our patterns of land usage, emissions of pollutants and toxic chemicals, deforestation, and habitat destruction are all taking a toll. Potentially worst of all, climate change could dramatically affect every life form on Earth—a problem directly attributable to industrial era technology. Ironically enough, climate change could ultimately lead to the collapse of the very technological civilization that produced it. Clearly, Dessauer's bright future has yet to materialize.

The evidence, as I will argue, points strongly toward a pessimistic and possibly catastrophic outcome for both humanity and the planet. This is consistent with the pantechnical view of an ongoing universal process of technification under conditions of surplus energy, one that is ultimately far beyond the control of humanity. Empirical evidence can provide many clues about our metaphysics, along with an important justification for our views. Dessauer made substantial and striking progress in his articulation of a metaphysics of technology, but the actual course of events has impressed upon us the need to rethink the subject on a deeper level.

JUENGER: POET, WRITER, PHILOSOPHER

As important as Dessauer's book was, it had little immediate impact. Technicians and engineers had no use for such ideas, and professional philosophers had not yet taken up the subject. Heidegger published his primary work *Being and Time* in 1927, the same year as Dessauer, but at this date was not yet concerned with technology per se; it would be nearly a decade before he began to articulate his ideas on the subject and almost 20 years before he publicly lectured on it. Rather, Heidegger at that time focused almost exclusively on the "question of Being." He did, however, develop a philosophy of tools, tool usage, and of the flattening and homogenizing process known as "the They" (*das Man*) that presaged his later work in the philosophy of technology.[7] Unfortunately, Heidegger's difficult and complex language and his contorted presentation technique detracted substantially from the book's value.[8]

Jose Ortega y Gasset was among the first formal philosophers to examine technology, beginning in the mid-1930s. But his was more of a pragmatic analysis than a metaphysical search for a technological essence. He did touch on some relevant themes, as in the following passage from his essay "Man the Technician," in which he connects the human and the technological:

> [M]an begins where technology begins. . . . This is why I have previously emphasized that the meaning and the final cause of technology lie outside itself, namely in the use man makes of the unoccupied energies it sets free. The mission of technology consists in releasing man for the task of being himself. (1941: 117–118)

In noted contrast to Dessauer—and more in line with Scheler and Whitehead—Ortega was very skeptical of the technological "promise":

> [T]echnology, for all its being a practically unlimited capacity, will irretrievably empty the lives of those who are resolved to stake everything on their faith in it and it alone. . . . Just because of its promise of unlimited possibilities, technology is an empty form, like the most formalistic logic, and is unable to determine the content of life. That is why our time, being the most intensely technical, is also the emptiest in all human history. (151)

But it is a mild critique and the essay only marginally a metaphysical inquiry.

It was at this time, circa 1939, that Juenger undertook his analysis of technology. As the younger brother of the prominent writer Ernst Juenger, Friedrich—born 1898—was exposed to the literary and intellectual world of Hannover from an early age.[9] Politically aware since his teenage years, he was appalled at the outcome of World War I and particularly the onerous burden placed on Germany at Versailles. Friedrich was a nationalist but opposed the rise of the National Socialists in part because of their glorification of and faith in modern technology.[10]

In the late 1930s, Juenger drafted an extended manuscript that he called "*Illusionen der Technik*"—"Illusions of Technology." But with his nation at war, publication was delayed. Eventually he arranged a first printing with Klostermann's of Frankfurt, in 1944, as "*Über die Perfektion der Technik*." Unfortunately a bomb strike destroyed nearly the entire run; only a few rare copies remain. Finally in 1946 the small booklet appeared once again, with the streamlined title *Die Perfektion der Technik*.

The book name is potentially deceptive, at least for English readers. A literal translation, "The Perfection of Technology," does not do justice to Juenger's vision. German *perfektion* means something like purification and a completion or final realization. As technology approaches *perfektion*, it realizes itself by subsuming and ultimately displacing the organic, the natural, and the truly human. It would be a monumental failing of humanity to allow such a thing to occur, according to Juenger; hence the more appropriate English title, *The Failure of Technology*.

As commentator Frederick Wilhelmsen explains in the introduction to the book, Juenger seeks and (arguably) attains a properly metaphysical goal: "he has shown us the essence of technology" (vii). This essence, though, is not something abstract or ethereal like Dessauer's Platonic forms; it is concrete and tangible. Technology's essence resides in its *actions*, what it *does*, and how it affects humanity and the Earth. There are classically metaphysical themes throughout the book, but they are defined pragmatically, in terms of the effects that they have. And it is precisely these effects—almost uniformly negative—that serve to confirm and justify Juenger's larger conception of the essence of technology.

At the highest level, technology is defined by two essential qualities: automatism and an all-embracing universalism. On the first point, Juenger sees technology as an autonomous force in the world, one that has grown beyond our ability to control. This Frankenstinian monster rose from an innocent beginning—the simple need to lighten our burden of manual work—to become a dominating force in modern times. Technology started out in the form of crude but functional tools, which, over time, became increasingly precise and effective. As their capabilities progressed, our tools became mandatory accompaniments of human labor. Technology spread ever further, evolving into machines—at first powered by men and animals, then engine powered, and ultimately becoming fully self-sustaining and automatic. As automation spread and industry grew, more humans and more of the planet were drawn into the technological system of production. But as he rightly observes, in order to produce more, the system must *consume* more—more of humanity and more of nature. The result is ecological plunder and human alienation.

As an autonomous force, technology is best represented in the various automatic machines and automated factories, Juenger believed. Here we see its nature most clearly: a machine that "lives." But of course it is not truly alive, but rather something else—something hideous and monstrous, a "dead life," a form of the living dead. And this dead life has come to dominate us all. "The machine . . . gives the impression that something lifeless penetrates into, and permeates, life. This is what the observer senses and what evokes in him ideas of age, coldness, death" (36). The man who succumbs to mechanism becomes himself a walking dead man, "[one] of those vampire-like beings, endowed with a semblance of life, who may be justly called human automatons" (119). The vast cloud of dead life known as modern technology has settled over the civilized world, saturating and smothering it—like the poison gas that crept into the trenches of World War I.

The autonomy of technology is perhaps its defining characteristic. "Technological progress," he says, "is synonymous with an increase of all kinds of automations" (33). Indeed, "we are surrounded by an automatism towards which all branches of technology are developing." Since it runs automatically, the technological system functions as an independent entity in the world, complete with its own plan and its own ends. At one level it is driven by purely natural laws and causal, mechanistic forces; but on another it functions purposefully and teleologically; "the causal and the teleological viewpoints [are] equally valid" (67) as applied to technology. Juenger is succinct on this point: "technology pursues its own ends" (74). These ends may be circumscribed under a single heading: absolute mechanization of the world: "technology . . . aims at complete automation (*einem perfekten Automatismus zustrebt*)" (83).

And it is succeeding. We know this because it has penetrated all aspects of society. This is an astonishing fact, given that technology began as so many diverse and simple tools, and now has become a monolithic and integrated phenomenon. Universalism reigns in the technical realm. Juenger writes:

> It may seem strange that this titanic modern industrial system with its human organization that tries to engulf everything, and whose power we encounter at every step, should have grown from seemingly unconnected trials and errors, from widely scattered inventions, from decidedly humble beginnings. But the convergence of these inventions is only an expression of the convergence of a way of thinking which is absolutely uniform, no matter what its point of origin. Wherever this thought goes to work, its every manifestation contributes to the mechanical arts all over the world. (71)

Among other outcomes, it drives social order: "Technical thinking, imbued with an unlimited drive for power (*unbegrenztes Machtstreben*), acts imperiously and recklessly. Full of unshakable faith in organization, it promotes and expands organization in all directions" (80). It dominates science ("science becomes the servant of technology" [91]), politics ("technology usurps the state" [104]), and higher education ("the university becomes a technical training center and servant of technical progress" [100]). It drives forward relentlessly, growing and expanding its reach: "technical progress attacks everything that is at rest, everything that possesses permanence and stability, everything that does not lend itself to technical progress or withdraw from it. Technical progress attacks everything that denies it the resources . . . which it craves to devour" (90).

As expected, the effects are universally detrimental. Regarding humanity, it results in an alienated, depressed, and detached human being, one who is a virtual slave to the clockwork demands of the system. The alleged benefits of technology accrue primarily to the elite few, the managers, who personally profit from its excess production and manufacturing efficiencies; for the masses it dispenses only poverty. It promotes both physical and psychological illness, via unhealthy living and working conditions, poor-quality "industrial" food, and modern diseases like diabetes and cancer—indeed, asks Juenger, what is cancer but "autonomous growth" within the human body, a process that mirrors the cancerous growth of the technological system? Our future, he predicts, is one of "exhaustion, apathy, and dull depression" (178).

But technology also damages the nonhuman world of nature. Here Juenger was a true philosophical pioneer. He makes extensive comments on the ecological destruction wrought by technology and thus offers an early form of an environmental ethic—and in fact seems to argue for intrinsic value in nature, a notable accomplishment.[11] As noted earlier, the industrial system produces more only by consuming more—more of nature and its declining natural resources. "It is a ruthless destruction, the like of which the earth has never before seen. A more and more ruthless destruction of resources is the characteristic of our technology. Only by this destruction can it exist and spread" (23). "Since technology presupposes destruction," he adds, "it cannot be fitted into any healthy economic system." It is intrinsically dysfunctional and unsustainable.

Pollution is unavoidable: "Technology darkens the air with smoke, poisons the water, destroys the plants and animals. It brings about a state in which nature has to be 'preserved'" (24). This veritable "pillage of the earth . . . carries within itself the seeds of its own destruction" (85). The scientist, armed with potent technological instruments, "begins to squeeze and torture nature" (108) in search of her secrets. As technology's influence grows, it will threaten the entire globe: "Since technology['s] progress spells the progressive pillage of the earth, it is obvious that in a state of perfection it will practice the most complete and most intensive exploitation on a planetary scale" (177). But nature will not stand idly by; it will respond in kind. "[N]ature answers the conquest of technology by a counter-invasion of its own; as we destroy it, it destroys us with the elemental forces we think we have captured" (133).

For all this gloom and woe, Juenger does not surrender to the technological Moloch—but neither does he offer much of an action plan. The complete technification of the world will not happen, he says optimistically, because human nature inherently surpasses pure mechanism, and because such a global system would collapse under its own excesses before it could monopolize the planet. He calls for neither a revolution against technology nor any retrograde, back-to-the-land social movement. "Nothing is further from my mind than the romantic rejection of technology" (152). Rather, he asks simply for something vague and obscure, and clearly Heideggerian: "watchfulness (*Wachsamkeit*)," he says, "must be practiced by every man of spirit who wants to go on feeling that he is more than a mere cog in a gigantic machine" (153). *Watchfulness*—as if this were a sufficient response to the life-destroying power he has just spelled out.

It is an oddly passive conclusion to such a forceful analysis. Yet Juenger closes the book with his convictions intact: "Technology's striving for power is unbroken," he reminds us (204). "Modern man wakes up to the fact that the elemental forces he has enslaved in his machinery are turning against him with ever growing, viciously destructive force" (187). Pay attention, he seems to say to his fellow man; watch the system, see what it will do, be aware. Only when humanity is sufficiently awakened can we hope to achieve salvation.

Clearly Juenger's metaphysical system is largely implicit. He makes no formal attempt at such analysis. The negative consequences of technology are the prime concern, and they themselves provide an outline of the technological essence. *The Failure of Technology* is undoubtedly a work of lasting importance. In many respects it anticipates the later work of Ellul—with its emphasis on technological autonomy, universalism, self-augmentation, monolithic character, and insatiable growth. In fact, one can make a case that it is more closely aligned with Ellul than with Heidegger, who apparently was its most direct beneficiary.

The influence on Heidegger, however, cannot be ignored. We know that Juenger and Heidegger exchanged letters beginning in 1941 or 1942, and we have some evidence that Heidegger read at least a draft copy of Juenger's book.[12] Hence it is virtually certain that Juenger influenced Heidegger's thinking on technology—perhaps most critically, in the idea that technology posed a threat to humanity, that it was something profoundly dangerous.[13] Not only this, but we can plausibly infer that Juenger's mild solution of "watchfulness" became translated into Heidegger's equally mild notion of "keeping watch over": "Everything, then, depends upon this: that we ponder this [technological] arising and that, recollecting, we watch over it" (Heidegger 1977: 32). Juenger arguably provided Heidegger with the essentials of both the problem and the solution.

As with Dessauer, it is instructive to view Juenger's work from the pantechnical perspective that I advocate. If in fact technology is a universal process of unfolding and ordering, Juenger was then correct to grasp its general character as that of an unstoppable force that is universally present. And because the Pantechnikon functions with the power of natural law, it indeed seems to operate "automatically" and thereby becomes manifest in autonomous machines and industrial processes. Furthermore, it is entirely appropriate to describe the process as teleological, given that the cosmos is biased toward order and strives relentlessly to that end.

But also in Juenger we have one of the first elaborated critiques of the dangers of technology. As mentioned, others had already highlighted the threat in a general way, including Scheler, Whitehead, and Ortega. One could also mention prior critical works by Weber (1905), Veblen (1906), Spengler (1931), and Mumford (1934)—though again, these were relatively mild criticisms and relatively undeveloped in their analyses of the dangers. Equally striking, though in an opposing way, is the weakness of Juenger's "watchfulness" response. That the cosmos as a Pantechnikon presents a danger to humanity is clear; it is a kind of tsunami of complexification, one that both swept us into power and now threatens to sweep us away, using our own tools to do so. One does not deal with a tsunami by watching it. One plans in advance, prepares countermeasures, moves to higher ground—*moves out of the way.* But Juenger will have none of this; no "romantic rejection of technology" for him. From the pantechnical standpoint, Juenger's response to the crisis is wholly inadequate.

Regarding the threats to nature, Juenger was prescient. He presses us to reconsider the dangers of our own time. If anything, he underestimated the risks, given the very real possibility of such present day technocatastrophes as nanoreplicators or a world-altering singularity. Compared to such nightmare scenarios, global climate change is all but inconsequential.

And yet one may ask: If pantechnical progression is a natural process, and indeed is nature *as such*, and if the cosmos strives toward "the better," how can it also be detrimental to nature? The Pantechnikon brought forth our planetary ecosystem, with its tremendous breadth and diversity of life; why would it destroy that? The answer, I suggest, relates to the very process of complexification. Higher levels of structure order and organize lower levels—sometimes benignly, sometimes destructively, but always with a loss of autonomy for the lower orders and a gain for the highest.

We can imagine different scenarios of future evolution of the Earth—some in which nature is benignly harnessed for the sake of a higher order of complexity, and others in which the biosphere is swallowed whole in a kind of retrograde devolutionary act. The future may mark an improvement from the perspective of the universe as a whole, but in none of the possible scenarios does it get 'better' for existing structures. After nearly 14 billion years of evolution, the Pantechnikon has produced both a living Earth and an exemplary creature, *Homo sapiens*. From our perspective, the global system is optimally evolved. *It cannot get better; it can only get worse.* Our primary duty is not to the universe but to ourselves and to the present planetary system that we depend on. This must be the focus of our future actions. To act otherwise could open the way to either enslavement or extermination.

HEIDEGGER: EXISTENTIALIST, ORACLE, SAGE

Any road to a viable metaphysic of technology must pass through a tiny village in southwest Germany. It was there, in a small cottage in Todtnauberg, a few kilometers from Freiburg, that Heidegger conceived many of his groundbreaking ideas—including those that would be published in 1954 as *Die Frage nach der Technik*, or 'The Question Concerning Technology' (QCT).

When his first book, *Being and Time*, came out in 1927, it was hailed as a brilliant work, at least within the German-speaking world. Not so in Anglo-American circles. Dummett (1978: 437) remarks that, in the 1940s, "Heidegger was perceived only as a figure of fun, too absurd to be taken seriously." This view continued throughout the 1950s and 1960s, but interest picked up in the 1970s as he neared the end of his life—he died in 1976 at age 86. From the 1980s on, though, his reputation accelerated; the number of scholarly English-language books about him has risen dramatically, from around 60 published in the 1970s to more than 200 in the 1990s to more than 300 since the year 2000. And yet there is scarcely a philosopher in history more difficult to comprehend.[14]

Heidegger's early work on Being is promoted today by some as his most important accomplishment and even as one of the peaks of 20th-century philosophy. Graham Harman, for example, states that "Heidegger deserves to be called the greatest philosopher of the past century" strictly on the basis of his early achievements. Of his later work on technology, however, Harman is appalled: it is "sadly monotonous," little better than "a gloomy drama in which every invention merely strips the mystery from the world and turns all things into a manipulable stockpile of present-at-hand slag." Generally speaking, Heidegger "is horribly overrated as a philosopher of technology" (2009: 112).

There is some justification for such a view, as I will explain. For my part, though, I am inclined to reverse this pronouncement. As I see it, Heidegger's early writings on Being are tedious beyond measure and his accomplishments there meager. His philosophy of technology, at least as articulated in QCT, is also excessively pedantic but does contain sufficient insight to merit discussion and is a significant milestone on the path toward an understanding of this most important of contemporary phenomena. I tend to agree with John Caputo, who stated that Heidegger's "critique of technology is the most powerful part of his work, the part where everything he has to say comes to a head" (1988: 542). Not the best of the 20th century but rather the best that Heidegger has to offer.

Given the intense interest in the man in recent years, it is clearly unnecessary to provide yet another extended discussion of his approach to technology. Works such as Zimmerman (1990), Dreyfus and Spinosa (2003), Malpas (2006), Cerbone (2008), Ihde (2010), Watts (2011), and Wisnewski (2013) all do an adequate job. I do, however, need to give a concise overview of the main themes—both for the benefit of the uninitiated reader and because I think there is a logical framework to the essay that has not yet been covered by other commentators. Most importantly, an overview allows us to better assess Heidegger vis-à-vis the pantechnical framework.

QCT contains virtually the whole of Heidegger's outlook on technology. It is a modest-sized essay, about 8,800 words in original. I suggest that his scheme is best viewed through four cardinal terms: Revealing (*Entbergen*), Bringing-forth (*Hervorbringen*), Ordering (*bestellen*), and Enframing (*Gestell*). Like the points on a compass, they map out the terrain. They provide a steady point of reference as we traverse the often murky landscape.

Taking each of the four in turn: 'Revealing' refers to the process by which things come into being. Every concrete thing that appears in the world comes about by some process of change. Animals are born, buildings are constructed, statues are sculpted, mountains rise up. In some sense, all extant things were hidden away in the background of the cosmos, unseen, before someone or something revealed them to the world. In revealing, the latent becomes actual, the implicit explicit. Heidegger also uses words like "presencing" (*Anwesen*), "disclosing," and "unconcealment" as synonymous for this process—all involve bringing into the light of day something that has previously been hidden. Once revealed, things possess an actuality or reality in themselves; they are now an aspect of "truth," they are true facts of existence. Revealing is thus equated with truth. Heidegger deploys the Greek *alêtheia* (uncovered, exposed) to express this notion of truth.

There are two forms of revealing: Bringing-forth and Ordering. Bringing-forth is the original mode of revealing; for virtually all of cosmic history, it was the *only* mode. Heidegger sees it as the process of "making" in a straightforward sense, and thus he equates it to the Greek *poiêsis*—recall the discussion in Chapter 2 on Aristotle's theory of technê.

There are, in turn, three forms of Bringing-forth: simple technology, such as handicrafts; art; and nature. The first two are obviously within the human realm, as they relate to objects, artifacts, and works of fine art that we conceive and construct. Heidegger thus classifies them as forms of technê—essentially agreeing with Aristotle that technê is, by definition, a human construction. The third form of Bringing-forth, nature, encompasses the entire natural world: plants, animals, clouds, stars, and planets are all examples of objects brought forth by the universe. Not being man-made, such things obviously do not qualify as technê. There being no counterpart to technê in the natural realm, they are simply "nature."

The pleasant intonation of "Bringing-forth" is deliberate. It is portrayed as a gentle, benign, innocent process. It is almost as if the veil of the cosmos quietly parts, and there on the stage before us, something new appears. Bringing-forth is unthreatening, noncoercive, a danger to no one. In fact it is an unalloyed blessing—identical to creation itself.

Not so for the other form of revealing. Ordering is, as one may infer, something forceful and harsh. It is the imposing of a structure, a wrestling of unruly nature into a systematic mold. Heidegger calls it a "challenging revealing" (*herausfordernden Entbergens*). Unlike the diversity of forms of Bringing-forth, there is only one mode of Ordering, namely, modern technology. Being man-made, modern technology is of course also a form of technê. Technê thus cuts across the boundaries, having both benign (Bringing-forth) and forceful (Ordering) manifestations.

The essence (*Wesen*) of modern technology is naturally the focus of attention. Heidegger names this essence as *Gestell*, normally translated as 'Enframing' but occasionally as terms like 'Framework,' 'posure,' or 'construct.' *Gestell* is commonly described as a structuring device, something like a bookrack or a system of shelves. It takes an otherwise disordered and chaotic collection of things and puts them in order. Enframing is both the essence of modern technology and an Ordering, which is a challenging form of revealing.

These are the guiding concepts of the work. With these in place we are well situated to look at his specific claims. Heidegger begins by debunking one of the most common myths of technology: that it is neutral, that it is just a tool, that its goodness or malevolence lies only in how it is used.

> Everywhere we remain unfree and chained to technology, whether we passionately affirm or deny it. But we are given over to it in the worst possible way when we regard it as something neutral; for this conception of it . . . makes us utterly blind to the essence of technology. (1977: 4)

Modern technology is a completely new phenomenon; it has its own imperatives and own demands. It carries its own set of values, which may be utterly opposed to those of humans or nature. By viewing technology as simply neutral, we completely miss this fact and thereby become vulnerable to manipulation.

Modern technology is fixated on energy—its production, exploitation, and use. Being so, it seeks to draw everything in the world, including humans, into the sphere that Heidegger calls "standing reserve" (*Bestand*). By this he means things taken merely as sources of energy, ready at hand, awaiting orders for use. As such, technology both plunders the natural world and dehumanizes mankind—and not by accident but by its very nature. This is how Enframing operates in the world.

Modern technology, he repeatedly emphasizes, *is* Enframing—the process by which technology structures the world into standing reserves of energy. It places demands on people and nature to restructure themselves according to the technological imperative. It orders, and it compels.

As a manifestation of Enframing, technology is able, Heidegger says, to hide its true nature. It appears to us in so many benign forms: medicine, transportation, information, heat, light, entertainment. It grows food, produces useful consumer goods, creates profits, makes life easier. It displays itself in innocence. We come to see it as merely a tool, to be used as we wish, for good or for evil. Yet as hard as we might try, we inevitably look past its essence; technology "keeps itself everywhere concealed to the last" (22).

This hidden nature of modern technology, combined with its insidious power to convert and debase everything it touches into a standing reserve of energy, makes it dangerous. But not an ordinary danger. Chainsaws and handguns are dangerous, but technology per se is an order of magnitude beyond. It is "*danger as such*"—that is, the very essence of danger. It presents a profound and even existential threat to the world: "it is the supreme danger (*die höchste Gefahr*)." Heidegger elaborates:

> As soon as what is disclosed no longer concerns man even as object, but does so, rather, exclusively as standing-reserve, and man in the midst of objectlessness is nothing but the orderer of the standing-reserve, then he comes to the very brink of a precipitous fall; that is, he comes to the point where he himself will have to be taken as standing-reserve. Meanwhile man, precisely as the one so threatened, exalts himself to the posture of lord of the earth. (1977: 27)

We deceive ourselves into thinking we are the masters of our technology. But its true nature, Enframing, is hidden from us. We see technology as merely neutral. It is not. In reality, *it* dictates the terms of engagement. It orders, we obey.

The power of modern technology and its hidden nature are such that it imposes a fate upon humanity; it "holds complete sway over man." And yet that fate is not sealed. In grasping the essence of technology, we are opened to a "freeing claim"—an escape route, as it were. We must see the danger for what it is, and then consciously and deliberately observe the unfolding of the technical process. Enframing is consequently both a danger and the source of our enlightenment; not only is it the ultimate danger but also, ironically enough, the "saving power" (*das Rettende*). As Heidegger says, we must "begin to pay heed to the coming to presence of technology," and in doing so we are "holding always before our eyes the extreme danger." We must "watch over it"—explicitly recalling, of course, Juenger's "watchfulness."

And then what? We don't know. Heidegger suggests that by deeply absorbing the truly monstrous nature of modern technology, we will somehow find a way to coexist with it. He offers up the fine arts as one such means, as a way of tapping into a pure, shining form of creativity that will aid our salvation. But this is only a shot in the dark—a dark time in which we have few viable alternatives. Otherwise we give ourselves over to the essence of technology, allowing it to achieve a kind of total domination of revealing, that is, of *all truth*: "the frenziedness of technology may entrench itself everywhere to such an extent that someday, throughout everything technological, the essence of technology may come to presence in the coming-to-pass of truth" (1977: 35). Technology may become the final truth of the world, the ultimate fact of existence.

ASSESSING HEIDEGGER

Such, in brief, is Heidegger's analysis. It is an impressive accomplishment. But how valuable is it? How much wisdom and true insight does it contain? We notice a few things immediately. First, it is a pure metaphysical analysis; no data, no statistics, no calculations, no experiments, not even citations. Such works are always difficult to assess; we must rely on the plausibility, coherence, and intuitive strength of the claims. But there is also the linkage to the real world. Metaphysics is not fiction. There must be some tangible consequences, some practical implications, if it is to be meaningful.

A number of questions arise regarding this work. First, Heidegger treats modern technology as a collective whole and never distinguishes among the vast array of devices and kinds. Granting that modern technology must have some essence or essential nature, why is there only a single essence encompassing the whole diverse concept?

Second, and relatedly, what is "modern" technology? At one point Heidegger refers to "modern machine-powered technology" (*die moderne Kraftsmaschinentechnik*), which presumably came about with the advent of the steam engine and coal-fired furnace, circa 1700. He then mentions the idea that modern technology "is based on modern physics" (1977: 13–14). And then we also have the inference that, due to Enframing, all modern technology makes energy demands. So here we have three characteristics—Which is definitive? Is any one sufficient, or must all three be present? More to the point: Is every product of modern industrial processes an instance of "modern technology"? A bicycle? A child's toy? A paperclip? A bar of soap? If these things are *not* modern technology, and yet they are neither art nor craft, *what are they?* Recall that all of human technê must fall under one of the three designations, according to Heidegger.

Hence we have a central issue: where to draw the line for "modern" technology. This is both a present-day and a historical issue. We would like to know (a) which products today count as modern and (b) when in the past they first appeared.

Third, modern technology's essence, Enframing, seems to enter the world, whenever it does, as something entirely new. People commonly believe that something "incomparably different" (*unvergleichbar andere*) appeared with modern technology. But this is an illusion, says Heidegger. Even though the devices are new, the essence is not. This is so because Enframing is more than the essence of modern technology; it is a means by which Being (*Sein*) reveals itself. We know this from the short affiliated essay "The Turning," in which Heidegger states that "Enframing [is] one among Being's modes of coming to presence" (37). This is consistent with his prior claim that Enframing is one form of Revealing—after all, what is "revealed" but Being itself, in a particular manifestation? With the coming of modern technological machines, circa 1700, "Being . . . has adapted itself into Enframing." In other words, the *appearance* of the mode is new, but the mode itself, like Being, is eternal.

Even granting this, we still have a problem: Why did the Enframing mode stay hidden for so long? Why did it take nearly 14 billion years of universal evolution for it to appear? What was so cosmically significant about Thomas Newcomen, in 1712, boiling some water in a metal container to drive a piston? Or James Watt in 1769, when he invented the double-action steam engine? By all conventional accounts, these were just two more logical steps on a very long road, reaching back at least to the ancient aeolipile of Hero, circa 60 AD.

But for Heidegger, the steam engine was an event of profound ontological importance—a previously unseen mode of Being appears for the first time ever. The only comparable events were when early humans first created simple stone hand tools (more than 2 million years ago) and primitive artistic renderings (some 30,000 years ago). Prior to these events, there was only nature as the mode by which Being revealed itself. But recall: These three—nature, stone tools, simple art—are all of a kind, all Bringing-forth; hence their appearance marked no world-changing break in the cosmic order. Modern technology *did* mark such a break; for the first time *ever*, a mode of revealing made itself known—a mode that had been lying in wait, silent, for eons.

We may be excused for finding this implausible. Any purported sharp break in the ontological order should call for focused attention and for sustained justification. When we survey the natural world, we generally see continuity throughout; sharp breaks are rare or nonexistent. We do see cases of rapid change, but of rapid *continuous* change, and this is not a break. To salvage Heidegger's scheme, we would have to postulate some component of Enframing in all technology, and indeed in all revealing. Then we could allow "modern" technology to be defined as that which demonstrates a preponderance of the universally present Enframing mode. This is both more satisfying and more consistent with reality; an ancient campfire, for example, seems to "set upon" nature in such a way as to demand energy resources. The same holds with primitive hunting tools, which would seem to challenge nature (wildlife) to supply energy (food). This was not Heidegger's view, but perhaps it should have been.

Other questions arise: Why does Enframing have the nature of a challenging revealing, one which demands everything be taken as an energy resource? Heidegger seems to take this as a brute fact of existence. But even granting this, why is it so dangerous? Doesn't humanity need and rely upon sources of energy to sustain itself? The answer here seems to have two parts: First, Enframing debases everything—including humanity—by treating things only as standing reserve. A debased existence more than offsets the superficial benefits of more energy consumption. Second, Enframing hides not only itself but also the other, the "good" revealing, Bringing-forth. It thereby monopolizes all truth, or rather becomes the *only* truth—and this is the ultimate danger.

And then there is the question of the saving power: Why does it reside in the very source of danger, in Enframing? This is highly counterintuitive; one would expect to find the saving power in some oppositional entity, like nature or wilderness. Apparently the answer lies in the claim that Enframing opens a kind of window onto Being itself. The very power and danger of Enframing indicates that it is an exceptionally clear window, a direct access to the heart of existence. We need this because modern man has forgotten Being, and only by returning to it can we understand our essential human nature (*Dasein*) and thereby restore our dignity. Rather like staring at the sun, we need to peer directly at Being and the process of revealing, but without succumbing to the danger—quite a challenge that Heidegger has lain at our feet.

There are several other issues we could press upon: If Enframing is the essence of Ordering, what is the essence of Bringing-forth? Why is technê strictly a human endeavor? What marks it out for special metaphysical status over the technical acts of other animals? Why are there precisely 2 forms of revealing, and not 3, 4, or 20? Why is it impossible for nature to exert a challenging revealing? What is the role of the mind, how and where does it appear, and why does it have no evident bearing on the question of technology? Is it really appropriate to put so much emphasis on etymology as a source of clues regarding the nature of Being? It seems that Heidegger has produced more questions than answers.

Perhaps we ought not be overly demanding. The real question is not whether he has created a flawless and airtight metaphysical scheme—which is certainly impossible—but whether he has captured a significant degree of insight and wisdom. If we claim he has, we then need to justify that claim. And, if there is valid insight but we can now, with 60 years of hindsight, see the flaws, then we ought to be able to improve upon his system.

What, then, are his key insights? I would propose the following: (1) Technology is something that acts through humans but is ultimately beyond human control; (2) technology is intimately connected to Being, in the sense of an association with the fundamental workings of the universe; (3) the Greeks had significant insight into the phenomena of both technê and logos; (4) nature and human technê are on a continuum; (5) technology is essentially about power and energy; (6) technology acts in a quasi-deterministic fashion; and (7) technology poses a significant danger to humanity. All of these are consistent with a pantechnical metaphysic.

Where does he fall short? From my standpoint, the following corrections must be made:

(1) Modern technology is not fundamentally different from premodern or traditional technology, and it represents no breach of the cosmic order. There is no new mode of Being that makes itself known there. Any attempt to identify a breakpoint or decisively "modern" technologies will fail.
(2) Technology is does not appear solely through human action. Animals, plants, and all life forms participate in technê. And more than this, all of nature produces structures via a process that is continuous with human technê.
(3) The "essence" of technology is the action of a universal Logos on the energetic substance of material reality via a process that we may call technê. It is the deployment of matter and energy to create higher levels of complexity and order wherever this is possible—namely, where matter is fluid, in some form, and energy abundant. To call this "Enframing" is an adequate and acceptable first approximation—but only an approximation.
(4) Environmental well-being is of vital importance. Heidegger gives scarcely any indication of ecological awareness. This is a notable omission, especially considering that Juenger was so attuned to the problem. The impact of technology on nature must be a component of our ethical outlook, both for instrumental reasons and because an enlightened ethic will see intrinsic value in the nonhuman world. Empirically speaking, advanced technological societies inevitably damage their local environments and disrupt the global ecosystem; this fact must figure into our metaphysical analysis.
(5) The role of mind is decisive and cannot be ignored. The universal push for order and complexity is a manifestation of intelligence, and this intelligence is built into the objects that it has brought forth. The order and complexity of a system or object is a measure of the intelligence intrinsic to it—that is, to the intensity of its mind. Modern technological systems embody high degrees of intelligence, function intelligently, and thus exert an 'intellectual,' or psychological, effect back upon their users.
(6) There is no "saving power" in technology. Heidegger was an optimist. Further promotion of advanced technologies only heightens the risks to us and to the planet. Yes, it is impressive to see these new technologies in operation and exciting to possess them ourselves; it is truly a kind of insight into Being. But it is also like looking down into an active volcano—bad policy. Even worse: Our technophiles are presently drilling deep holes down into the magma, creating the risk of newer and larger volcanoes. They are tapping into power sources that they cannot control. The wise choice in such a situation would

be—*stop drilling*. That is the real saving power: the power to stop. And if it is true that the eruption will some day come of its own accord, so be it. But only a foolish society would hasten that day.

"ONLY A GOD . . ."

In 1966, Heidegger gave a rare interview to the German periodical *Der Spiegel*.[15] Among other topics, he elaborated on his views on technology, with the benefit of some 12 years' experience since the initial publication of QCT. Remarking on the political ramifications of the growing technological phenomenon, he said:

> During the last 30 years it has become clearer that the global movement of modern technology (*die planetarische Bewegung der neuzeitlichen Technik*) is a power whose magnitude in determining our history can scarcely be overestimated. It is for me today a decisive question as to how any political system—and which one—can be adapted to this technological age. I have no answer to this question. I am not convinced that it is democracy. (Heidegger [1976] 1981: 55)

Present political systems, he argues, fail to confront the problem of technology precisely because they believe it to be something safely under human control. "In my opinion this is not possible," he says. "Technology in its essence is something that humanity, of its own accord, cannot master." Technology continually presses upon us; "[it] increasingly dislodges man and uproots him from the earth." It has spread throughout society, to the point where "all our relationships have become merely technical ones."

For Heidegger, this fact is a cause for growing pessimism. Neither man nor philosophy can help. Heidegger admits that he himself can do nothing. In the most famous line of the interview, he declares, "Only a god can save us (*Nur noch ein Gott kann uns retten*)." This indicates something of the desperation and the gravity of the situation. But save us from what?—he does not say. When pressed for an answer by the interviewer, he simply replies, "I cannot help you." He has neither a concrete sense of where we are heading nor a notion of how to prevent it. The saving power is nowhere to be seen. One might have hoped for better.

How, then, are we to assess Heidegger? He expressed numerous important insights into technology but surprisingly little in the way of originality. I have focused here on the link to Dessauer and Juenger, but Heidegger drew from other thinkers, most notably Nietzsche. Though he made only a few passing critical comments on technology,[16] Nietzsche articulated the concept of the will to power (*der Wille zur Macht*) as a universal quality of existence, and this had a major influence on Heidegger. Common opinion notwithstanding, it is not only humans who express or embody such a will, in Nietzsche's system;

all life forms, and in fact all of nature, manifest it.[17] In *Beyond Good and Evil*, he defines all physical forces "unequivocally" as expressions of will to power: "The world seen from within . . . would be 'will to power' and nothing else" (sec. 36). This view is repeated in his writings published as *The Will to Power*. He wrote, with emphasis: "*This world is the will to power—and nothing more!* And you yourselves are also this will to power—and nothing more!" (sec. 1067).[18] And again: "[T]he innermost essence of Being (*Wesen des Seins*) is will to power" (sec. 693). Heidegger was well aware of all these passages. If Being is the will to power, and if the essence of technology is equated with or is a mode of Being, we can then see a basis for Heidegger's three-way connection among technology, power/energy, and Being.

But Nietzsche had only an indirect influence on the technology question. Of Heidegger's seven key insights listed earlier, the first six can be explicitly found in Dessauer's *Philosophie der Technik,* which came out 27 years prior to QCT. The seventh—that technology is profoundly dangerous—is the central theme of Juenger's book. These ideas, combined with the related writings of Scheler (1915), Spengler (1931), Jaspers (1931, 1949), and perhaps even the likes of Mumford (1934) and Orwell (1937), cover much of the ground that Heidegger claims as his own. Heidegger's philosophical sophistication clearly exceeds that of his two main predecessors, of course, and his terminology is certainly more complex, but in his case this is no virtue. He made the same points, less clearly, and despite the benefit of writing after the catastrophe of World War II. Still, he was obviously able to impress world opinion to a far greater degree than the others were; had it not been for him, the philosophical discourse would assuredly have been substantially different.

In the end, it must be said that Heidegger was not a very good philosopher of technology, and frankly not a very good philosopher at all. Anyone who can declare that intelligibility is suicide for philosophy scarcely deserves serious consideration. And yet, for a variety of reasons, he was and remains a figure of some importance. Amid the obscurity, he did express valid and important insights on the problem of technology. For this fact alone we owe him at least a nod of recognition.

And so we have paid our due. We made a stop at that small cottage in the hills of southern Germany. We stayed a bit. We had a cup of tea. We looked, we listened, and now we move on.

NOTES

1. For a brief overview of these individuals, see Mitcham (1994: 20–29).
2. I opt for the usual English spelling in place of the German "Jünger."
3. Little known in the English world, there exists only a partial translation in Mitcham and Mackey (1972), under the title "Technology in Its Proper Sphere."
4. And see my discussion at the end of Chapter 10.

5. Recall that both *Philosophie der Technik* and *Sein und Zeit* came out in 1927, and thus there was a simultaneous emphasis on *Dasein*. Dessauer, however, used it in reference to the philosophy of technology, whereas Heidegger did not; '*Dasein*' appears only twice in his seminal essay "The Question Concerning Technology."
6. I examine some of these problems in Chapter 11.
7. See Ihde (1979) or Dreyfus (1984) for an elaboration of the connections.
8. In English, *Being and Time* is a 400-page monument to obscurity. By all accounts it is scarcely better in German.
9. Both Juengers had some influence on Heidegger. Oddly, though, he himself suggested that Ernst had the larger effect on his work on technology rather than Friedrich—explicitly citing Ernst's 1932 book *Der Arbiter*. This is quite strange, given that Ernst had no interest in metaphysics, was entirely political in outlook, and adopted a protechnology stance. Thus the influence can only have been in a negative sense, as a reaction—which was precisely the position taken by Friedrich. Zimmerman (1990: 46–57ff) tries to substantiate the case for the importance of Ernst's role but falls short.
10. At the Berlin Auto Show of February 1939, Josef Goebbels gave a short but revealing speech on "the meaning of technology" in a National Socialist context. He said, "We live in an age of technology. The fast pace of our century affects all areas of life. Hardly any aspect can escape its powerful influence. Thus a danger arises that modern technology will make men soulless. National Socialism never rejected or opposed technology. Instead, one of its primary aims was to deliberately reaffirm it, to ensoul it, to discipline it, and to place it at the service of our people and our culture. . . . [National Socialism] has revealed a new romanticism in the outcomes of modern inventions and technology. . . . [We] understood how to take the soulless framework of technology and fill it with the rhythm and energy of our time" (Goebbels 1939: 105).
11. On this count, Juenger was some 10 years ahead of the generally recognized pioneer in this area, Aldo Leopold, and his landmark book, *A Sand County Almanac*.
12. See Morat (2007: 224–245).
13. This idea did not exist in Heidegger's opus *Being and Time*. Clearly something in those 22 years—between the publication of *Being and Time* in 1927 and Heidegger's first lectures on technology in 1949—convinced him that a great danger existed in modern technology.
14. One can well understand and sympathize with the remark by Wisnewski (2013: xiii): "Heidegger . . . is often dismissed as obscure—even deliberately so—or as writing just plain nonsense." Steiner (1979: 4–6) mentions those who believe "his writings are a thicket of impenetrable verbiage" and that "the questions he poses are sham questions"; for these critics, "Heidegger is a prolix charlatan and poisoner of good sense." Blackburn (2000: 43) recalls Ryle's concern about "an alarming tendency toward unintelligibility" in Heidegger, adding, "this is the tendency that blossomed." A few years later, he lamented "the purgatory of trying to read Heidegger" (Blackburn 2005: 78). Edwards (2004: 9) calls him the "greatest catastrophe" in the history of philosophy; "until fairly recently, Heidegger was not taken seriously by philosophers in Great Britain and the United States." He then expresses dismay at the "tide of unreason" that is bringing him back into prominence. And he relates in passing the complaint of Oxford philosopher Anthony Quinton regarding Heidegger's "ponderous and rubbishy woolgathering."

   Surprisingly, Heidegger himself would likely have taken such words as complimentary. He rather notoriously stated the following in one of his own works: "Making itself intelligible is suicide for philosophy" (1999: 307). This suggests a program of deliberate obfuscation—an unpardonable sin, in my opinion. But not to worry, says Blackburn; "outside a very few pockets Heidegger is not really an influence in Anglo-American professional philosophy" (2000: 48). The whole problem of Heidegger is compounded by the fact that his approach seems to have been taken as a recipe for success in contemporary continental philosophy. It runs something like this: Pick a vaguely philosophical topic, write in dense, obscure prose, make sagely pronouncements . . . and maybe someday a cottage industry will spring up, "explaining" your work to the masses. (The French certainly learned this lesson.)

15. It was published 10 years later, on May 31, 1976, upon Heidegger's death—per his stipulation.
16. To be discussed in Chapter 7.
17. Undoubtedly Nietzsche owes this general insight to Schopenhauer; see *World as Will and Idea,* especially Volume 1, Book 2. Earlier echoes can be found in Spinoza's *conatus* doctrine.
18. This in the context of a world that he describes as "a monster of energy."

6 Borgmann's 'Critical' Metaphysics

Heidegger and the other German metaphysical thinkers of the early to mid-20th century lived long enough to experience rapid technological change and a major restructuring of society. All were born before the year 1900 and thus were alive to witness several dramatic advances: human flight, the private automobile, electrification, mass production, motion pictures, television, nuclear power, early computers. Within a single person's lifetime, someone could remember the Wright brothers' flight in 1903, and then, just 66 years later, watch a televised landing of men on the moon.

In a sense, this was probably a more dramatic change than a comparable time span to the present day. Someone alive today might recall from childhood World War II or the atomic bombs dropped on Japan, but the overall magnitude of change is arguably less significant. After all, life in the Western industrial nations of 65 or 70 years ago was not much different, in many respects, than it is today: cars, single-family homes, factories and 9-to-5 office jobs, going to school, jet airplanes, radio, and television. Even considering the vaunted information revolution—personal computers, the Internet, smart phones, and so on—the changes in day-to-day life are arguably less severe than in the past. Has the pace of change slowed? Or has it shifted to new ground? Are we still on the road to a technological Armageddon, as Juenger, Heidegger, Jaspers, Mumford, Ellul, and others warned? How did the metaphysical picture of things change as we moved into the latter half of the past century? And what light can a pantechnical analysis shed on this?

To reiterate a point I made at the beginning of the present work, there was a marked change in postwar philosophy. It became increasingly dominated by formal, analytical thinking; it became more exact, more precise, more scientific. It thereby purchased its legitimacy, but at a price: metaphysical speculation was diminished, deemphasized, and finally relegated to mere history of philosophy. Metaphysical investigation, along with political philosophy and much of applied ethics, got shunted down a dead-end corridor. Scientific, technological philosophy, on the other hand, was perfectly suited to the postwar boom in modern industrial society. The field dealt primarily with linguistics, definitions, logic, and arcane argumentation. Philosophical journals came to resemble those of mathematics. Philosophy itself then reflected and supported the pervading mechanistic, objectivist, industrialist, technological worldview.

And yet, as I stated, metaphysical thinking did not vanish. Philosophers and others still sought to understand the core of technology: its essential nature, how it works, and how we ought to respond. In other words, they employed what I have called a synthetic metaphysical approach. Ellul made a major contribution to this effort with his *Technological Society* of 1954; a treatment of his ideas follows in Part II. His thinking, combined with lingering neo-Marxist concerns about alienation, and the evidently growing capacity of technological devices, all promoted further critical discussion. By the 1960s—when Ellul's book first appeared in English translation—the challenging spirit of the time was well suited to skeptical analyses, broadly along synthetic metaphysical lines.

Thus, for example, we find in 1964 the publication of such works as Blauner's *Alienation and Freedom* and Marcuse's *One-Dimensional Man.* Both of these tackle the problem of technology from a Marxist perspective, though in a psychosocial context; neither presses deeply into the issue of technology per se. Blauner laments that the modern industrial worker is systematically alienated, and he identifies "the worker's relation to technology as a major condition of alienation" (1964: 19). In the modern system, the worker "becomes reduced to a mechanical device." Sounding vaguely Heideggerian, Blauner adds that the industrial worker "approaches most completely the condition of *thingness*, the essence of alienation." For his part, Marcuse focuses more on the technological phenomenon, identifying "technology as form of social control and domination" (1964: 158). In modern society, "technology . . . provides the great rationalization of the un-freedom of man and demonstrates the 'technical' impossibility of being autonomous, of determining one's own life." Anticipating my own terminology, Marcuse remarks that "the Logos of technics has been made into the Logos of servitude. The liberating force of technology . . . turns into a fetter of liberation" (159).

The following year, biologist René Dubos emphasized an important and little-acknowledged point, namely that human beings are not biologically well adapted to modern technological society. Our genetic and psychological structure was formed during a 2-million-year period of foraging in small groups amid a vast, pristine wilderness. But today we live in huge artificial environments, disconnected from natural processes and in the company of thousands or millions of fellow humans. Unfortunately,

> biological man remains fundamentally the same as he was when he emerged from his animal past. Outwardly, man makes adjustments to the new conditions of life; inwardly, however, he has so far failed to make true adaptations to them, and this discrepancy creates physiological and psychological conflicts which threaten to become increasing traumatic. (1968: 232)

This is a major hurdle because our physiological structure is more or less hardwired; life under nonevolutionary conditions will necessarily subject us to permanent and unavoidable stresses. One solution might be to guide technological development in a more benign and human-friendly direction. Unfortunately this is not possible, says Dubos, because technology is becoming self-directing. Like an unruly teenager, it increasingly goes where it will, independent of our wishes. Consistent with Heidegger and Ellul, Dubos observes that

> science and [modern technology] now often function as forces independent of human goals. In many cases . . . knowledge creates concepts that man cannot restate in terms of his experience. . . . All too often, knowledge and technology pursue a course which is not guided by a pre-determined social philosophy. (239)

As an autonomous force, technology will inevitably act in its own best interest, which is often in opposition to that of humanity. As our living conditions continue to diverge from our evolutionary past, the stresses can only increase.

By the late 1960s and early 1970s, other critical voices joined in the dialogue. In an important series of articles, Henryk Skolimowski articulated a progressively harsher critique of the technological phenomenon, including some of the earliest philosophical analyses of the adverse impact on nature.[1] Theodore Roszak wrote a number of high-profile books that touched on the problem of technology, including *The Making of a Counter Culture* (1969) and *Where the Wasteland Ends* (1972). Writing as a historian and psychologist, his focus was on the budding "technocracy" and its adverse effect on the citizenry, primarily the young. But he made no attempt to analyze the technical phenomenon itself and supplied nothing of a metaphysical analysis. A somewhat more incisive study came in the work of Ivan Illich, particularly in his books *Tools for Conviviality* (1973) and *Energy and Equity* (1974). But again the emphasis was on sociological effects rather than technology per se.

For all this commentary, formal philosophical treatment was scarce. With few exceptions—Marcuse, Habermas, Skolimowski—professional academics saw little need to address technology; and of the exceptional voices, only Skolimowski foresaw a role for metaphysics. Mitcham and Mackey's anthology *Philosophy and Technology* (1972) was a milestone work, pulling together for the first time a variety of current perspectives from professional philosophers. Toward the end of the decade came Hans Jonas's *Das Prinzip Verantwortung* (1979), in English as *The Imperative of Responsibility* (1984), which provided an ethical analysis of the problems posed by technology, and Don Ihde's two early monographs, *Technics and Praxis* (1979) and *Existential Technics* (1983).

But the standout philosophical book of that time was certainly Albert Borgmann's 1984 work, *Technology and the Character of Contemporary Life*. It is to this book that I now turn.

BORGMANN ON TECHNOLOGY

Without question, Borgmann is a leading figure in philosophy of technology of the past few decades. Born in Freiburg, Germany, in 1937, he moved to the United States in 1970 and began teaching at the University of Montana, where he remains to this day. To date he has written some half dozen books, of which *Technology and the Character of Contemporary Life* (TCCL) is the most prominent. This is widely cited in the literature and was the subject of a dedicated anthology in 2000.[2] For present purposes, Borgmann makes for a good point of elaboration. He is a fair representative of the field, and his writing brings out some important contrasts with my pantechnical approach. He is particularly relevant because of his purportedly antagonistic stance toward modern technology—an impression that turns out to be misleading.

His main ideas can be roughly sketched in brief. Our modern worldview is dominated by an outlook that he calls the *device paradigm*. A 'device' is a technical object that delivers a commodified product (heat, food, entertainment) in a manner that is inscrutable to the typical user. Television, a stereo system, central heating, and a digital wristwatch all typically fit into this category. They supply the requested product upon demand—a product that is engineered, standardized, and widespread. And they supply it via a complex technological process that is hidden or removed from the user. The user obtains what he wants, but without knowledge of the means by which the product was delivered. It simply 'appears,' on demand, as if by magic.

In contrast to the device is the 'thing,' specifically a 'focal thing.' This object—also a technical product—delivers its product in different way: directly, uniquely, and via effort and involvement. For example, in contrast to the 'device' of central heating we have a 'focal' fireplace or wood stove. In contrast to fast food or a TV dinner we have homemade or homegrown food, cooked from scratch. A guitar, for another example, is a focal thing; it is fully transparent, we see directly how it works, and in the hands of skilled musician it provides a unique musical experience. The musical event is a special and deeply felt occurrence. Performer and audience gather together, on some sort of occasion, to share something intimate and personal. On the other hand, a stereo set—or, today, an iPod—delivers uniform music (meaning, identical each time it is played) at any time, day or night. There is no uniqueness, no specialness. Music there is a commodity. The guitar performance, as a focal event, and the guitar itself, as a focal thing, provide something very different—an *experience.* Both are music, but one is commodified and abstracted from the hidden technological (electronic or electromechanical) base, and the other is intimately and visibly connected to the means.

Under sway of the device paradigm, modern society progressively converts focal things and events into devices. More and more of the goods and services we procure are commodified, and devices take up an ever-larger share of our daily existence. This is problematic, says Borgmann, because devices are *shallow*; we see only the surface layer, the results, and not the whole process. We are detached from our products of consumption. Devices exist in the "foreground" of modern life and as such are relatively superficial. Focal things, by contrast, have real substance and real depth. They exist in the background, as a kind of solid foundation to life. They promote interaction, involvement, responsibility, civility. Borgmann's program is clear: to move away from or minimize the role of devices in our lives and to restore the place of focal things.

It is a helpful analysis, as far as it goes. In terms of metaphysics, though, there is not much synthesis. Beyond the concept of the device, we do not know what technology is; we do not know what it does to us or to the planet; and we have only the mildest of action plans—though perhaps fitting to his analysis of the situation. There are numerous examples but little in the way of useful empirical data, which might have helped to substantiate his claims. And there are deeper, more fundamental philosophical problems with his approach, ones that bear directly on the nature of technology.

But first, let me cite some points of agreement. I think Borgmann has a correct intuition about many aspects of the present situation. He opens the book with the observation that most people see the problems of technology as something extrinsic to technology itself—that is, as problems of politics, social justice, environment, and so on. This, he says, is "a serious mistake." Rather, there is something systematic about modern technological society, something intrinsic to it, that is at the root of the problem: "I propose to show that there is a characteristic and constraining pattern to the entire fabric of our lives" (1984: 3). "It is," he adds, "concrete in its manifestation, closest to our existence, and pervasive in its extent"—all valid observations of the world conceived as a Pantechnikon. Even more to the point: "All of reality is patterned after the paradigm, and in this case we can say that the paradigm has acquired an ontological dimension" (104). I could not agree more—though Borgmann hardly takes his own words seriously enough.

This pattern—the device paradigm—dominates our modern world and hence is of greatest consequence. In fact, it reveals "a fatally debilitating tendency" (4) of modern life. Furthermore, because ours is a technological society through and through, "a penetrating inquiry of technology must inevitably be a social critique" (10). Social power accrues to our corporate heads and political leaders, but their autonomy is deceptive: "We think of executives of large firms and of nationally prominent politicians as powerful. But their range of options is very constrained. Their power is contingent on their adherence to the technological paradigm" (63). In a technological society, it seems, even the masters are slaves.

Points of convergence continue in Part 3, "The Reform of Technology." Truly effective actions, says Borgmann, must alter the very structure of the device paradigm; "a fruitful reform of technology must be one *of* the device paradigm and cannot allow itself to be confined *within* the framework of technology" (155). Effective reform requires a central concept or ideal, and he proposes "nature in its pristine state"—that is, wilderness.

"Wilderness . . . constitutes a focal concern and a fruitful counterforce to technology." Nature is deep; as 'the real world,' it reminds us of our authentic engagement with things on a focal level. And by respecting wilderness, we acknowledge that some parts of nature are off limits to technological exploitation. Nature has intrinsic value, whereas technology views things only in instrumental terms. Nature thus commands our respect and perhaps even our reverence, attitudes that are utterly foreign to the technological stance. Also, "in the wilderness, time and space are restored to us" (191), in the sense that events there are allowed to unfold at their own pace and according to their own inherent rhythms. In immersing ourselves in free and unrestrained nature, we *become* free; "the liberation of nature is inseparable from human liberation" (193).[3]

But for all these valid and important insights, Borgmann repeatedly fails to draw meaningful conclusions or to use them to construct a deeper metaphysical scheme. First there is the basic ambiguity in his paradigm concept. He never clearly defines devices or focal things, preferring to give examples and illustrations. Conditions and context matter, it is true, but this blurs the whole concept considerably. Store-bought "commercial" wine is a device, but homemade wine is a focal thing. A mechanical clock, such as a cuckoo clock, is a tangible and transparent focal thing, but an electromechanical or digital clock is not. Many related questions arise: Must we see the mechanism and the process behind things in order for them to be focal? Is it a question of the knowledge and experience of the user—and thus in part subjective? Is focality more of an attitude or is it something ontological? Is it possible to produce a focal computer? Or a focal automobile? Is a bicycle focal or device? What about a music box? And 'deviceness' would seem to exist on a continuum: an acoustic guitar is focal, but yet its components—the strings, for example—are highly engineered industrial products; are the strings devices and yet the guitar focal? An acoustic guitar, furthermore, is less of a device—and more focal—than an electric guitar, presumably. But is a live electric guitar performance focal or device? A vinyl record is a device, but surely less of a device than a CD, which in turn is less of a device than the memory chips in an iPod. A difficult proposition, to untangle all these threads.

One option might be to adopt a less binary and more analog approach. Perhaps both deviceness and focality exist on continuums, such that any given object or event has more or less of each. I have in mind a two-dimensional graph, with focality on one axis and deviceness on the other. Then we might locate each entity on this graph, with more or less of each aspect. This would not solve all problems, but it might better reflect the continuity within and among things. But perhaps it is not Borgmann's intention to create a consistent metaphysical schema. Perhaps we ought simply take his system normatively. As he himself said, several years later, "The chief function of the device paradigm . . . is to provide a perspective on . . . ethical issues" (2000: 346). Apparently his program is primarily one of ethics, not metaphysics.

Second, he falls into the trap of the technological dichotomy: of believing that "modern technology" is a distinct and unique phenomenon relative to traditional technology. Heidegger, as I explained, also went afoul of this distinction, and it introduced irresolvable difficulties into his ontology. Borgmann is equally explicit in his stance: "The concern of the present study [TCCL] is modern technology" (1984: 35). He adds that "we have tentatively and formally defined [modern] technology as the characteristic way in which we today take up with the world."[4] But the same ugly dilemma arises: What is "modern"? Borgmann gives us two suggestions. Just after the above quotation, he writes of the need to "return to the period where the pattern was first articulated. This is to go back to the founding event of the modern era, to the Enlightenment and to its first beginnings in the early seventeenth century." The Enlightenment is a vague term relating to the rise of philosophical skepticism and rationalism; its starting point is ambiguous, though some would point to Descartes's *Discourse on Method* (1637) or *Meditations* (1641) as the initiating events. Others might reach back a bit further, citing Bacon's *Novum Organum* of 1620. Evidently this is Borgmann's time frame.

But what was the first true device? The early to mid-1600s saw few dramatic new technological advances. Apparently the early modern era was the start of device paradigm *thinking*; actual products would take some time. This is supported by Borgmann's other suggestion, namely, that the steam engine was the decisive invention. He says, "Modern technology begins with the introduction of the steam engine in mines, factories, and for transportation" (57). But what does he mean by "steam engine"? Is he referring to Papin's "steam digester" of 1679? Or maybe Savery's nonpiston engine of 1698? Perhaps he means Newcomen's piston-driven steam engine of 1712? (This was in fact the first to be used in mines, circa 1715.) Or could he be thinking of Watt's famous steam engine—which was not even invented until 1769? His reference to "transportation" means the steam locomotive; but this was not operational until 1804 and wasn't put into widespread use as a transportation system until the 1830s and 1840s. Was it one of these? Or all of them? Does it matter?

Were these early steam engines even "devices" at all? In a sense, they were simply fancy stoves, and a stove is one of Borgmann's prototype focal things. Do they become devices when put to use? When mass produced? When used for profit? When put on wheels? More difficult questions.

The point is clear: No one machine or no single invention can signify a truly metaphysical break in the technological order. Borgmann can only be referring to the "steam engine era," a broad swath of time over which steam power gradually made incursions into several areas of life. This was not an ontological event but a slow and progressive shift in machine power, accompanied by a realization of the potentials opened up by new sources of energy.

Thus, we must be very careful in using the well-worn phrase 'modern technology' as if it were a thing in itself. I have no issue with declaring modernity to begin with the steam engine era, nor even with calling these 'modern' technologies. But I am concerned about making a fundamental, metaphysical distinction between modern technology and its predecessors. There is no sharp breakpoint but only a smooth progression along certain lines of development.

Even so, we can understand his motivation. If one is concerned about aspects of present-day technology, and yet everyone realizes that humans cannot do without some technology, then there is strong pressure to declare modern technology bad and traditional technology good—and thus to make fundamental distinctions. It is better and more accurate to view all of technology as a continuous progression toward order and toward greater manipulation and control of energy. Such a process has both positive and negative consequences for humanity and nature, and as the technology grows in sophistication, the balance shifts from *primarily* positive to *primarily* negative. This view better aligns with our empirical data and is more consistent with the historical philosophical opinion—as I will describe later on.

ON DETERMINISM AND SUBSTANTIVISM

Some further points of disagreement: Early in TCCL, Borgmann sketches out a structure of the dominant categories of theories of technology. He identifies three rivals to his own "paradigmatic" approach: *substantivism*, *instrumentalism*, and *pluralism*. In the substantivist view, "technology appears as a force in its own right, one that shapes today's societies and values from the ground up and has no serious rivals" (9). This approach he characterizes as embodying "technological value determinism" and as a position of "antitechnology" because it generally views technology as a "pernicious force." It aims at "a comprehensive elucidation of our world by reducing its perplexing features and changes to one force or principle. That principle, technology, serves to explain everything, but it remains itself entirely unexplained and obscure."

The chief advocate of this view, says Borgmann, is Jacques Ellul.[5] For Ellul, *technique* (his word for technology) is seen as "the autonomous and irresistible power that enslaves everything from science to art, from labor to leisure, from economics to politics." Technique strives relentlessly toward a specific end, namely *efficiency*, in all that it does. But efficiency is a "systematically incomplete concept" for Borgmann, because it has no external goals and no external standards by which to measure things. As a self-contained totality, technique is its own end, and thus efficiency is rendered meaningless. In sum, argues Borgmann, "Modern technique, a power in its own right, is put forward by Ellul as its own unexplained explanation." This whole theoretical approach is "easily dismissed," adds Borgmann, because it reduces to a mere "demonizing of technology" (10). Any talk of substantivism is "misleading" (41); indeed, "the rule of technology is not the reign of a substantive force people would bear with resentment or resistance" (105).

Obviously this is opposed to the present view. Pantechnical metaphysics is clearly a substantivist approach. As a universal process, technology is in fact "a force in its own right"; it may be true that it has "no serious rivals," but it *does* have rivals, namely the other forces at work in the universe, including those in the human sphere of action. It is a form of determinism, though this requires further explanation. And it is, in one sense, 'anti' technology, in that technology is rightly seen as a threat to human and planetary well-being—'pernicious,' if you like. But on the other hand, from the high perspective of the Pantechnikon, one cannot essentially be 'anti' technology at all; technology is creation itself. The ethical norms of the pantechnical outlook—or indeed of any nonsuicidal worldview—require self-preservation of humanity and nature, and this demands an extreme restriction on many forms of contemporary technology. Thus one may still be opposed to 'modern' technology, broadly defined, without introducing the ontological distinctions that Heidegger and Borgmann do.

Borgmann wants to make an *a priori* rejection of substantivism, but he has no grounds to do so. He offers no counterarguments. His main concern seems to be with Ellul's unique form of this view and in particular with its circularity. Here he has a valid point. Ellul—as I will show—does in fact rely fully on technique as an explanatory force, but without providing anything at all in the way of an underlying metaphysics. Ellul presents a devastating critique of the effect of technique on human society, one that has become increasingly confirmed by events of the past few decades. But efficiency is truly an ambiguous end, and it offers little explanatory power. The pantechnical outlook remedies both these concerns: 'striving toward order' via the expropriation of surplus energy is the preferred and more explicit end, and technology is thereby seen to have a concrete metaphysical interpretation.

The instrumentalist viewpoint, by contrast, sees technology simply as so many tools or instruments, put to use by humanity for human ends. It is anthropocentric and value neutral. The only relevant tools are manmade ones, and the tools themselves are intrinsically neutral, neither good nor evil; their value comes only from the mode of their use. Borgmann rightly declares this a "shortsighted" outlook; he flatly states that "instrumentalism does not constitute a proper theory of technology" (11). This is self-evidently true from the pantechnical perspective.

The other competing outlook, pluralism, is perhaps the dominant one among professional philosophers. This viewpoint tends to be empirically driven, focusing on specific technologies and particular applications rather than on capital-T Technology as a comprehensive phenomenon. Confronted with such comprehensive theories, the pluralist can only reply with "counterexamples, unresolved problems, and disregarded evidence." Technology is too diverse and too complex, according to the pluralist, to accommodate any efforts at grand systematization. He sees "no striking overall pattern" in technology, only specific and concrete instances. For the pluralist, technology has no essential nature and poses no problem in itself. Pluralism thus reduces to "a learned reflection of ambiguity." I find no disagreement with this assessment.

Setting these issues aside, there are more fundamental conflicts. In constructing a critique of 'modern technology,' Borgmann has set himself up as a kind of critic of technology—adopting a vaguely 'antitechnology' stance, to use the term loosely. On the one hand this seems clear, and yet repeatedly throughout his book, Borgmann comes off as a technological apologist or even as an advocate. The ambiguity is rooted in his dichotomous outlook: devices are bad, focal things are good. Thus one is justified, on his view, in being antidevice and profocality. What this means precisely is unclear, except that his list of nominally focal things—music (acoustic), poetry, art, running, the culture of the table, a stove or fireplace, cathedrals, Greek temples—all existed prior to his "modern era" of the steam engine. One would presume, therefore, that Borgmann might oppose much of modern technology—or perhaps all of it—on the basis of its intrinsic predominance of deviceness.

But this is not so. Technology, it seems, has issued a "promise" to humanity. This promise dates to the beginning of the Enlightenment, specifically to Bacon and Descartes. Bacon's *New Organon* exclaims that "we are laying the foundation . . . of human progress and empowerment" ([1620] 2000: 13). This empowerment was founded on the three decisive technologies of the day—printing, gunpowder, and the compass. "In fact these three things have changed the face and condition of things all over the globe" (100). By these means "we extend the power and empire of the human race itself over the universe of things" and, correspondingly, "the empire of man over things lies solely in the arts [i.e., technology] and sciences." Indeed, man must "recover the right over nature which belongs to him by God's gift" (101)—referring to the mandate of Genesis.

Descartes, for his part, took an equally assertive and hopeful view. He realized that his theoretical discoveries in science and physics could be quite "useful," and he noted that we must "do all in our power to secure the general welfare of mankind." His "practical philosophy" yielded knowledge of the "power and action of fire, water, air, the stars, the heavens and all other bodies in our environment," and therefore that "we could use this knowledge . . . for all the purposes for which it is appropriate, and thus make ourselves, as it were, the lords and masters of nature." Such knowledge would allow "the invention of innumerable devices which would facilitate our enjoyment of the fruits of the earth and all the goods we find there" (1985: 142–143)—so he wrote in 1637. Through technology we can bring the blessings of the Earth to humanity, for the betterment of all. This is the promise.

So conceived, the promise is legitimate, according to Borgmann. Technology—*modern* technology—carries with it the great promise for humanity, something that remains undiminished. One need only recall the pain and hardship of early, premodern days to realize what a blessing modern technology truly is. "The splendor of the promise of technology appears bright to this very day when we remember how recently misery and deprivation have been shaping human life, especially in the newly settled West of this country" (36). Thereafter follow several pages elaborating the miseries of premodern Western life. Borgmann is emphatic in his endorsement of the technological blessings. He baldly states, "Without modern technology, the liberal program of freedom, equality, and self-realization is unrealizable" (34). How such a definitive and unconditional position is to be justified, he does not say.

Thus, according to Borgmann, an enlightened society *needs* modern technology. And yet the manner in which we have it—the device paradigm—fails to realize the promise. Thus our program must be one not of reining in modern technology, for this would render our higher aspirations impossible, but rather one of "recovering the promise" in the best possible way. We must "disentangle" the promise from the device paradigm in such a way as to "respect the legitimacy of the promise and to guard the indispensible and admirable accomplishments of technology" (153)—resulting in his program of reform, the third part of his book.

And yet we are struck by the naiveté of his plan. "Medical technology provides healing and wholeness" (246), but with no recognition that the very technological system that provides this medical care has itself produced many of the same ailments it offers to cure. Likewise, "media technology allows us to consider all things and to be enlightened about the world in an intelligent and compassionate way"—and yet we are faced with a barrage of propaganda, issued by large media conglomerates, with a vested interest in a particular outlook. Who today would argue that the Internet has brought any meaningful sense of enlightenment, intelligence, or compassion to society at large? We have a surfeit of information, to be sure, but wisdom is in shorter supply than ever.

Borgmann's reform program would simply "prune back the excesses of [modern] technology and restrict it to a supporting role" (247). Coming to the fore are his focal practices and things, entities which, he says, are neither pretechnological nor antitechnology. "Rather, they unfold their significance in an affirmative and intelligent acceptance of technology." But once again, this is true only in a trivial and self-evident sense: We cannot live without some technology—whether it is called 'focal' or not is incidental.

And yet the banalities continue. Focal things help us realize "the wide-ranging and effortless way in which technology provides a context of security, comfort, and enlightenment" (248). So conceived, "technology can fulfill the promise of a new kind of freedom and richness." Technology is not the problem, it is the *solution.* Without it, we lose all possibility for enlightenment: "The ultimate calamity would be the complete destruction of technology; it would be the eradication of all hope" (249). Such statements are pointless trivialities, having no bearing on the philosophical issues at hand. Borgmann is either confused about what to do or too timid to lay out a stronger program of action.

ON THE QUESTION OF NATURE: A "NEW MATURITY"

Perhaps Borgmann's greatest weakness occurs in his treatment of the environment. He dedicates one chapter to the topic, "The Challenge of Nature." Wilderness has been ravaged by technological society, and yet the solution to this problem "is not a flight from technology but the realization that nature in its wildness attains new and positive significance within the technological setting" (182). In other words, nature needs to get used to modern life. Technological man is here to stay, and his priorities are determinate. Any system of environmental ethics—such as Aldo Leopold's notion that nature has intrinsic value and that its integrity is the final criterion of right and wrong action—is either "too tattered" or "so flexible" as to be useless. The idea that technology "ruthlessly invades" nature is "no longer adequate"; "the advanced technological setting is no longer characterized by violence of machinery" (189)—a difficult sell to anyone who has witnessed a forest clear-cut, a mountaintop removal, urban sprawl, mass topsoil erosion, an ecosystem devastated by invasive species, or any of a number of other abuses.

After all, technology is here for our benefit, says Borgmann: "Without technology we could not venture safely or comfortably into the wilderness" (193). To accept such a standpoint shows a "new maturity," and furthermore its acceptance "is required of us." "Through technology," he says, "we learn to respect the wilderness" (194)—as if there were no other way. Lest we be concerned about ecological destruction, the device paradigm has the advantage of letting us "deal technologically with the physical limits to growth"—in other words, technology will come to the rescue. The spread of the device paradigm has produced "a high level of ecological consciousness" in recent years, something that would have been "impossible" without its extension. Indeed, "the global ecological situation is more promising socially and physically than most critics had dared to hope a decade ago" (147). One wonders to which globe he is referring.

All this comes to a head in the issue of global climate change, potentially the largest technological catastrophe in history. Borgmann's view on the subject is telling. First, TCCL has no mention of it at all. Granted, 1984 was a bit early in the climate debate, but it would not have been impossible for a perceptive scholar to pick it up. For example, the *New York Times* carried one of its first major stories on global warming as early as June 1977.[6] Another story followed the next month.[7] Three more articles appeared in 1979, and then came a clear pronouncement in 1981: "Study finds warming trends that could raise sea levels; Climate of Earth is found warming" (August 22). Nearly a dozen articles ran in the NYT between then and the end of 1983—plenty of opportunities for Borgmann to address the issue.[8]

Even years later, Borgmann's comments remained sparse. For example, a lengthy "reply to critics," published 16 years after TCCL, offered all of two passing comments: "Global warming . . . remain[s] a grave challenge"

(2000: 341) and "concerted reaction to global warming is still tentative" (362). More brief references came in his *Real American Ethics* (2006) and "Does the Crisis of Global Warming Require a Spiritual Solution?" (2008), but it wasn't until 2012 that Borgmann published a dedicated essay on the topic: "Science, Ethics, and Technology and the Challenge of Global Warming." This short piece, about eight pages in length, is oddly inconclusive. He begins with the observation that all three projects—science, ethics, and technology—"have failed to meet the challenge of global warming" (2012: 169). In order to press deeper, we need to examine the theoretical underpinnings of each. We have such theories for science and ethics but not technology, and this is the missing piece that we need. Borgmann then embarks on a discussion of the relevant scientific theory and then the ethical theory (Aristotle, Kant, Mill, etc.) and then, finally, technology . . . but by this time global warming is nowhere to be seen. We are offered some trite closing observations ("The paradigm of instantaneous, ubiquitous, and attractive availability is a crucial feature of the rule of technology"; "All of us realize that we have to pay our dues to the machinery of technology"), and then the article ends. Borgmann's final appeal is for us to "shift from consumption to engagement" and "from indolence to responsibility" (177). A mild program, at best—though fully consistent with that of TCCL.

My environmental rebuke here is aimed at Borgmann, but it holds for most all philosophers of technology, unfortunately. Global climate change—like the environment generally—is almost nonexistent as a *philosophical* issue of technology. Technology philosophers seem to believe that concern about nature falls under a different academic category, such as environmental ethics, and thus they see no need to address it. Recent anthologies in philosophy of technology may include a section that nominally deals with the environment, but the readings are either narrowly targeted (e.g., GMO foods, biotechnology) or are classics in environmental ethics that have little to say about technology. The dedicated technology philosophers rarely tackle environmental issues directly.

We can conjecture why this is so. I suspect at least two pragmatic issues are at play. First, on a practical level, global climate change via greenhouse gas emissions seems to be inextricably connected to advanced industrial society. There appears to be no way to continue with our high-tech existence and yet reduce carbon emissions to a benign level. Theoretically we could embark on a worldwide campaign for nuclear power, but this will not happen, due to fears of nuclear weapons development, concerns about radioactive waste, and finite (nonrenewable) limits on uranium fuel. Or we could promote mass solar panel installation, mass wind farms, mass geothermal plants, and so on; of these only solar panels could conceivably meet world demand, but even this would require massive expenditures on power storage (e.g., batteries) and distribution networks. Power-hungry nations like the U.S. would have to erect huge solar panel systems; in our case, some 40,000 square miles would be required to meet near-future demand, an area roughly one third the size of the state of Nevada. Any such proposal would evoke furious environmental and land-use battles regarding the paneling over—and thus destruction—of huge areas of the American southwest. Additionally, we have the vested interests of the fossil fuel industry; they would spend vast amounts of money to oppose any large-scale shift toward alternative energy sources. The most rational solution, after all, is to reduce energy use. But this necessarily entails a lower standard of living, and this is politically impossible.

Thus, in global climate change, technology has produced a problem for which there is likely to be no easy solution short of a major retrenchment of our industrial way of life—that is, of technology itself. This, however, is a 'radical' solution. Philosophers of any stripe are not very good at radical solutions. As a discipline, we tend to be increasingly conservative, concerned more about prestige, image, finances, and political correctness than with truth, justice, and wisdom. Our salary, our promotions, and the interests of wealthy donors all weigh heavily on us. Technological society has provided the resources needed to fund our universities and to create multibillion-dollar endowments, and thus it is hard to put forth fundamental critiques of the very system that has produced such beneficence. We philosophers, who should be the harsh voice of reason and wisdom, have instead found it "unnecessary" to challenge technological society. We find it easier to conduct oblique analyses of "actual technologies, actual devices, actual machines." Through our "deliberate choice and reasoned arguments," we focus on minutiae, tangential issues, and multiply rehashed microstudies. Critiquing technology at a fundamental level, like critiquing wealth or power, is a losing strategy. It is one best avoided.

Of course, I cannot be sure how much of a factor this is in the thinking of other philosophers of technology; they certainly do not admit it in public. For his part, Borgmann is actually rather explicit: his convenience and material well-being are a function of modern technology, and he is not about to give that up—even at the cost of largely ignoring a looming disaster like climate change. No doubt it applies to others as well, though less explicitly. It lurks in the background, casting a shadow over the thinking of many an academic. We personally, and our beloved institutions, profit so mightily from technological largess that we have a very hard time seeing technology as anything other than a beneficent force in the world—or at worst, as an unruly teenager who just needs a bit of discipline. "He's really a *good* boy," we say.

The second issue relates to the question of rational reform of technology. Borgmann explicitly calls for a reform program that carries with it the unstated but obvious assumption that rational reform is actually possible. On specific and small-scale issues, this is reasonable. Technological problems are addressed and resolved on a regular basis, via something like a rational process of reform—that is, identifying the problem, examining alternative solutions, and implementing a fix. But global problems require global coordination, and this has proven exceptionally difficult.

Despite a large scientific consensus on the nature of the climate-change problem, for example, the nations of the world have shown themselves to be utterly incapable of rational action. Total global emissions currently exceed 50 billion tons of carbon dioxide equivalent per year. This must be reduced to approximately 44 billon tons per year by 2020 in order to have a reasonable chance of limiting global temperature increases to 2° Celsius. Instead, we are on track to hit 58 billion tons by 2020.[9] The U.N. has been conducting annual climate change conferences since 1995. Despite the data and growing consensus, these conferences have been a string of utter failures—no binding commitments by the major nations of the world to take enforceable action to reduce emissions. Thus, even in the face of a global catastrophe, the prospect for rational reform is slim. And if reform fails here, we have little reason to believe that it will succeed with the larger and more complex problem of technology generally.

The orthodox philosopher of technology may reply: "This is not a problem of technology, but one of political will, or social tradeoffs, or ethics—environmental or otherwise." And yet: The problem has a known and explicit cause—*advanced technological society.* The root cause is modern industrial technology. If our philosopher responds by saying, "The problem isn't technology, but rather too many people on the planet," we have a reply: *It is technology itself that has created the conditions by which there are too many people on the planet.* It was only with the technological advances of the Industrial Revolution—specifically, new energy sources, new medical techniques, and increased food production—that we experienced the upward, exponential acceleration of the global population. An appeal to the population issue does not solve the problem, it only displaces it.

Why is rational reform of our technological system so difficult, even in the face of something as serious as climate change? Because human rationality is fundamentally limited, whereas technology is not. Our innate abilities to reason, anticipate danger, and plan ahead were formed over 2 million years of hunter-gatherer existence. The dangers were generally clear and transparent, even if the exact causal mechanisms were not known. Relevant time scales were short; 'long-range planning' typically involved not more than the coming 12 months—with its natural cycles of food supplies and seasonal challenges. Consequences of our actions were usually immediate, or within a short time span. They were clear to all involved.

Today we face vastly more complex problems, and on much longer time scales, than humanity has evolved to cope with. Of course, some of us can anticipate long-term and subtle dangers. But mass society is inevitably driven by the mass, and the rationality level of the mass is disturbingly low. It is so low, in fact, as to effectively operate *irrationally*—by which I mean, with the shortest of short-term thinking, and responding only to the most obvious and immediate threats. Truly rational action must be preemptive. It involves anticipating threats and acting before they become manifest, in order to avoid or mitigate harm. The larger and more complex the society, the less able it is to function rationally.

Technology, then, is a compound danger. It is the cause of the problem—global carbon emissions, in this case—and it has produced a society that is functionally irrational and thus incapable of solving the very problem it has created. Technology creates the danger and then makes us unable to collectively respond to it. It virtually guarantees that we and the planet will feel the full impact of climate change and global warming. With the present state of industrial society, we will not take serious action until it becomes an obvious threat. But when that time comes, it will be far too late. The disaster will be upon us, and upon all life on Earth. At that point, our only action will be a scramble for survival, under the most brutal and cutthroat conditions imaginable. Future generations will not forgive us, but perhaps they will understand our functional impotence in the face of a dominating technological system.

But we can press a bit deeper still. Why is technology a compound danger? Because it is an expansionist enterprise. It is manifest as the embodiment of change, progress, and growth. Technology drives ever forward, relentlessly, with an insatiable thirst for order and complexity. As long as energy exists in superabundance, technology will put it to use. Order and complexity will grow. This is in its very nature, because, as I argued earlier, it is born of universal expansion. The growth of order on Earth is fundamentally connected to the growing, expanding cosmos and its concomitant energy differentials that produce all order in the universe. Technology is growth incarnate; it therefore drives forward the growth of the human species and the complexity of its society. Technology gives rise to a new being, a social being, that exists on a plane far above that of the individual human. We rational individuals, enlightened as we may be, are unable to affect this vast social being. It is a new creature on this planet. It has no evolutionary history, save through us. By cosmic standards, it represents a new rationality. By our standards, it is functionally irrational. In furthering its own development, it threatens our very existence. In aiding its emergence, we risk collective suicide. If present conditions continue, disaster is inevitable.

RECENT THOUGHTS

One might have thought that, over the years, Borgmann would have reflected on these and related issues, and therefore modified his views. As of 2000, at least, that had not happened. His "Reply to Critics" reiterates many of the same rosy and facile observations that plagued TCCL. The piece opens with a veritable roll call of good news: "the threat of nuclear holocaust has all but disappeared, environmental problems have been reduced if not eliminated, the culture of consumption is approaching a sustainable modus operandi, the rate of population growth is declining, democracy has been spreading" (2000: 341). Suffice to say that he is either wrong or misleading on every count. A few lines later he states that the good life of global democracy and material abundance "is at least conceivable now and likely reachable." The United States, as the sole global superpower, is "the model of the kind of open and enterprising democracy that is most hospitable to full employment and vigorous economic growth." Overall, the "general rise of freedom, security, and prosperity" imposes on us an "obligation . . . to extend these blessings as far and as rapidly as possible." One wishes this were satire.

On the positive side, Borgmann rightly observes that, in the intervening 16 years, "the rule of technology has been enormously expanded and solidified" (342). And he correctly understands that conventional philosophy has utterly failed to grasp the significance of this fact. Analytic philosophy is "culturally barren," and its predominance in the universities "has had a deadening effect" on academia. And we know why: Analytic philosophy is *technological* philosophy, and thus it is intrinsically incapable of standing at a sufficient distance to conduct an adequate critique.

But his essential philosophical views are unchanged. He continues with his dichotomous thinking, emphasizing yet again the uniqueness of *modern* technology. He repeats his "rejection of determinism," continuing to hold a superficial view of human freedom. As mentioned earlier, the chief function of the device paradigm is "to provide a perspective on . . . ethical issues" (346) rather than on metaphysics. His program is still one of reform but, remarkably, of a variety even less decisive than in TCCL. This is so because, sadly, "the cultural climate of today is not hospitable to ambitious reforms" (360). Therefore it is only "refinements and adjustments that are needed, it seems, not fundamental reform." He does, however, hold out "some reformatory zeal" in three areas, namely social justice, environment, and "quality of life." But because "technology aids and undergirds liberal democracy," we need not worry ourselves about any far-reaching action plan. Indeed, "the device paradigm requires neither intrinsic reform nor global replacement" (364). The system is managing itself just fine.

One important update in his "Reply" relates to information technology, which had dramatically accelerated between 1984 and 2000. Borgmann observes that it is "currently the prominent and most influential version of the device paradigm" and is "sweeping everything before it." A revealing comment follows. The power and scope of information technology is so impressive, he says, "that one is sorely tempted to embrace the determinist and substantivist implications of this most recent technology" (352). Precisely. And in fact one *would* embrace a determinist and substantivist view if one were not predisposed to exclude it. The empirical data certainly support such a stance, as I will explain.

Yet even years later, Borgmann still had not changed his opinion. In a published interview of 2007, he referred to technological determinism as a view that is "surely wrong" (2007: 10). But once again, he concedes that "the straightforward application of [technological] libertarianism to the technological culture is bizarre: millions of people freely and responsibly make the same deplorable use of [technology]; they make the same unfortunate decisions." Almost as if . . . a kind of determinism held sway.

In spite of all this, we need not worry, says Borgmann. "Computer technology deserves admiration" for its "astounding creativity," "diligence," and "intelligence." Generally speaking, the impact of computerization on our economy and society has been "on the whole for better than worse." The entire integrated system works for our benefit, and we ought to be grateful, not critical: "The machineries of life in a technological society have been configured into an interlocking system, and if we want to hold on to anything like the physical welfare and cognitive scope it affords, its basic machinery must be kept intact" (2000: 365)—this from the reputed "critic" of modern technology.[10]

I emphasize all these points for precisely this reason—Borgmann has been cast by himself and others as a prominent technology critic, and his stance clearly shows what passes for a contemporary critique of the technical phenomenon. The milder-than-mild reform program, interspersed with gushing praise and notes of resignation, are all that present-day philosophers can muster. From a survey of the philosophical literature, one could be excused for thinking that technology posed no problems whatsoever—apart, perhaps, from very individual and specific applications that are best addressed on a case-by-case basis, probably by someone else. This is a striking and fundamental shift in outlook from the view offered by philosopher Webster Hood, who stated the following—in 1972: "That technology represents a problem of major importance requiring analysis and interpretation needs no argument. [Technology] is the controlling power in our age, affecting and shaping virtually all aspects of human existence" (1972: 347).

How can it be that some four decades ago, before anyone ever heard of cell phones, home computers, or the Internet, technology was self-evidently a philosophical problem of first rate; and yet today, with the world facing technological disaster on several fronts, it merits no discussion at all? *How can this be*—with the global climate in a feverish state, human population heading toward 10 billion, and technology pressing ever-deeper into everyone's daily existence? Were Hood and the other likeminded skeptics—Skolimowski, Roszak, Illich, Mumford, Schumacher, Dubos, Ellul, Heidegger, and philosophers from centuries past—simply mistaken? Were all those thinkers simply projecting their fears and worries onto modern technology unjustifiably? Have recent events borne out the error of their ways?

Decidedly not. Something else must be at work. It is almost as if Heidegger's prediction were coming true: that the spread of technology has caused us to become blind to its essence, such that we take it either in an instrumental, neutral sense, as a purely social construct, or as something only incidentally problematic. The technological system seems to function, on several counts, as a self-sustaining and self-purifying process. Anything and anyone that might block its path or disrupt its progression is shunted out of the way. The predominant philosophical views are, perforce, those that are agreeable to the system. This is not magic, not a grand conspiracy, and not some mystical occult power. The system works very rationally and very methodically in its own self-interest. In fact it *defines* rationality in a technological context. Those thinkers who see in technology no fundamental problem or see only minor and incidental issues are inevitably viewed as the "most rational," "most authoritative," and thus most worthy of consideration. And who among academic thinkers today does not want to be viewed as preeminently rational and authoritative?

THE QUESTION OF THEOLOGY

I close this chapter on Borgmann with a final observation—perhaps incidental, perhaps not—regarding his religious outlook. He is unusual among present-day philosophers in that he openly acknowledges and writes about his Christian faith. Even his secular works contain scattered references. TCCL has several comments on divinity, Christianity, and the sacred. His "Reply" quotes the Bible (359) and closes with references to "the Kingdom of God" (368), natural theology, and St. Augustine. Granted, Borgmann never presses his views on the reader, and his Christianity is not intrinsic to his philosophical analysis. But still, his religious inclination certainly supports and may even partly explain his stance on technology.

I see at least three reasons for concern here, within the technological context. First, inherent in all monotheistic views is the notion of human free will. This is taken for granted because the concepts of salvation and damnation require it.[11] And if we are free, we have ultimate authority over technology. If it dominates us, this can be due only to the frailty of human will. There can be no technological determinism because there is no determinism, period. God gave us free will, and it can never be usurped by mere machinery.

Second, technology cannot be an all-powerful force in the world, because God is this force. God made or allowed the technological system to emerge. He controls it, fully and completely. If technology ever threatens to overwhelm humanity, God will step in and save us. Or, if he does not, it was his will. Either way, there is no problem. With God looking out for us, we need never fear technology.

Third, all monotheistic religions demand a large degree of faith. Faith is, by definition, belief in that for which you have no rational justification. It is, quite literally, the counterpoint of reason—in plain words, the embodiment of irrationality. Now, to be sure, all of us have some small regions of our psyche that operate irrationally. If we follow Plato, our appetites, wants, and desires function exactly this way.[12] Normally this is not a problem. But if irrationality comes to take too large a portion of our mental lives, we are inclined to make mistakes—possibly very serious ones.

And the more irrational the religion, the greater the risk. Certain religions, such as Taoism or Buddhism, seem to operate with a minimal quota of faith. But something like the Christian story, which, from a factual perspective, is a pile of absurdities, demands faith to excess.[13] Devout Christians must, of necessity, allow unreason to play a large role in their existence, and this cannot but adversely affect one's philosophical thinking. In Borgmann's case, we can see how this might—and obviously this is conjecture—alter his outlook on the metaphysics of technology. Certain lines of thinking are closed off before the analysis has even begun. Even Ellul, who was a Catholic theologian as well as social philosopher, allowed Christian concepts to creep into his analysis, albeit to an apparently minimal degree, given that he argued for a strongly substantivist and determinist position.[14] On the other hand, Ellul does have a kind of faith in human ability and free will to overcome even the apparently insuperable power of technique. And he suggests that divine intervention may be one of our only hopes (1964: xxx).

Such issues are, of course, of little or no concern for the Christian. But for myself and others who cannot accept the Christian story, this raises a flag of caution. It does not invalidate Borgmann's analysis, but it does give one pause regarding the fundamental intuitions that lie at the root of his thinking.

Borgmann is surely not alone in this regard. For example, Don Ihde's early work with Paul Tillich and his dissertation on Paul Ricoeur suggest that he too was—and may still be—a man of religious faith. This fact may likewise bear on his explicitly antiessentialist and antisubstantivist views. Obviously religiosity is not a requirement on this count, as it is certain that many nonreligious philosophers also hold antisubstantivist opinions. Normally we do not know a given philosopher's view on such matters. But where we do know, it is legitimate to speculate on the role such beliefs play in that person's philosophical system.

Borgmann, then, is an instructive case study for present purposes. He is a prominent and representative member of mainstream philosophy of technology of the past three decades. He offers many important insights and yet provides a notable counterpoint to my own thesis of the Pantechnikon. His main work, *Technology and the Character of Contemporary Life*, came out nearly contemporaneously with Ihde's early works, with Haraway's *Cyborg Manifesto*, and not long before Feenberg's initial writings on technology—just to name a few other representatives. I will address the views of Ihde and Feenberg in later chapters, but my objective here was to focus on the nominally metaphysical aspects of Borgmann's thinking.

With this I close my sketch of the theoretical or conceptual aspect of the metaphysics of technology. I have attempted to present a variety of details regarding both the origins of technology conceived as a universal phenomenon and also its essential nature or intrinsic properties. All this constitutes a necessary but not sufficient, aspect of a synthetic metaphysical approach. The other aspect is the pragmatic dimension; it addresses what technology does, how it functions in the actual world, and how humanity can or should respond to it. This aspect, I suggest, is best begun with an examination of several historical perspectives that saw technology as a danger or threat. Earlier philosophers, when investigating technical phenomena, have traditionally focused on effects and consequences on human beings and on society generally. But these pragmatic considerations carry with them an implicitly critical metaphysics, one of an entity that is detrimental to human well-being—despite the nominally obvious benefits.

My conjecture here is that many past philosophers caught a glimpse of the Pantechnikon but were not in a position to articulate a drawn-out metaphysical worldview. This is natural enough—faced with only a crudely developed technological system, they felt little need for a decisive response. Instead they concentrated on the particular adverse effects, leaving the metaphysics implicit. But many of their insights are still valid; they stand as early warnings. And their observations and analyses are both more coherent and more consistent when viewed from the standpoint of a pantechnical ontology.

NOTES

1. A further discussion of his views occurs in Part II.
2. *Technology and the Good Life?*, Higgs et al. (eds.).
3. Cited in reference to Rodman (1977).
4. He refers here to an earlier statement in the book: "The [device] pattern . . . inheres in the dominant way in which we in the modern era have been taking up with the world; and that characteristic approach to reality I call (modern) technology" (3). Later he reiterates: "To avoid misunderstanding, let me repeat that my concern is with modern technology and its character. . . . I will simply speak of technology when I mean modern technology" (12).
5. Though later he cites Heidegger as an "extreme proponent" of it (40).
6. "Climate peril may force limits on coal and oil" (June 3, 1977).
7. "Scientists fear heavy coal use may bring adverse shift in climate" (July 25, 1977).
8. It wasn't until 1987 or 1988 that global warming became prominent in the news.
9. Under the "business as usual" model. See *Emission Gap Report* 2012, UNEP. Projections out to 2050 are even worse; the OECD baseline projection is 80 billion tons, versus 20 billion required to minimize temperature rise.
10. The same attitude continues in his 2007 interview, in which he refers to "the unquestionably beneficial and indispensible structures of technology" (12).
11. Predestination notwithstanding.
12. See *Republic,* Book IV, 439d.
13. Particularly striking are those Christians who accept and embrace the absurdities, because it provides a "test of faith." It certainly is a test—and it's not one of faith. I emphasize that this concern does not afflict Christianity alone. Ranking only slightly higher on the scale of absurdity is its allied religion, Judaism. And Islam certainly has its fair share of such problems. Hence the fairly clear distinction between philosophy and theology.
14. To the point where one must question the orthodoxy of Ellul's beliefs.

Part II

Praxis

7 Classical Critiques
Glimpses of the Pantechnikon

In *Nicomachean Ethics*, Aristotle reminds us of one of Plato's favorite questions: Are we on our way from or to first principles?[1] That is, are we starting with some intuited, reasoned, or otherwise given metaphysics that then lead us to certain conclusions about the world? Or do we start from factual knowledge about the world, which then suggests certain metaphysical views? Plato often took the former route, preferring to begin with grand metaphysical visions and then inferring practical conclusions. Aristotle opted for the latter: "We must begin with things familiar to us," he said, "[f]or the facts are the starting point" (1095b). Pragmatic man that he was, Aristotle saw no value in obscure theoretical entities that had little relevance to real life. Metaphysics was important, but it had to have consequences—tangible effects that impacted our daily lives. Simply because these higher concepts were 'metaphysical,' or rather 'first principles,' in no way should suggest that they are detached from reality. Quite the contrary: they are the *most* real of philosophical entities and thus have profound practical implications.

There are virtues in both approaches. For my part, I like to think of it as a dialogue—empirical data supporting metaphysics, and metaphysical principles helping to explain data. One starts with a general idea, perhaps stimulated by something heard or read, or even just a hunch, and then seeks to explain it via a general theory and supported by real-world data. As data are examined, the theory is modified. As the theory takes shape, certain results are suggested and sought for. Even *what counts as facts* is affected by the metaphysical theory or worldview that one adopts, and so the interaction can be quite complex. But granting this, there is evidently a kind of coevolutionary process that occurs in many cases, and I think it best describes my own approach. In previous chapters I have emphasized the first principles of technology, with a minimum of empiricism. From here on, I will attempt to elaborate on the pragmatic aspect of the matter, looking specifically at what it does to us and to nature—without losing track of the larger metaphysical picture that it must ultimately support.

The outlook, unfortunately, is not promising. The metaphysics of the Pantechnikon portrays a universal process of the formation of order and complexity wherever energy is present in excess and the physical conditions allow. Man-made technology is the latest and most complex phase of this process here on Earth. We do not, however, control such a process; rather, we are the means by which it is manifest. As long as abundant free energy is available, technology will advance—regardless of our wishes. Or more precisely, regardless of the *circumstances*; our wishes are not relevant, largely being a product of technological thinking. No matter the situation, we will always wish for more technology.

For millennia, the advance of technology was not a major concern; it was relatively benign and of limited potency. In the modern era, however, our tools began to exceed our grasp. The technological system is now accelerating in scope and power, threatening to cause extreme damage to humanity, nature, and possibly to itself. Stresses are increasing, and there are many signs that technology is causing severe hardship. For humanity, this takes the form of physical, psychological, and moral harm. For the environment, it comes in a variety of ways: species extinction, deforestation, pollution, and global climate change, to name a few. In all cases, we have considerable empirical data to support the claim that technology is at the root of these problems. They are furthermore intrinsic to technology, such that we will be unable to fully alleviate them without either eradicating the technological source or implementing a yet more complex technological solution. Unfortunately, the notion of a technological fix to a technological problem is bound to fail when the problem is technology itself.

To further compound the difficulties, we have entrenched interests that unflinchingly support the advancement of technology and oppose all contrary efforts. Additionally, technological thinking has so pervaded common culture—at least in the industrialized nations—that it is extremely difficult to get a fair hearing in the court of public opinion. The average person sees and thinks technologically; he cannot conceive of non- or "anti-" technological solutions. Thus, from both above and below, as it were, protechnological forces drive us ever onward in spite of the evident hazards. But this is precisely what we would expect in a pantechnical universe.

The challenge is difficult. But we must persevere—our quality of life, if not our very survival, is at stake. Let us give the data a fair reading. And let us begin by listening to voices of the past, the great thinkers in our own philosophical tradition, with an open mind. Their insights, ancient or modern, are perhaps more valuable and more valid today than when they were first spoken. The great philosophers were every bit as clever and a good deal more insightful than most of us today. Life was more transparent then and their minds less beclouded. Thus, even though the immediate danger was slight, they could sense a process that was heading into dark waters. They could see then, from a much greater distance, things that we today look right past, simply because they are—here.

AN UNACKNOWLEDGED TRADITION

I propose, then, to begin with a look at the history of technology criticism. By some accountings, this falls under the heading of 'antitechnology.' On the one hand this is understandable, insofar as the critics are finding fault with the technology of their day. They are indeed critical and condemnatory—sometimes harshly and viscerally so, other times more objectively. On the other hand, they are clearly not condemning *all* technology, as the label seems to imply. No thinker has ever suggested that humanity utterly do without—something that is impossible in any case. They are, more correctly, opposed to those aspects or modes of technology that they find particularly harmful.

Furthermore, in the present context and from the standpoint of the Pantechnikon, this 'antitech' label is even less relevant. Technology is here identified as the universal process of creation, as the realization (technê) of a universal order (Logos). From this perspective, one can no more be antitechnology than one can be antiexistence; it is a logical impossibility. Hence I reject the antitechnology label as being vague, imprecise, ambiguous, and defamatory.[2] The past thinkers were critics, yes; opponents, yes; but antitechnologists, no.

There are several reasons this historical study is important. First, it is largely unknown, even among the experts; I elaborate on this point in what follows. Second, many people know of *some* of these critics, perhaps even most, but few understand the full picture. It is a virtue in itself to compile something approaching a complete list of such writings, even if in condensed form. Third, it demonstrates the longstanding and sophisticated nature of severe doubts about technological progress. This is no recent phenomenon, churned out by a handful of cranks. It has a long and noble pedigree that cannot be ignored. Fourth, the history of philosophy of technology is almost exclusively a *critical* history. It is not as if there were two streams of thought, pro and con, and I choose to present only one; for all practical purposes, and from a philosophical standpoint, there is only one outlook, and it is largely negative. No philosophers have written in praise of technology, and none have responded to the critiques of its detractors. This fact alone should give us pause.

There are further reasons. If there is validity in the pantechnical metaphysical picture, then it obviously has existed throughout human history and thus was amenable to comment—even though, in the human realm, technology was present only at a low level of intensity. Thus, on my view, we should expect that past philosophers caught only a glimpse of the Pantechnikon and commented accordingly. I would predict that they would observe certain salient features of technology: its suprahuman power, its ability to alter and even determine social conditions, its psychological impact on people, its fixation on energy and power, its unstoppability. In fact, all these points have been raised in the past.

If they did not offer a fully articulated vision of the Pantechnikon, there were reasons for this. For most of them, the human technical sphere was not advanced enough to clearly see the effects. Technology posed no immediate danger and seemed to present no special philosophical problem. Additionally, many of them lacked the appropriate intellectual concepts to formulate their views more explicitly. They, like all of us, had only the philosophical tools of their time, and this imposed certain limitations. Hence they largely bypassed theoretical metaphysical explanations in favor of practical consequences and other pragmatic considerations. But even here, it was, for the most part, a series of isolated observations. There is no attempt to place a critique in a larger context and no use of data to confirm theoretical views. My objective is to remedy both of these shortcomings.

Finally, what follows is not merely a sequential listing of views. It is an attempt to examine past comments in light of a new, pantechnical perspective. This not only brings new insight into prior ideas, but it reciprocally illuminates the concept of the Pantechnikon itself. There is thus a double benefit to such a review in the present context.

My first point, however, requires further discussion. It is surprising how little knowledge of past critiques is displayed by contemporary philosophers. This is clearly part of a broader, *technological* trend of hypermodernism in philosophy, wherein 'the old' is typically disparaged as obsolete, irrelevant, and uninteresting. Just as old technology is (generally) useless technology, so too, many say, old philosophy is useless philosophy. This is all the more true with a putatively contemporary issue such as the philosophy of technology.

For example, Cooper (1995) examines the conflict between the "liberation" (pro-) and "enslavement" (anti-) camps regarding technology's role in society. The enslavement camp is allotted a mere two-page section, covering a handful of vaguely antitechnology sentiments. Cooper cites by name the following individuals: Marx (no critic of technology but a proponent of technological determinism), Alasdair MacIntyre (a moral and political philosopher with no interest in technology), Marcuse (a valid critic), Heidegger—and no more.[3] All technology concerns are grouped into three categories: (1) technology out of control ("the Frankenstein thesis"), (2) the control of the many by the few, or the "antidemocratic" thesis, and (3) "epistemic enslavement" of the human psyche. Each category is given just one or two paragraphs of elaboration. Granted that Cooper's essay is not a lengthy piece, but to suggest that this is anything close to an adequate study of the "enslaving" aspects of technology is a gross understatement—particularly as the lead essay in the anthology in which it appears.

More revealing is the monograph by Dusek (2006). Nominally an objective study for use as a college text, this book tries to present a balanced view of all relevant issues. Technology critics are corralled into Chapter 11, titled "Antitechnology: Romanticism, Luddism, and the Ecology Movement." Subheadings in the chapter cover the three title areas plus deep ecology, ecofeminism, overpopulation, and sustainability. Even prior to getting into the text, one senses that this will be a less-than-compelling presentation of technology criticism.

The first section, on romanticism, mentions by name Rousseau (a true critic), Schelling (a noncritic), Whitehead, Carlyle, and Dickens, though with few or no details on their specific critiques. Dusek then covers the poets—Wordsworth, Blake, Ruskin, Emerson: interesting observations but irrelevant to a philosophical analysis. Section two on the Luddites recalls that historical drama, again, of some interest but little philosophical value. The section closes with a few words on author, activist, and psychotherapist Chellis Glendinning, a pop critic of modern technology. The ecology section covers, in rapid sequence, Haeckel and Spencer (noncritics), "the Nazis," ecosystem models, Kuhn, Whitehead, and various obscure ecologists. Then comes deep ecology, with a very brief look at Naess and Spinoza. The ecofeminism section cites Warren, Salleh, and a few others. Overpopulation is then addressed, with yet more references to "the Nazis," followed by quick mention of Malthus, Marx, Mao, Ehrlich, Hardin, and Commoner—not a *philosophical* technology critic among them. The closing section quotes the Brundtland Report of 1987 and the Native American "seven generations" test. One can be excused for losing the technology thread here.

In fact it is worse than sheer inadequacy. Such a chapter leaves an impression on the reader—the student—that there is no cogent antitechnology movement throughout all of history, that technology critics have nothing meaningful to say, and that the position is more or less equated to either naive romanticism, stereotypical Nazism, or fringe environmentalism. This is so far from the truth that Dusek commits a veritable sin against his own subject matter. More generally, it is indicative of the low state of scholarly comprehension of intelligent opposition to technology.

To be fair, Dusek is not alone. Other recent anthologies and sourcebooks fare little better. The collection edited by Scharff and Dusek (2003) includes just one historical critique, a short essay by Rousseau. There is a reasonable sampling of recent critics, from Heidegger and Ellul to Winner, Schumacher, Mumford, and Jonas, but they are not presented in any systematic way.[4] Johnson and Wetmore (2009) include 34 readings, with but two by critics—Winner and Bill Joy. Kaplan (2009) covers Heidegger, Marcuse, Jonas, and Winner—but Ellul is nowhere to be found, and there are no historical critiques. Finally, Hanks (2010) includes entries by Heidegger, Ellul, Winner, Marcuse, Mumford, and Jonas, along with a few issue-specific critiques and a quaint homily by Wendell Berry on why he is "not going to buy a computer." Overall, the absence of historical pieces is perhaps understandable, given the pressure on such anthologies to be "current and timely." Understandable—but then there is *no time at all* for the important historical insights and the long-term trends that they reveal.

The most recent short shrift appears in Tabachnick's book *The Great Reversal: How We Let Technology Take Control of the Planet* (2013). Nominally a critical philosophical study and one that promises to be severe, this book again presents a woefully inadequate depiction of the contrarian viewpoint. His chapter on "responses" includes a four-page section on "antitechnology" views, which ends up being mostly on Heidegger, and a two-page section on neo-Luddism that merely touches on a handful of topics, ranging from radical environmentalism to *Fight Club* and Ted Kaczynski to Kirkpatrick Sale. The remainder of the chapter skips lightly through the usual suspects: the Romantic poets, Oswald Spengler, "the Nazis," and more Heidegger. All in all, another vapid portrayal.

Without doubt, part of the explanation for the poor presentation of technology criticism is the belief by authors and editors that there is *not much to it*—that the view represented by Heidegger, Ellul, and others amounts to mere "demonizing of technology," as Borgmann put it, and thus is "easily dismissed." Substantivism and the corresponding technological determinism are no longer credible theories, they say. We need not waste our time even refuting such views; a wave of the philosophical hand will suffice. Smith (1994: 38–39) states the standard view concisely; he "lament[s] the stubborn tenacity of technological determinism," calling it a "variety of superstition," somewhat akin to creationism. Betraying his confusion on the issue, he asks, "How can something so demonstrably wrong-headed continue to sway adherents?" "Demonstrable"—but undemonstrated.

Let us, then, take a look at those wrong-headed views of the past—views in which some of our greatest thinkers were horribly misled by their own fears and apprehensions, thinkers who were unable to see beyond the tips of their philosophical noses, who fell straightaway into the obvious foolishness of technological determinism and substantivism, and who failed to see that technology is always developed by humans to serve strictly human interests and always operates with the best of intentions, at least in principle if not in fact.

Or perhaps this: Let us get a sense of their true thinking—their *skeptical* thinking—with an eye to the metaphysics underlying the technical phenomenon. Let us place their views in a larger sociohistorical context. There is much to be learned.

APPREHENSION IN THE ANCIENT WORLD

The essence of technology is necessarily connected with its character as we experience it. Most plausibly, that essence underlies our experiences and, in a loose sense, explains them. As explained in Part I, a pantechnical metaphysics is consistent with the original Greek notions of technê and logos. The dialectical nature of metaphysics implies that our experience of technology gives us clues to its metaphysical nature, and intellectual speculations on its metaphysics can be confirmed by observations in the real world. If the metaphysics of the Pantechnikon are broadly correct, it must then have certain definable and objective consequences—consequences that will be noted by perceptive observers. In Chapter 1, I showed that Heraclitus, Plato, Aristotle, and the Stoics articulated various elements of a technê-logos worldview, one that was evidently consistent with their observations of the natural world. For them, a cosmos guided by a logos- or nous-like rationality and informed by a creative, artistic, or technê-like formative process offered the best and most coherent metaphysical picture.

But they lived in a time of relatively benign and unobtrusive technology. The tools and devices of that society were simple, straightforward, and transparent in nature, even if difficult to master. The technician was not yet in a position of social status; the intellectuals of the Greek world looked with disdain on those whose occupation involved the design and construction of the "merely useful" goods. A modern technological mindset had obviously not yet taken hold. Hence there were few compelling reasons to examine the phenomenon critically, let alone metaphysically, as a particular manifestation of more universal processes.

Rather, it was more often the opposite. The simplicity, immediacy, and directness of the tools, combined with their obvious advantages, served to hide the larger issues from the view of most observers. Once one has chosen or been compelled to a life of agriculture, for example, one clearly needs functional and reliable tools. When one is under attack, one needs the strongest and surest possible defense. Apart from deliberate misuse, basic tools have no obvious side effects. They seem to have no downsides and only benefits to the users. Who could reasonably question them?

And yet even then, in the ancient world, some saw in rudimentary tools and technology the power to restructure and thus disrupt existing society. Such men viewed even the simplest techniques as things with the potential to degrade human relations with the world and with each other—often in ways that are obscure or unpredictable.

The Greeks were among the first to view technology with a skeptical eye. They initiated a long and impressive chain of critiques that grew in intensity as the technological phenomenon progressed. Such a historical overview is therefore important because it demonstrates not only a persistent sense of concern about technology over the past two millennia, but also its growing urgency parallels the growth in power and pervasiveness of technology itself. And most importantly, these critiques should help us to assess my various metaphysical claims.

This historical study naturally focuses on Western philosophy and tradition, but I will begin with a few comments from ancient China—and with the central text of Taoism, the *Tao Te Ching*. This collection of short sayings and aphorisms has been historically attributed to the venerable sage Lao Tzu, who is said to have been an older contemporary of Confucius—placing him in the early part of the sixth century BC. The title means, roughly, 'the book of the Way (*tao*), and of Virtue (*te*).' The Way, or Tao, is the ultimate creative force of the cosmos and is something every good Taoist wishes to align himself with. Achieving a life of harmony with the Tao is the path of virtue.

Taoism advocates a simple life, free from the luxuries and complexities of civilization. Luxuries—those things that are 'hard to come by'—are more than unnecessary; they are virtual obstacles to achieving a harmonious life: "Goods hard to come by / serve to hinder [man's] progress" (I.12).[5] The wise man, therefore, works to free himself from desire for artificialities and human contrivances. He seeks the simple life.

Artificial things are a result of man's cleverness and hubris. The way of the Tao is simple and natural, and thus it requires no complex analytic thinking and few human constructions or creations. Cleverness leads to innovation, which results in the luxuries and artifices that corrupt society and obstruct the free and open life:

> Exterminate ingenuity, discard profit / and there will be no more thieves and bandits. (I.19)
>
> When cleverness emerges / there is great hypocrisy. (I.18)
>
> Woe to him who willfully innovates. (I.16)
>
> Exterminate learning and there will no longer be worries. (I.20)

"Learning," or technical knowledge, is singled out as particularly important. It very clearly leads one down the wrong path, away from the Tao:

> In the pursuit of learning one knows more every day;
> In the pursuit of the Tao one does less every day.
> One does less and less until one does nothing at all,
> And when one does nothing at all, there is nothing that is undone. (II.48)

Learning produces tools, technology, and this is a source of great concern for the well-run society. Tools alter and disrupt both the social and natural orders. As tools become more highly refined, the state becomes more unstable and more vulnerable; it enters a darkened time, a condition of benightenment:

> The more sharpened tools the people have,
> The more benighted the state. (II.57)

Paraphrasing this stanza, we might say: *The sharper the tools, the darker the times.* Here we have a prescient, poetic, ominous vision of technology—from well over 2,000 years ago.

Given this warning, the wise man knows when enough is enough. He knows when to stop:

> One ought to know that it is time to stop.
> Knowing when to stop, one can be free from danger. (I.32)
>
> Know when to stop
> And you will meet with no danger.
> You can then endure. (II.44)

Today we have tools that would have been, for Lao Tzu, inconceivably potent. Clearly our era is one of very "sharp" tools. This suggests that it is correspondingly a very dark time. The danger, Lao Tzu would say, is great. Our very endurance is at risk. We must know when to stop.

Another relevant critique is found in the work of Chuang Tzu, a quasi-Taoist philosopher who lived some 100 years later. His main book, also known as the *Chuang Tzu*, is another classic of the orient. Chapter 12 tells of a Tao-inspired old man, a gardener, who declines to use even simple technology because of its tendency to corrupt the spirit.

As the story goes, a passing Confucian, Tzu-kung, comes across the gardener, who had carved a narrow passage down to the bottom of a well, in order to draw water by hand with his pitcher. It was a slow and arduous process. Wanting to be helpful, Tzu-kung says to the man: "Suppose you had a machine which in one day would irrigate a hundred fields. You would have plentiful results to show for very little effort. Wouldn't you want one of them?" "How does it work?" asks the gardener. Tzu-kung proceeds to describe a lever system with a weight on one end and a bucket on a rope leading down into the well—a simple device known as a well-sweep. The old gardener then scrunches up his face, laughs, and says:

> I have heard from my teacher [Lao Tzu] that those who possess machines must end up becoming mechanical in their affairs. And those who are mechanical in their affairs must end up becoming mechanical in their hearts and minds. If a mechanical heart is situated in one's chest, then one can't prepare oneself to receive pure simplicity. If one can't be prepared to receive pure simplicity, then the spirit becomes unsettled, and Tao has no place to enter. It's not that I wasn't aware of such machines, but that it would be disgraceful to end up that way.[6]

The gardener then sends Tzu-kung on his way, confused and embarrassed.

It is a striking parable, especially given, as I said earlier, the self-evident benefits of simple technology and apparent lack of drawbacks. One wonders: Can the use of even a simple lever system really cause one to become "mechanical in one's heart and mind?" Evidently Chuang Tzu thought so. He could see that a fixation on tools and devices draws one toward earthly, material things and away from the life of simplicity, contemplation, and spirituality. Such a "disgrace" must have been commonplace, or he would likely have not been compelled to comment.

ATHENIAN CRITIQUES

Meanwhile, on the other side of the world, the Greek philosophers were embarking on a mission of inquiry that would create Western civilization and Western philosophy as we know it. Their cultural achievements were numerous and well known: art, architecture, literature, and philosophy, among others. This group, consisting of only some 30,000 Athenian citizens, created one of the peaks of human existence, using, by our standards, the most primitive of tools. As I explained in Chapter 2, technê was but one of many aspects of everyday life, the 'art or skill' by which things come into being. It was furthermore connected with the concept of logos, either conceptually (Plato) or by definition (Aristotle). I quoted from *Gorgias* the passage in which Plato notes that every mode of technê is concerned with a particular kind of logos; this includes rhetoric, medicine, physical training, and so on. Each art deals with its own specific underlying order: "All of [the various technê] are concerned with a particular logos, namely, the one that corresponds to each particular technê" (450b). Plato thus establishes, for the first time, a fundamental connection between technê and logos—a connection that would soon be formalized by Aristotle.

But is there anything of concern here? Plato seems to think so. One telling passage comes from *Phaedrus.* Here Plato recounts the ancient Egyptian story of the god Theuth (or Thoth), who was said to have delivered all the great skills and arts to the people of Egypt—some 4,000 years before Plato's time. These included arithmetic, geometry, astronomy, and games such as checkers and dice. The most important gift, however, was writing. According to the story, Theuth visited an Egyptian king, Thamus, to brag a bit about the great things he had given the people and to suggest that Thamus disperse these wonderful skills throughout the land. But wise and skeptical man that he was, Thamus had a few questions first; for him, precaution was the order of the day. He went down the list, one by one, and insisted upon hearing the case for each: Were they really an improvement in the condition of man?

Each technê had its own advantages and drawbacks, and few were problem free. But Theuth was certain that writing, of all the *technai*, would be praised by Thamus: "This invention, O king, will make the Egyptians wise and will improve their memories; for it is an elixir of memory and wisdom that I have discovered." But he was surprised by Thamus's reply:

> Most ingenious Theuth, one man has the ability to beget technê, but the ability to judge of their usefulness or harmfulness to their users belongs to others; and now you, who are the father of writing, have been led by your affection to ascribe to them a power the opposite of that which they really possess. For this invention will produce forgetfulness in the minds of those who learn to use it, because they will not practice their memory. Their trust in writing, produced by external characters that are no part of themselves, will discourage the use of their own memory within them. You have invented an elixir not of memory, but of reminding; and you offer your pupils the appearance of wisdom, not true wisdom, for they will read many things without instruction and will therefore seem to know many things, when they are, for the most part, ignorant and hard to get along with, since they are not wise, but only appear wise. (275a)

Plato is not relaying this story to criticize it; he clearly agrees.

We need to be clear on the warning posed in this short passage. The concern was not simply that the written word could be misused or misappropriated—that is obvious enough. Rather, there are fundamental, inherent qualities in writing that make it not only deficient but outright deceptive. Writing is inadequate because it is static; frozen in time, it is unable to respond to the needs of the reader. It cannot answer questions. It cannot explain itself. It must be interpreted. Hence the author is unable to transmit anything of real value: "He who thinks, then, that he has left behind him any technê of writing, and he who receives it in the belief that anything in writing will be clear and certain, would be an utterly simple person." In fact, the only real value of writing is for *the writer*, to remind himself of his own ideas.

Truth can only be conveyed in a dynamic, interpersonal manner, between living human beings. But writing is, in a sense, dead; it is a mere skeleton of living thoughts and hence offers nothing of substance. Worse—it gives the *illusion* of substance, causing the reader to think he has gained something of greater value than he really has. It is a deceptive technology and thus carries with it a danger. Here then, at the very outset of Western civilization, we find in Plato the establishment of some basic themes in technology criticism: the deceptiveness, the inherent flaws, and the unanticipated consequences.

Another revealing passage occurs in *Republic*. Book II is a detailed inquiry into the nature of justice. At the beginning of this book Plato relates the story of the Ring of Gyges. As we recall, it is a fantastic, science fiction–like story about a poor shepherd who peers into a chasm in the ground and finds a giant corpse wearing only a gold ring. He takes the ring and later discovers that, by turning it on his finger, he can disappear and reappear at will. Invisibility gives the shepherd unmatched power, and he quickly dispatches the local king, thereby winning the kingdom.[7]

For Plato, this story is a lesson in justice and the inherently corrupt nature of mankind: Given sufficiently powerful tools, humans will exploit, plunder, and kill. Tools, if not actually producing immorality and corruption in us, at the very least give license to our most malevolent natures. In either case the result is the same. The implication is that we cannot be trusted with powerful tools and thus ought, rationally, to limit their potency and spread.

These concerns seem not to have impressed Aristotle. This is perhaps to be expected, seeing that he was much more the technician—in the broader sense—than his mentor. His fascination with logic, definition, categorization, and precision marks the foundation of our analytic, 'technological' tradition in philosophy.

Aristotle was a philosophical technician but not a technologist. In fact the Greeks generally, and despite their prodigious accomplishments, seem to have successfully held technology at bay—either by will or good luck. There was a notable lack of major technological breakthroughs during Athens' golden age, and their available technologies were relatively crude: primarily, four metals (gold, silver, bronze, iron)[8], ceramics, domesticated animals, oil lamps, and candles. They had wheels and simple wheeled vehicles, seagoing ships, and simple crossbows, but the inventions of Archimedes did not come until the post-Aristotelian period. There were new applications, of course, but for the most part the basic technologies of classical Greece had been known for centuries.

In any case, Greek intellectuals clearly viewed the practical, technical arts with disdain. Ellul observed that, for them, "technical research was considered unworthy of the intellect" (1964: 28–29). They "held contemplation to be the goal of intellectual activity, refused the use of power, [and] respected natural things." He added that "the Greeks were suspicious of technical activity because it represented an aspect of brute force and implied a want of moderation."

We find evidence for this in both Plato and Aristotle. Later in *Gorgias*, Plato briefly discusses the role of the "engineer" (*mêchanopoión*, lit. "maker of machines"). Socrates contrasts the Sophists' penchant for robust self-praise with that of oarsmen or engineers, neither of whom go around praising themselves even though their work is vital to human well-being. Should they attempt to do so, the Sophists would be appalled. "You [Callicles] despise him and his technê, and you would call him 'engineer' as a term of abuse" (512c). This was evidently the common view among the Greek intelligentsia.

In *Symposium*, Plato elaborates on the virtue of spiritual expertise. One who is wise in such matters is a "man of the spirit," but he whose profession is one of technê is "merely a mechanic" (*banausos*, 'artisan'). Again we see the generally disparaging tone. Finally, in *Laws* we find a caution against becoming a craftsman. "No citizen," says Plato, "should enter the ranks of the workers whose vocation lies in the arts and crafts (*technêmata*)" (846d). A citizen's primary task is keeping the polis in good working order, and he cannot do this and also be a craftsman. As before, the manual workers and technicians come across as second-class citizens.

Aristotle, in *Politics*, casts a bit more light on precisely why the artisans were denigrated. Apparently over the centuries it became custom for the lower classes—the slaves and foreigners—to do the bulk of the manual labor, which included many of the practical *technai*. Such people are not worthy of full inclusion in the ideal state, he says. "The best form of polis will not admit [the artisan class (*banausoi*)] to citizenship" (1278a8). Manual workers are distracted by their material tasks and cannot focus on the best style of life, that of contemplation: "for no man can practice excellence who is living the life of a mechanic or laborer" (1278a21). Somewhat later he reiterates the point: "there is no room for excellence in any of [the inferior] employments, whether they be artisans, or traders, or laborers" (1319a27).

Finally, near the end of *Politics*, in a discussion on childhood education, Aristotle condemns those who press their children into physical training at the expense of the more intellectual pursuits; "parents who devote their children to gymnastics while they neglect their necessary education, in reality make them into mechanics" (1338b32)—that is, into people who are merely useful and one dimensional. Thus we find a fairly consistent and fairly negative depiction of the technical artisan among our two leading sources.

The Greeks, then, did not fear technology; rather, they simply refused it dominance in their society. This refusal, said Ellul, "was the result, perfectly mastered and perfectly measured, of a certain conception of life. It represented an apex of civilization and intelligence." Greek rejection of technology

> was a deliberate, positive activity involving self-mastery, recognition of destiny, and the application of a given conception of life. Only the most modest techniques were permitted—those which would respond directly to material needs in such a way that these needs did not get the upper hand.
>
> In Greece a conscious effort was made to economize on means and to reduce the sphere of influence of technique. . . . The great preoccupation of the Greeks was balance, harmony, and moderation; hence, they fiercely resisted the unrestrained force inherent in technique, and rejected it because of its potentialities. (29)

This seems to have been the last period in human history when technology could be restrained, if only temporarily, by an act of will.

CHRISTIANITY

Three hundred years after classical Greece, into the early Christian era, we find an ambiguous stance toward technology. On the one hand we have the Old Testament mandate of "dominion" over the Earth, and the implied suggestion that tools are the means to this domination. On the Christian view, nature is no longer a sacred realm of enspirited things; it is now seen as something "given into the hand" of mankind, for his use, at his pleasure.

At the same time, however, we find Biblical passages suggestive of a cautious attitude toward technology and even of an association with sin. The forbidden fruit of the Tree of Knowledge surely represents technical as well as moral knowledge. Notably, the first act that Adam and Eve undertake upon eating of this fruit is a technical act: the sewing of garments.[9] Their murderous son, Cain, took upon himself the role of a technician and builder—of cities.[10] And the grandest technological accomplishment of the Old Testament, the Tower of Babel, was condemned by God as a sign of hubris.[11] Apart from these faintly negative associations, though, technology largely avoided reprimand for the next 1,500 years.

INTO THE RENAISSANCE

In the second millennium AD, technology underwent a slow but inexorable acceleration. Between the years 1000 and 1500 we find several notable advances, including improved metal horse gear, gunpowder, the compass, new architectural techniques (as in the great European cathedrals), the modern clock, the cannon, oceangoing commercial ships, the printing press, and the musket. Advances in agriculture and commerce, directly attributable to improved technology, led to a growing population and the first large, high-density cities. The conditions were set for a variety of problems, including poverty, growing social inequality, and health and disease concerns. Almost on cue, the world witnessed its first widespread catastrophe—the plague—in the mid-1300s.[12]

By the 1500s it was becoming more apparent that technology cut with a double-edged blade, capable of tearing into the fabric of mankind. It was a threat both to man's moral integrity and to the well-ordered worldview constructed by the Church. Writing in 1530, Henry Agrippa observed that the "worldly arts and sciences"—'arts' meaning the practical arts, or technology—posed precisely these threats:

> There is nothing more pernicious, nothing more destructive to the well-being of men, or to the salvation of our souls, than the Arts and Sciences themselves. And therefore quite contrary to what has been hitherto practiced, my opinion is, that these Arts and Sciences are so far from being to be extolled with such high applauses and panegyrics, that they are rather for the most part to be dispraised and vilified: And that indeed there is none which does not merit just cause of reproof and censure.

The power of technology could be abused in myriad ways, heightening the prospect for disaster. The most evil of men could wreak ever-greater havoc, and the mighty power of reason could corrupt even the most well meaning. As Agrippa said, "there is nothing more deadly than to be, as it were, rationally mad."

Jean-Jacques Rousseau picked up the essence of Agrippa's message, secularized it, and built it into an entire philosophical outlook. In his first essay, *A Discourse on the Arts and Sciences*, Rousseau constructed a shockingly novel argument: that science and technology have, since the beginning of history, led not to enlightenment and virtue but rather to moral corruption and social decay. This is because they have a unique power to entice humanity into an unnatural social bondage, which in turn yields a host of ills. The arts and sciences "fling garlands of flowers over the chains which weigh [us] down" and "stifle in men's breasts that sense of original liberty for which they seem to have been born" ([1750] 1993: 5). The temptations of technology cause men "to love their own slavery" and hence to trade away their native freedom.

The progressive advance of technology over the ages has invoked a corresponding and proportionate decline in our humanity. Notably, this was not some phenomenon limited to Rousseau's own era: "Will it be said that this is a misfortune particular to the present age? No, gentlemen, the evils resulting from our vain curiosity are as old as the world. . . . Our minds have been corrupted in proportion as the arts and sciences have improved" (8). He then lists several historical examples. In each case an original, robust, and free society proceeds down a path of cultural and technological development—which, when achieved, leads to decay and collapse. Rousseau cites the Egyptians, Athenians, Romans, and Ottomans as evidence. In contrast, he says, consider the austere but robust societies of Scythia, early Persia, Sparta, the Germans, and the Swiss—simple, proud, noble, unfettered and unbothered by the trappings of a technological society. There is no victory, it seems, when it comes to the sciences and arts; to succeed is to fail.

Rousseau's thinking is unique in two important ways. He is the first to see that science and technology lead to a comprehensive restructuring of society, affecting political and economic aspects as much as our morals and social norms. He understood the causes and effects of a truly *technological* society: its homogenizing effects, its tendency to transform mankind into a herd-like animal, to press us into becoming something artificial, something false.[13] In building an artificial culture, mankind remakes himself as an artificial being. We cannot technologize the world without technologizing ourselves—this is Rousseau's moral.

Second, his essay provides, for the first time, something of an analysis of the situation. He observes, first of all, that technology arises from and is sustained by our weaknesses and failings, not our strengths. It is a kind of compensation for human limitations, a way of moving beyond our ordinary frailties—frailties obtained, of course, over millennia of evolution and in symbiotic relationship with nature. "The arts and sciences owe their birth to our vices; we should be less doubtful of their advantages, if they had sprung from our virtues" (15). Born of vice, they embody vice and reproduce it in their objects of attention.

Rousseau's critique is specific. Scientific inquiry results in error and misjudgment as often as it produces truth.[14] Through its labor-saving devices, technology serves to induce idleness and boredom in people; we no longer perform real, meaningful work but instead the meaningless, detached labors of virtual slaves to the system. The arts and sciences inevitably produce a prodigious and wasteful luxury of goods that corrodes morality and eviscerates virtue. Luxury also produces effeminate and cowardly men, weak of character and servile in disposition. Finally, the technological society places disproportionate value on "talent" rather than virtue and on trivial intellectualization rather than meaningful action in the world: "The question is no longer whether a man is honest, but whether he is clever. Rewards are lavished on wit and ingenuity, while virtue is left unhonored. There are a thousand prizes for fine discourses, and none for good actions" (23–24).

Rousseau's final plea: "give us back our ignorance, innocence, and poverty, which alone can make us happy" (27).

By the late 18th century, even the academic philosophers were taking note of the technological phenomenon and viewing it with an appropriate skepticism. Among the more notable was Kant. In his *Groundwork of the Metaphysics of Morals* (1785), he expounds on the contrast between reason and instinct as a means to happiness. Surprisingly, he observes that if happiness is taken as our objective in life, instinct is superior to reason.[15] This is because reason and its byproducts—the technical arts and sciences—more often than not lead us *away* from happiness:

> We find that the more a cultivated reason purposely occupies itself with the enjoyment of life and with happiness, so much the further does one get away from true satisfaction. . . . And from this there arises in many [people] . . . a certain degree of *misology,* that is, hatred of reason; for, after calculating all the advantages they draw—I do not say from [merely] the invention of all the arts of common luxury, but even from the sciences—they find that they have in fact only brought more trouble upon themselves instead of gaining in happiness. ([1785] 1996: 51)

An insightful observation, though predated by similar reflections of Rousseau. Kant apparently did not further elaborate on this idea, but comparable thoughts were expressed by Goethe, Schelling, and Schopenhauer—a series of German thinkers that would climax with Heidegger.

All the while, the technological pace was quickening. The Industrial Revolution was officially underway as of the mid-1700s. Steam engines of various sorts were becoming prevalent, as were coke-fired furnaces. Power weaving looms appeared around 1785, inspiring the Luddite revolt. By the early 1800s, steam locomotives appeared. New products, like the camera and the refrigerator, were coming on the scene. And the electric telegraph was just around the corner.

Amid these developments, Scottish philosopher and essayist Thomas Carlyle expressed his deep concerns about the spread of technology and mechanical ways of thinking. His important 1829 article "Signs of the Times" offers one of the earliest English-language critiques. Aligning himself with Rousseau, Carlyle laments the moral decay and complexity of contemporary life, seeing in it signs of social and spiritual degeneration. This age, he proclaims, "is the Age of Machinery," both inwardly and outwardly. Outward in the sense not only of the physical machinery but also in the mechanization of procedures in all walks of life. Every endeavor, from science and business to politics and the arts, had a formalized process by which things were now to be done. "Philosophy, science, art, literature, all depend on machinery," he wrote ([1829] 2007: 36). There were rules and laws to follow, periodicals to publish, meetings to attend, corporations to form, and royal societies to join. Everything, in other words, was becoming bureaucratized.

Inwardly and correspondingly was the mechanization of the mind and spirit. Echoing the ancient wisdom of Chuang Tzu, Carlyle exclaimed that "Men are grown mechanical in head and in heart, as well as in hand." They no longer rely on individual inspiration and effort but rather seek out the latest tools and devices to aid them in their work: "for Mechanism of one sort or other, do they hope and struggle. Their whole efforts, attachments, opinions, turn on mechanism, and are of a mechanical character." "Our true Deity," he said, "is Mechanism" (43). The purely intellectual activities, such as metaphysics and morality, fall by the wayside—they being relatively immune to mechanization. But if this is the case, it is of no use: "what cannot be investigated and understood mechanically, cannot be investigated and understood at all" (38). Thus, says Carlyle, have we abandoned the inner world and placed all our hopes in the outer.

The common mindset had so altered by this time that mechanical thinking had come to seem both natural and, worse, inevitable—"as if it could never have been otherwise." But of course, this is not so, says Carlyle. The whole prior history of mankind and all its accomplishments were achieved without the dominance of mechanistic thinking. The geniuses of science, literature, the arts, and philosophy required no procedures or processes, no royal guilds, no monthly periodicals, to achieve greatness. Thus it was, and thus it will always be, he says hopefully.

Carlyle was a religious man, and as a devout Christian he was committed to the same two propositions as Borgmann: God, not technology, is the presiding power in the universe; and human beings will always have the chance for redemption. For Carlyle, those who see Mechanism as God are the pagans, and they can be converted. The human spirit can never be broken or permanently enslaved; we need only have "faith in the imperishable dignity of man," and in the "high vocation" that man has been called to. The reigning "spiritual maladies" are but "chains of our own forging" (48). Through education and growth of knowledge, mankind will come to see its errors; thus we must press ever forward: "for not in turning back, not in resisting, but only in resolutely struggling forward, does our life consist."

Carlyle's naïve optimism comes to a resounding conclusion. We can recover the wisdom of our forefathers, he says, break out of the glass bell that surrounds us, and be free once again. Mechanism "is not always to be our hard taskmaster, but one day our pliant, all-ministering servant." We need not despair because "a new and brighter spiritual era is slowly evolving itself for all man." Perhaps.

Despite his religious-themed conclusion, Carlyle's essay marks an important advance in technology criticism. His rationale notwithstanding, he is the first to see technology as a God-like being, one that has tremendous power in the world and that commands worship and adulation—"our true Deity is Mechanism." In the framework of the Pantechnikon, there is some justification for such a view; technology possesses an almost god-like power in its ability to transform the world. And it is the embodiment of a kind of cosmic wisdom, the Logos, which is progressively revealing itself as the universe evolves. Carlyle, of course, overlays these insights with the trappings of Christianity and thus draws erroneous conclusions, but perhaps he cannot be faulted for that; Christian irrationality still held sway over much of the intellectual world in the early 19th century.

Furthermore, Carlyle was the first to describe an expanded notion of technology: as process, procedure, organization, and structure. Technology was no longer simply tools and machines but all organized means to particular ends. This conception anticipates Ellul's definition, but precedes it by some 120 years. It was a remarkable insight and points to a more penetrating look at the essence of technology—though without a metaphysical structure to support it.

Carlyle's essay was also unique in that it drew one of the earliest extended critical replies: an article by an American mathematician and lawyer, Timothy Walker. His piece, "In Defense of Mechanical Philosophy" ([1832] 1982), marks perhaps the first sustained attempt in the modern era to defend technology against its critics. Even in the intervening 180 years, in fact, there have been surprisingly few such defenses; technology apologists generally either ignore the critics, offer ad hominem fallacies, or give simplistic off-the-cuff dismissals. Sustained, philosophical replies are rare. In Chapter 10, I will examine a few particularly relevant defenses.

CRITICS IN THE INDUSTRIAL AGE

About 25 years later and an ocean away, Henry Thoreau lived the call for simplicity that was inaugurated by Rousseau. The basis for his two-year experiment in primitive living, documented in *Walden*, was his view of modern society as a degrading and counterproductive mode of existence, all but absorbing a man's life—"he has no time to be anything but a machine" ([1854] 1992: 6). Much preferred is the condition of "voluntary poverty," which allows one to live a "life of simplicity, independence, magnanimity, and trust" (13). These are classic themes that pervade his work.

Technological society turns against us, said Thoreau, and consumes us. It is thus more accurate to say that we serve technology rather than the other way around; "men have become the tools of their tools" (33). The system is now the master and we the slave. Even the animals own us: "I am wont to think that men are not so much the keepers of herds as herds are keepers of men" (49). The culmination of this view—"We do not ride on the railroad; it rides upon us" (82)—perfectly captures his mindset.

Despite the poetry and force of Thoreau's words, he offered little in terms of systematic or philosophical analysis. His focus was on the absorption of humanity's life energy, our enslavement in a larger technological system, and our subsequent loss of autonomy—in other words, the practical effect on the ordinary man. But yet behind his words is a looming presence, something that presses ahead unstoppably; technology, to Thoreau, advances with a locomotive force that threatens us all.

Not that some did not try to derail it. British essayist Samuel Butler recognized, by 1863, that humanity was facing an almost irresistible power. He believed that technology was relentlessly encroaching on us, threatening even to surpass us, its creators, in the flow of world events. Decisive in Butler's thinking was Darwin's theory of evolution—*Origins of Species* being published just four years earlier. Hence the title of his short, striking, but little-known piece: "Darwin among the Machines." It is a milestone in the history of technology critique.

In essence, Butler outlines an evolutionary theory of machinery. The concepts of natural selection and survival of the fittest applied not only to the sphere of life but also to that newest of kingdoms, the Kingdom of the Machines.[16] Machines come into being, modify their forms, grow in power and ability, consume energy, emit wastes, and otherwise function analogously to living organisms. Thus it is reasonable, thought Butler, to apply Darwinian theory to them as well. When we do this, we realize that they are stronger, faster, and better than us in many ways—even to the point of developing nascent mental attributes. Crucially, they evolve much, much faster. Looking to the future, then, we can foresee a day when machines will become the superior race and we are little more than their servants and domestic pets: "man will have become to the machine what the horse and dog are to man." It would be, says Butler, a kind of irreversible enslavement.

> Day by day, the machines are gaining ground upon us; day by day we are becoming more subservient to them; more men are daily bound down as slaves to tend them, more men are daily devoting the energies of their whole lives to the development of mechanical life. The upshot is simply a question of time, but that the time will come when the machines will hold the real supremacy over the world and its inhabitants is what no person of a truly philosophic mind can for a moment question. ([1863] 1968: 212)

In the face of such a threat, the only rational course of action is to go on the offensive, to attack the system before it overwhelms us. Butler marks out a radical proposal, in the most eloquent of language:

> Our opinion is that war to the death should be instantly proclaimed against them. Every machine of every sort should be destroyed by the well-wisher of his species. Let there be no exceptions made, no quarter shown; let us at once go back to the primeval condition of the race. If it be urged that this is impossible under the present condition of human affairs, this at once proves that the mischief is already done, that our servitude has commenced in good earnest, that we have raised a race of beings whom it is beyond our power to destroy, and that we are not only enslaved but are absolutely acquiescent in our bondage.

A follow-up essay, "Mechanical Creation" (1865), describes in prescient detail the process by which the machine society would surpass mankind. Butler considers the following question: How can it be that machines could attain dominance without our knowledge or intervention? Three reasons:

> Firstly, man is committed hopelessly to the machines. He cannot stop. If he would . . . bring up his children with a fair prospect of their thriving, he must go on improving the machines. . . . Secondly, man's interests may not be really opposed by his becoming the lower creature; the interests of the two races may continue in the same direction. . . . [T]he insensate mass [of people] will readily acquiesce in any arrangement which gives them cheaper comforts without yielding to unreasonable jealousy. . . . [Thirdly,] the change will be so slow and subtle that man's sense of what is due to himself will never be rudely violated at any given moment; and custom will deaden our sense to the noiseless and imperceptible aggressions of our own creations. ([1865] 1968: 236–237)

As long as the machines need us to survive, we will survive. We will be bred for specific purposes, much as pets and farm animals are. This is our future, such as it is. At least we need not worry about being crushed by the machines; "they will, as far as we can see, find us always . . . serviceable." It would hardly suit them "to exterminate us," he says optimistically. Of course, if they were to develop true autonomy and the ability of self-reproduction, then the rules will have changed. But that prospect, for Butler, was in the distant future, thousands of years hence. If only that were so.

Butler, then, was the first true technological pessimist. Given that we cannot stop the machines—*as of 1863*—the damage is already done, the process of enslavement is in motion, and we blindly consent to it. We see in his brief writings an early anticipation of several later themes, many to be developed in detail by Ellul. But from our present-day perspective, we are struck by how much Butler underestimated the situation. Technology advances vastly quicker than even he imagined, and its power and scope already exceed anything he foresaw. Its autonomy is a likely condition of the near future, not millennia hence. And he had no sense of the widespread environmental destruction that could be wrought by an overpopulated technological society. There is a notable irony here: Readers must have thought Butler half mad, but as it turned out, he was not mad enough.

By this time the second phase of the Industrial Revolution had begun. High-quality steel was being produced, and new devices like bicycles and gasoline engines were coming on the market. On the downside, the ugly grinding factory existence that Dickens described so well was in full swing. The capitalists had fully realized the benefit of modern technology and were putting it to maximum use. And it was around this time that the consummated "marriage of science and technology" (White 1967: 1204) occurred, allowing the acceleration of both.

Karl Marx was obviously impressed by the ability of capitalism to make use of industrial technology—and not in a good sense. As he saw things, technology was not intrinsically problematic; only when misused for a selfish profit motive did it become a social evil. But the spread of industry by the mid-1800s suggested to him, even more than to Carlyle and Butler before, a process that was fundamentally reshaping and reordering modern society. As such, Marx is widely but controversially viewed as the founder of the thesis of technological determinism. I will therefore defer my discussion of his views to the chapter on that subject.

For present purposes, the last major figure of the 19th century was Nietzsche. Most of his relevant comments come from his so-called middle period, roughly 1877 through 1882. Typically, they are not in the form of extended elaborations but rather appear as aphorisms or other short passages. As with many of his themes, we are obliged to take isolated comments and reflections and then piece together, in pointillist fashion, a coherent picture.

We find an early and relatively neutral comment in *Human, All Too Human* (1878–80), where Nietzsche remarks on the question of political power in socialism. Political power is like the powers of nature—dangerous if not carefully controlled. He compares it to the power of steam, when "pressed into service by man as god of the machine"; if the design is faulty, it explodes. To adequately manage such energy, we must consider how to make it "an instrument for the attainment of [our] objectives" (vol. I, sec. 446). Like Marx, he could see that this instrumentalist view cut both ways:

> Mankind mercilessly employs every individual as material for heating its great machines: but what then is the purpose of the machines if all individuals (that is to say, mankind) are of no other use than as material for maintaining them? Machines that are an end in themselves—is that the *umana commedia* [human comedy]? (sec. 585)

If machines are "an end in themselves," then the system works for its own benefit and not for ours. Technology progresses simply for the sake of more technology and no other reason.

Later in the same work, Nietzsche offers more cautionary notes. Of concern was the growing tendency of modern technology to promote large-scale organizations and to make of men a metaphorical cog in the bureaucratic machine. Indeed, machinery teaches us to think in precisely this way:

> *The machine as teacher*—The machine of itself teaches the mutual cooperation of hordes of men in operations where each man has to do only one thing: it provides the model for the party apparatus and the conduct of warfare. On the other hand, it does not teach individual autocracy: it makes of many *one* machine and of every individual an instrument to *one* end. Its most generalized effect is to teach the utility of centralization. (sec. 218)

Loss of autonomy is evidently a bad thing in itself. But this centralizing tendency, in conjunction with the ideological inculcation, also has a depressing effect on men's minds. It provokes a reaction:

> *Reaction against machine-culture*—The machine, itself a product of the highest intellectual energies, sets in motion in those who serve it almost nothing but the lower, non-intellectual energies. It thereby releases a vast quantity of energy in general that would otherwise be dormant, it is true; but it provides no instigation to enhancement, to improvement, to becoming an artist. It makes men *active* and *uniform*—but in the long run this engenders a counter-effect, a despairing boredom of soul, which teaches them to long for idleness in all its varieties. (sec. 220)

Here we see one of the first observations in history of an explicitly adverse psychological effect of technology.[17] Where all this leads, we do not know: "*Premises of the machine age*—The printing press, the machine, the railway, the telegraph are premises whose thousand-year conclusion no one has yet dared to draw" (sec. 278).[18]

In another passage, Nietzsche observes that industrial mass production depersonalizes the items of consumption, thus demeaning society in a fundamental way. Here he echoes Marx's discussion of alienation, though focusing on the consumer and society rather than on the worker:

> *To what extent the machine debases us*—The machine is impersonal, it deprives the piece of work of its pride, of the individual *goodness* and *faultiness* that adheres to all work not done by a machine—that is to say, of its little bit of humanity. In earlier times all purchasing from artisans was a *bestowing of a distinction on individuals,* and the things with which we surrounded ourselves were the insignia of these distinctions: household furniture and clothing thus became symbols of mutual esteem and personal solidarity, whereas we now seem to live in the midst of nothing but an anonymous and impersonal slavery. We must not purchase the alleviation of work at too high a price. (sec. 288)

Again, caution but not an extreme reaction.

Nietzsche's rhetoric heats up in *Daybreak* (1881), where he condemns the slavery intrinsic to the mass production system—and the men who allow themselves to be subjected to it. He derides the "factory slaves" for being happy and satisfied with their lot; rather, they ought to "feel it to be in general a *disgrace* to be thus used, and *used up,* as part of a machine" (sec. 206). Nothing about the "mechanical operation of a new society" could change "the disgrace of slavery into a virtue." In yielding to the machine, a man's inner value is discarded, and this is the greatest cost. Echoing Rousseau and Thoreau, Nietzsche praises "voluntary poverty" and "the free-heartedness of him without needs." To save themselves, the workers must leave not only their jobs but Europe itself, go abroad, and there, in a foreign culture, seek to recover their lost dignity; this is the ultimate "protest against the machine."

Such is the bulk of his commentary, and it offers a reasonably unified outlook. Machine society is degrading and psychically poisoning but evidently not an existential threat. It furthermore gives a false sense of pride and power, and thus we inevitably misuse it. He says as much in the late work *On the Genealogy of Morals* (1887) in a short passage that reflects a surprising glimmer of environmental awareness:

> Our entire modern way of life . . . has the appearance of sheer hubris and godlessness. . . . Our whole attitude toward nature, the way we violate her with the aid of machines and the heedless inventiveness of our technicians and engineers, is hubris. (Third essay, sec. 9)

Whether in the realm of nature or industry, technology seems to have an agenda. It is a self-established end and works against the deeper, truly human interests. McGinn (1980: 689) supplements this point: "[Nietzsche] would not have echoed the shibboleth: technology is neutral in itself; what matter are the uses to which it is put." Rather, he seems to have believed that technology embodies certain values and principles, and this fact makes it intrinsically unreformable. This is why Nietzsche calls for no reform program and no changes in use or ownership, as Marx did. Instead, in his one pointed recommendation, he states that humanity must literally abandon the techno-industrial system if we wish to save ourselves.

NOTES

1. Book I.4 (1095a33).
2. I will continue to sporadically use the "antitechnology" label as a convenient shorthand or in reference to original citations. This should not be taken to mean that I endorse the term as a meaningful concept.
3. Later in the piece he mentions two other legitimate critics, Spengler and Roszak.
4. Borgmann is included in the readings, as he is in most such anthologies, but as per Chapter 6, I consider him a noncritic.
5. All quotations taken from the Lau translation (Lao Tzu [1963]).
6. A modified translation can be found in Chuang-Tzu (1999: 82).
7. *Republic* (359d–360d). Knowledgeable readers will recognize this as the central plot line in Tolkien's *The Lord of the Rings*.
8. And, to a much lesser degree, lead.
9. "Then the eyes of both of them were opened, and they realized they were naked; so they sewed fig leaves together and made coverings for themselves" (Gen 3:6).
10. "Cain lay with his wife, and she became pregnant and gave birth to Enoch. Cain was then building a city, and he named it after his son Enoch" (Gen 4:17).

11. See Genesis 11:1–9.
12. There were earlier outbreaks of the plague, most notably during the reign of Ashdod, in Palestine, around 1000 BC, the Athenian plague circa 430 BC, and the Byzantine pandemic of the mid-500s AD. But these pale in comparison to the Black Death of the 14th century.
13. Rousseau is eloquent in his description of the oversocializing effect of the arts and sciences:

    > Before [technology] had molded our behavior, and taught our passions to speak an artificial language, our morals were rude but natural. . . . Human nature was not at bottom better then than now; but men found their security in the ease with which they could see through one another, and this advantage . . . prevented their having many vices.
    >
    > In our day, now that more subtle study and a more refined taste have reduced the 'art of pleasing' to a system, there prevails in modern manners a servile and deceptive conformity; so that one would think every mind has been cast in the same mold. . . .
    >
    > What a train of vices must attend this uncertainty!. . . Jealousy, suspicion, fear, coldness, reserve, hate, and fraud lie constantly concealed under that uniform and deceitful veil of politeness. (6–7)

    He has nothing but disdain for those "who seem, from morning till night, to have no other care than to oblige one another."
14. Thanks to Popper and Kuhn, we now know that it is worse than this: that *all* scientific theories are intrinsically false and that what passes for true is only the most workable or the least objectionable theory at our disposal.
15. Contra Aristotle. See *Nicomachean Ethics* (1098a1–b5).
16. Joining those of animal, plant, and mineral.
17. Chuang Tzu, Carlyle, and Hegel suggested such an outcome, but only obliquely.
18. Though Butler, as we know, did dare to draw such conclusions. Evidently Nietzsche was unaware of his work.

8 Recent Critiques
1900 to Present

In 1900, Georg Simmel published the first edition of his magnum opus, *The Philosophy of Money.* It is essentially a metaphysical study of the concept of value, centered on money as the means of exchange. Money is, in this sense, a mediating technology; in fact, it is the purest and most perfect technical expression of exchange. But it is still a technical device, and near the end of the work Simmel offers some relevant comments on technology generally.

With technology we make a common "metaphysical mistake": We transfer the value of the parts to the whole. Because individual technologies are beneficial, helpful, and benign, we tend to think the same of technology in itself. But this is wrong, says Simmel. The individual devices have been "autonomously developed (*der sie sich* in sich *entwickelt hat*)" from out of technology itself. These devices are valued as mere means and cannot, or ought not, be ends in themselves. The electric light is not what is valuable; rather, it is that which is illuminated. The telephone and telegraph are not valuable in themselves, but only as enablers of that which is said. Thus it is wrong to value technology as a whole; it can only be, at best, a collective of means that leads to intrinsically valuable ends.

For example, says Simmel, we surround ourselves with a preponderance of technical means in order to control nature. But these require effort to develop and more effort to maintain; and thus we spend much of our time servicing *them* rather than the ends. Hence, "the control of nature by technology is only possible at the price of being enslaved in it (*in ihr befangen zu sein*)." This is the great irony of the age: Machinery, which "was supposed to relieve man from his slave labor in relation to nature, has itself forced him to become a slave to it (*zu Sklaven eben an der Maschine selbst herabgedrückt*)."[1]

When it comes to economic production, we unfortunately suffer bondage from both the means of production and the technical products themselves; we get it 'coming and going,' as it were. Thus Simmel can claim that we are "slaves of the production process" as much as "slaves of the products." This complete domination by the means causes man to become "estranged from himself" (*sich selbst entfernt*). This, says Simmel, is the price we must pay for the "clamorous splendor of the scientific-technological age" in which we live.

Within a few years, related critiques began to arrive. Max Weber's *Protestant Ethic and the Spirit of Capitalism* appeared in 1905. Its focus on economic sociology prompted an examination of the technical means of "machine production" upon which the modern economy is founded. Through the economy, machine production rules all aspects of modern life.

> [The economic order] is now bound to the technical and economic conditions of machine production, which today determine the lives of all the individuals who are born into this mechanism, not only those directly concerned with economic acquisition, with irresistible force. ([1905] 1930: 181)

The technological mode of production thus imposes "an iron cage" upon us all. Weber hereby casts his lot with the technological determinists.

A similar complaint came from Thorstein Veblen the following year. In his essay of 1906, "The Place of Science in Modern Civilization," he spoke of the "cult of science" that had taken hold in America and throughout the Western world. Science's emphasis on materialism and factual knowledge has led to a condition of "race deterioration" and "discomfort on the whole" (1906: 588). Technology—defined in instrumental terms as "the employment of scientific knowledge for useful ends" (598)—has come to assume a primary role in both economic and cultural life:

> In the modern culture, industry, industrial processes, and industrial products have progressively gained upon humanity, until these creations of man's ingenuity have latterly come to take the dominant place in the cultural scheme; and it is not too much to say that they have become the chief force in shaping men's daily life, and therefore the chief factor in shaping men's habits of thought. (598)

In contrast with Weber, we see here a moderate form of determinism; technology is a dominant but not sole factor in social development and progress.

Veblen is struck by the "inhumanly ruthless fabric of technological processes" (605), something hardly suitable for a nominally civilized people. There are now "sciences" of everything, and "this opaque, materialistic interpretation of things pervades modern thinking" (608). And yet this is seen as a sign of progress, at least by those "whose habits of thought have been fully shaped by the machine process and scientific inquiry." The dominance of this worldview is "due chiefly to the ubiquitous presence of the machine technology and its creations in the life of modern communities"—a kind of technological universalism that would prove so important for Ellul.

Roughly one decade later, in 1915, an important essay appeared in Germany: *Das Ressentiment im Aufbau der Moralen*—in English translation, *Ressentiment.* The author, Max Scheler, was one of the leading early German phenomenologists. Concerned about the moral decline of humanity in the modern age, he drew from like-minded scholars, most notably Nietzsche. Scheler witnessed the rapid emergence of a mechanistic society that was the inevitable outcome of the scientific worldview that began in the 17th century with Bacon, Descartes, and Newton. Society was constructed in the image of the mechanical universe and increasingly functioned as a social machine. "The intellect can comprehend the universe only as a 'mechanical' universe, and this mechanical universe is then made the basic 'milieu' of universal life" (Scheler [1915] 1994: 134). In modeling something dynamic and living (society) after something mechanistic and dead (a machine), we simultaneously enervate the former and vitalize the latter.

The modern worldview "interprets life as such as an accident in a universal mechanical process." Life, even human life, seems an outsider, an intruder. Even the human body is a mechanism: the eyes become spectacles, the hands spades, all organs mere tools. "No wonder," remarks Scheler,

> that the mechanical civilization . . . is mistaken for the triumph, continuation, and extension of vital activity. Thus the infinite "progress" of mechanical civilization becomes the true "goal" of all vital activity. (141)

The resulting mechanized society is one in which humans become mere components in an increasingly depersonalized and dehumanized system. The fault lay not in the capitalist application of technology, as Marx thought, but in the structure of the technological-industrial complex itself:

> With the development of modern civilization . . . objects have become man's lord and master, and the machine has come to dominate life. The "objects" have progressively grown in vigor and intelligence, in size and beauty—while man, who created them, has more and more become a cog in his own machine. (141–142)

Regarding his analysis of the root cause of this situation, Scheler draws directly on Nietzsche. For Nietzsche, the great social crisis of the modern era derived from the Judeo-Christian "slave revolt" against the superior Greco-Roman culture of classical antiquity. Jews suffered oppression at the hands of the Romans and thus developed a deep, grudge-laden hatred, a *ressentiment*, that fueled their backlash. On Nietzsche's view, the Jews, led by Paul of Tarsus, developed a countermorality to undermine the Romans by targeting the source of their strength—their values.[2] Jewish *ressentiment* sustained the Judeo-Christian movement through the coming centuries. With the eventual collapse of the Roman Empire, the Judeo-Christian worldview acceded to power, maintaining it to the present day. Tragically, it brought along the lowly slave values, and these have been the chief source of our decay. For Nietzsche, this reading of history suggests a clear path forward: another moral revolution, a second inversion of values, that would overthrow the dominant Judeo-Christian belief system and recreate something like a modern form of the master morality of the Romans and Greeks.

By an analogous process, thought Scheler, the humanistic values of the Enlightenment had long held the materialistic, mechanistic urges of humanity in check. For centuries, the humanists—secular and spiritual—reigned over the mechanists, something the latter came to deeply resent. But with the growing power of science and technology and the concurrent decline of vital humanism, the mechanistic thinkers have come to the fore. They are now the dominant force, and they seek revenge. With their armory of mechanistic values, they now threaten to overwhelm traditional, noble human values.

> [I]t is not sufficiently clear that this generally acknowledged fact [of the triumph of mechanistic society] is due to a fundamental *subversion of values. Its source is ressentiment*, the victory of the value judgments of those who are vitally inferior, of the lowest, the pariahs of the human race! The entire mechanistic worldview, to the degree that it lays claim to metaphysical truth, is only the immense intellectual *symbol* of the slave revolt in morality. The rule of life over matter has weakened, the spirit (and the will above all) has lost much of its mastery over the automatism of life: that is the ultimate explanation for the expansion and development of the mechanistic worldview, and of the corresponding values which created it. (142)

This state of affairs is obviously no advance in the human condition but rather a kind of devolution of mankind, a real degradation in the state of human existence. Scheler specifically mentions such things as woman and child labor, disintegration of family life, overcrowded cities with poor living conditions, industrial toxins, overspecialization, and growing emphasis on money and materialistic values. He even offers a note of environmental concern, unusual for that time: "[necessary action extends to the] preservation of plant and animal life, and the woods, and to the protection of the landscape against the devastating tendencies of industrialism" (143). To the advocates of technological modernization—like Herbert Spencer, to cite his example—these are all temporary ills, "which will disappear with the still further development of industrialism." Scheler was justifiably skeptical. As we know, many of these problems have only worsened in the intervening century; one wonders how long humanity must wait for the promised amelioration to occur.

As things stand, though, we face a real and perilous decline. "The spirit of modern civilization," he said,

> does not constitute "progress" (as Spencer thought), but a *decline* in the evolution of mankind. It represents the rule of the weak over the strong, of the intelligent over the noble, the rule of mere quantity over quality. It is a phenomenon of decadence. (143)

In light of the present pantechnical analysis, we can say that Scheler correctly grasped the facts of the mechanistic worldview and its consequent values, and he could sense the impending dehumanization of mankind in an increasingly mechanistic society. He understood that social and technological *progress* can exist concurrently with human *regress*—and perhaps even demands it. This is a decisive step: that mechanistic society has now become detached, in a sense, from individual humans and flourishes at their expense. The technological society becomes a reality in itself, with its own good, its own values, and its own imperatives. It takes on a life of its own and, like all living entities, strives to flourish and grow, drawing from whatever resources are at hand and employing the most efficient means possible.

This, in fact, is a predictable consequence of the Pantechnikon. The technosphere, and the complex social being that it creates, draws increasing amounts of energy from its surroundings as it organizes those structures below it in the evolutionary hierarchy. Individual people, animals, plants, and nature generally are put to use in service of the higher and more complex system—this is the proper interpretation of Heidegger's Enframing. It is a true regress for humanity and nature because of our loss of autonomy, our increasingly undignified existence, and our generally degraded environment. The only things that flourish are technology, the technological society, and those who serve it. This is evolution at work—technology advances as other forms of existence decline.

THREE INTERWAR VIEWS: WHITEHEAD, SPENGLER, JASPERS

A. N. Whitehead was a mathematician and physicist long before he was a philosopher, and he thus was well situated to comment on the philosophy of science and technology. His first major philosophical book, *Science and the Modern World* (1925), begins with an understated but decisive observation: "The progress of civilization is not wholly a uniform drift towards better things" ([1925] 1967: 1)—in noted contrast to the received optimism of the day. The overriding theme of the book is that conventional mechanistic materialism, which presumes "an irreducible brute matter" that is "senseless, valueless, and purposeless" (17), is fundamentally defective in light of developments in atomic theory and quantum mechanics. The primordial aspect of reality is not the atom or subatomic particle but rather a "vibratory ebb and flow of an underlying energy, or activity" (35). This oscillating, periodic, "process" view of energy suggested to Whitehead that reality was far more integrated and interconnected than previously thought.[3] It also implied a continuity of existence, a mere variation in degree, between things lifeless and those alive and conscious. Reality was thus better modeled on the concept of the *organism* than on a system of discrete, interacting particles of inert matter: "The field is now open for the introduction of some new doctrine of organism which may take the place of materialism" (36).[4]

Nature is thus seen as dynamic, living, and creative. Mechanistic materialism is disproven by its central defect, namely, "that it provides none of the elements which compose the immediate psychological experiences of mankind" (73). Consciousness, feelings, emotions, mental images, free will, ethics—all these are unexplained mysteries within the conventional view. Whitehead concludes that "the only way of mitigating mechanism is by the discovery that it is not mechanism" (76)—that is, by abandoning the mechanistic view and replacing it with a process-oriented philosophy of the organism.

Science the Modern World closes, appropriately, with some philosophical reflections on the growing power of science and its attendant technology. Mechanized, competitive, capitalist civilization, Whitehead says, emphasizes the physical realm to the near-total exclusion of human values; modern society is "entirely devoid of consideration for the value of human life" (203). The science of political economy, for example, as exemplified by Adam Smith,

> did more harm than good. . . . [I]t riveted on men a certain set of abstractions which were disastrous in their influence on modern mentality. It dehumanized industry. This is only one example of a general danger inherent in modern science. (200)

A mechanistic way of thinking has become deeply engrained in the modern psyche, to the detriment of us all. The situation is grave, perhaps terminal: "it may be that civilization will never recover from the bad climate which enveloped the introduction of machinery" (203). Whitehead concludes:

> At the present moment a discussion is raging as to the future of civilization in the novel circumstances of rapid scientific and technological progress. . . . The point is that professionalism [i.e., technical specialization] has now been mated with progress. The world is now faced with a self-evolving system, which it cannot stop. . . . We must expect, therefore, that the future will disclose dangers. (204–207)

Here, for perhaps the first time, we see an unabashed determinism. Technology is an unstoppable, autonomous force advancing of its own intrinsic nature, contrary to the interests of humanity and civilization. Some 60 years earlier, Butler argued that we still had a chance to halt the machines, that it was still—barely—within our power to arrest their advance. For Whitehead, in 1925, that time had passed. The system will progress, it will evolve, and our very existence is imperiled.

Nonetheless, he suggests that there may be a path forward. Even if we cannot stop it, we can perhaps slow, delay, or redirect it in some meaningful way. This, at least, would mitigate the harms, allowing humanity and nature more time to adapt and to possibly work out other alternatives. One senses in Whitehead an argument not for the defeat of technology but rather for its deflection—for pushing this self-evolving system in the least harmful directions and shielding the most sensitive portions of the world from its harshest effects.

For Whitehead, however, these were just passing observations. He never developed a complete philosophy of technology. He simply stated the facts—technology is self-evolving, it cannot be stopped, and it will present us with dangers—and left it at that. His proposed change in worldview, to a process-based metaphysics, seems to have been his best prescription.

With the 1930s came global economic depression and further critical analyses. Oswald Spengler painted a particularly gloomy picture in his 1931 book *Man and Technics*. Noting the deep-seated skepticism about technology—"men have always regarded machines as the invention of the devil" ([1931] 1976: 85)—Spengler proceeds to dispel some common myths. We believe that our inventions serve only to benefit humanity, but we do not know and cannot know their true effects. Whether a given discovery is "useful or menacing, creative or distributive" is something unknowable. "The effect of a 'technical achievement of mankind' is never foreseen" (87)—and is increasingly likely to be for the worse. Furthermore, the idea that technology reduces human labor is likewise a fallacy: "The pace of discovery grows fantastic, and withal—it must be repeated—human labor is *not* saved thereby. The number of necessary hands grows with the number of machines . . . and our artificial life becomes more and more artificial" (88). Additionally, technology allegedly helps us to make better products, but the worker achieves no sense of satisfaction from this. In any large-scale undertaking, even ancient ones such as the construction of the pyramids, "the individual laborer comprehended neither the object nor the purpose of the enterprise as a whole" (89). Today the situation is far worse; a technical product's "real role in life . . . is entirely unknown to [the workers,] and in the creation of which, therefore, they have inwardly no share. A spiritual barrenness sets in and spreads, a chilling uniformity without height or depth."

Unsurprisingly, Spengler advocated a strong form of technological determinism. "In reality," he said, "it is out of the power either of heads or of hands to alter in any way the destiny of machine-technics" (90). Technology is driven by "inward spiritual necessities" and proceeds unceasingly "toward its fulfillment and end." We, who would cast ourselves as master of nature, have ceded control to technology. "The lord of the World is becoming the slave of the Machine, which is forcing him—forcing us all, whether we are aware of it or not—to follow its course" (90).

This has negative consequences not only for humanity but for the natural world as well. Astonishingly, Spengler even anticipates the danger of global climate change.

> *The mechanization of the world* has entered on a phase of highly dangerous over-tension. The picture of the earth, with its plants, animals, and men, has altered. In a few decades most of the great forests have

> gone, to be turned into newsprint, and climatic changes have been thereby set afoot which imperil the land-economy of whole populations. Innumerable animal species have been extinguished, or nearly so, like the bison; whole races of humanity have been brought almost to vanishing-point, like the North American Indian and the Australian [Aborigine]. All things organic are dying in the grip of organization. An artificial world is permeating and poisoning the natural. The Civilization itself has become a machine that does, or tries to do, everything in mechanical fashion. (93–94)

Prefiguring Heidegger, he remarks that we view all things in nature simply as resources for our use: a waterfall is hydroelectric power, cattle are meat, primitive cultures are new markets for industrial goods. He recognizes that the system is driven forward by its energy supplies—primarily coal, at that time. But he dismisses those who fear a pending energy shortage; technology will always find a way to fuel itself. "No stupid trifle like the absence of material would be able to hold up this gigantic evolution" (96).

Thus described, the question remains: What shall we do? A "weariness" of technology is spreading, says Spengler. The "born leader" is fleeing the Machine. Mankind is instinctively rebelling against the system. "In the consciousness of this unalterable state of things . . . men are so forlorn that it is mere human nature to revolt against the role for which the machine . . . earmarks most of them" (98). Things will not end well:

> This machine-technics will end with the Faustian civilization, and one day will lie in fragments, *forgotten*—our railways and steamships as dead as the Roman roads and the Chinese wall, our giant cities and skyscrapers in ruins like old Memphis and Babylon. The history of this technics is fast drawing to its inevitable close. It will be eaten up from within. (103)

A catastrophic future is unavoidable, and the foolish romantic alone thinks otherwise. "Only dreamers believe there is a way out." To be a realist is to be a pessimist; and he who looks to a better future is simply afraid of the awful truth: "Optimism is *cowardice*."

Spengler's compatriot and fellow philosopher Karl Jaspers, writing at about the same time, also came to negative conclusions about the technological era. Adopting a more objective and academic tone, Jaspers's book *Man in the Modern Age* (1931) cites a number of familiar complaints: technology as a "determinant of mass-life," booming population driven by technical advance, and workers as mere cogs in the industrial machinery. He places great emphasis on the detrimental effects of mass society and was among the first to employ such terms as "mass-rule," "mass-man," and "mass-order."

This, for Jaspers, is the key indicator of a dehumanized existence. The individual becomes lost in the collectivity; "he exists only as 'we'" ([1931]

1959: 49). The situation leads to an unending and irreconcilable battle between the system and the individual: it seeking to dominate, he striving for independence. "Tension between the universal life-apparatus and a truly human world is, therefore, inevitable. . . . Attempted mastery and attempted revolt will continue their reciprocal strike, each misunderstanding the other" (46–47).

We can see the effect of mass society in the characteristics of contemporary life: "public affairs become mere entertainment"; "a craving for novelty"; social life marked by "an unceasing activity"; and an "enthusiasm for the numberless and the vast." Lotteries and games become objects of public obsession. There is an overemphasis on youth and on "erotic exaltation." Children try to act as adults, and adults become childlike—and neither respects the other. The average man is leveled out and deracinated; he loses touch with his human essence.

Jaspers's critique continued years later, notably in his *Origin and Goal of History* (1949). The present era—the "Age of Technology"—is a time in which we have experienced a "catastrophic descent to poverty of spirit, of humanity, love and creative energy" ([1949] 1953: 96–97). There is an unavoidable feeling that "the spirit itself has been sucked into the technological process," which has come to dominate both industry and science. This fact "explains the secret lack of happiness in this world that is becoming ever more inhuman."

Technology, he continues, has "wrought a radical transformation in the day-by-day existence of man in his environment"—and the earth itself has been transformed "into a single great factory" (98). Because it bears so completely on all aspects of our existence, and because of its totalitarian character, "the importance of the irruption of modern technology and its consequences for absolutely all the problems of life cannot be overestimated" (99). An important insight: All problems are, at root, *technological* problems.

Jaspers attempts to provide a rough philosophical analysis of technology, but it is a scattered and generally unsuccessful effort. He offers a definition of technology and describes its two essential characteristics—understanding (i.e., rationality) and power. He introduces a decisive break in technological evolution around the year 1800, at which time we developed truly modern technology—though this move suffers from the same weaknesses that plagued Heidegger and others. He examines three "inseparable elements" of modern technology: science, the spirit of invention, and the organization of labor. He cites the pro and con views of Dessauer and Juenger, dismissing both. Ultimately he defends a third way, a middle path between demonization and praise: the idea that technology is fundamentally neutral. "Technology is *per se* neither good nor evil, but it can be used for either good or evil" (115). And again: "Because it does not set itself any goals, it is beyond or before all good and evil. It can serve the purposes of salvation or calamity. In itself it is neutral toward both of them" (119). Here we have a classic statement of the instrumentalist viewpoint.

Evidently, then, all the evils of technology are due to human failings and misuse. This would appear incongruous with his earlier statements and seems to be attributable mainly to Jaspers's lack of a deeper metaphysical analysis. Hence his other contradictions: Technology "has become an independent, impetuous force," and yet we must somehow regain mastery of it. The leading question of the day is "how man is to impose himself upon technology, which has become his master." And yet, we are "at the mercy of an uncanny process that is the inexorable and compelling outcome of man's own actions." He concludes the section by remarking that "technology is independent of what can be done with it; as an autonomous entity it is an empty power" (125)—odd statements that do not add up to a cohesive picture. But lacking a unified and incisive metaphysical framework, one can do little more than offer up such random observations.

A final work of the early 1930s bears mention: Lewis Mumford's detailed and ground-breaking study *Technics and Civilization* (1934). It is notable not only for its exhaustive academic treatment but also for a lingering skepticism throughout. Mumford never quite adopts a fully objective stance with his subject matter, and the negative consequences are never far from his mind. He cites the work of other technology critics like Weber, Veblen, and Spengler, writers who clearly influenced him. Perhaps it would be more correct to say that such thinkers planted the seeds of skepticism; much later, in the 1960s, Mumford would emerge as an unconditional critic himself.

ORWELL AND THE MACHINE SOCIETY

In the run-up to World War II, George Orwell was greatly concerned about the rapid spread of fascism throughout Europe. But the only alternative—socialism—was in bad shape, for two reasons. The first was its necessary emphasis on collectivism and the integrity of the social whole. For a small society, such as ancient Greece, this was not a problem because individuals maintained a direct and active role in the state of their society. But in a large population, socialism cannot function without highly defined and regimented social roles, a high degree of legislative constraint, and a concomitant diminishment of the involvement of the individual. In a mass society, the individual tends to become insignificant, irrelevant, and alienated.

Second, and because of the need for a high degree of order and coordination, modern socialism strongly embraces the machine culture. The need for efficient mass production and centralized control demands a technological society.[5] Despite Orwell's claim, it is not clear that socialism had a stronger inclination to the machine culture than fascism; arguably, the reverse was true. But still, for the Britons of the 1930s, they could clearly sense the growing mechanization of their lives, and they naturally connected it with their socialist political system.

Then as now, socialism placed a high priority on addressing human wants and needs. This would seem to be highly advantageous, and even a sign that a socialist democracy could respond to truly human concerns. But not so. The socialist state is a *well-intentioned* machine; but it is a machine nonetheless. And human existence in any machine entails a surrender of a substantial degree of dignity and autonomy. Even the most compassionate system of slavery is still slavery. Dehumanization is inherent in any mechanical society, no matter how gently implemented.

The actual effect on humanity, said Orwell, was unquestionably negative. The stereotypical socialist was a soft, sickly, dependent, weak-minded progressivist, bent on creating a safe, orderly, efficient, labor-saving, mechanized world—the type of world that Orwell despised. Alternatively, he sought an authentic, demechanized socialism, respectful of humanity and able to sustain the notions of liberty and justice without the dehumanizing aspects of collectivism.

This led him to address, in his nonfiction book *The Road to Wigan Pier* (1937), the dilemmas of life in a modern machine culture. It is a little-known but surprisingly well-articulated critique of the machine world, which, for him, is as repulsive as it is inevitable. Acknowledging past criticisms of technology, he nonetheless argues that "it is only in our own age, when mechanization has finally triumphed, that we can actually *feel* the tendency of the machine to make a fully human life impossible" (223). Everyone knows that "the machine is the enemy of life," and yet we are unable to grasp the implications of this fact.

Orwell is concerned that technology has made man soft and hence a degraded and pathetic figure. We all strive for the nobler virtues in humanity, but machine life makes it practically impossible: "The truth is that many of the qualities we admire in human beings can only function in opposition to some kind of disaster, pain or difficulty; but the tendency of mechanical progress is to eliminate disaster, pain, and difficulty" (226). He elaborates:

> In tying yourself to the ideal of mechanical efficiency, you tie yourself to the ideal of softness. But softness is repulsive; and thus all progress is seen to be a frantic struggle toward an objective which you hope and pray will never be reached. . . . Life has got to be lived largely in terms of *effort*. . . . The tendency of mechanical progress, then, is to frustrate the human need for effort and creation. (228–232)

The technological society, in meeting so many of our needs, has removed individual human initiative and thus personal satisfaction. The essentials of life are provided by "the system," not by individual or small-group action, as was the case throughout our evolutionary history. We forget how to provide for ourselves and thus become more or less dependent upon the system. We sense this dependency, and deep down we despise it, but we can do nothing about it.

Even to call it a dependency is an understatement, according to Orwell. Rather, it is a compulsion or even an addiction. The mere presence of technological options compels usage: "so long as the machine *is there*, one is under an obligation to use it" (231). Through a variety of social and psychological factors, technology entices and coerces us into engaging with it.

> The sensitive person's hostility to the machine is in one sense unrealistic, because of the obvious fact that the machine has come to stay. . . . The machine has got to be accepted, but it is probably better to accept it rather as one accepts a drug—that is, grudgingly and suspiciously. Like a drug, the machine is useful, dangerous, and habit-forming. The oftener one surrenders to it the tighter its grip becomes. . . . [T]he machine is here, and its corrupting effects are almost irresistible. One inveighs against it, but one goes on using it. (235–237)

These negative effects will only worsen because the system is advancing independently of human control. Reiterating the determinism of Butler, Whitehead, and others, Orwell remarks that technology is functionally autonomous: "there is a tendency for the mechanization of the world to proceed as it were automatically, whether we want it or not" (238). Like his predecessors, he neglects to analyze the root causes of this situation.

Not content with mere theorizing, Orwell seeks pragmatic solutions, and the situation demands positive action on the part of humanity. But his road to freedom is by no means an easy one. Revolt against technology would be opposed, first, on principle: "any attempt to check the development of the machine appears to us as an attack on knowledge and therefore a kind of blasphemy" (239). This is a true dilemma; any move to counter technological advance is typically seen as profoundly regressive and perhaps even irrational. This pseudo-argument has force only from a simplistic perspective in which modern technology is seen as the paragon of rationality. In fact, it is but one form—the form that promotes the growth of complex society and the social being. Small-scale societies, so-called primitive cultures, and individual humans are also embodiments of rationality, albeit in forms that function at dramatically different scales than modern technology. Technological rationality exerts itself more forcefully because it more effectively monopolizes the energetic resources of the world and because, I claim, it has a kind of evolutionary power behind it. But this does not mean that we must promote and serve it—particularly when doing so is counter to our own well-being. In fact, to defend technology under such conditions would be the height of *irrationality*—from the standpoint of the individual man or from that of nature. So who is the more rational: he who defends the interests of himself and his fellow human beings, or he who promotes a system that is destructive of his own long-term well-being?

Technology is very effective at promoting its own brand of rationality as the only acceptable form and thus as rationality per se. This creates a self-reinforcing but tautological argument wherein those who might oppose technology are, *by definition*, antirational. Any alternate form of thinking is not rational at all, on this view, and thus may be safely ignored, if not actively opposed—a nifty trick.

Furthermore, says Orwell, and more practically speaking, there is the realization that breaking free from the machine culture would mean living a hard life—something few are prepared to do. Orwell has heard the counter-argument: "Nobody seriously wants it."

> Now in a sense this is true. It amounts to saying, "We're soft—for God's sake let's stay soft!" which at least is realistic. As I have pointed out already, the machine has got us in its grip and to escape will be immensely difficult. . . . Clearly I do not, in a sense, "want" to return to a simpler, harder, probably agricultural way of life. In the same sense I don't "want" to cut down my drinking, to pay my debts, to take enough exercise, etc. But in another and more permanent sense I do want these things, and perhaps in the same sense I want a civilization in which "progress" is not definable as making the world safe for little fat men. (242)

The antitechnology critic is often charged with wishing upon humanity a reversion to the Stone Age; he is accused "of being a medievalist," whereupon the socialist will "begin to descant upon the horrors of the Middle Ages, leprosy, the Inquisition, etc." But as Orwell rightly observes,

> a dislike of the mechanized future does not imply the smallest reverence for any period of the past. . . . Press the point home, explain that you wish merely to aim at making life simpler and harder instead of softer and more complex, and the socialist will usually assume that you want to revert to a "state of nature"—meaning some stinking Paleolithic cave: as though there were nothing between a flint scraper and the steel mills of Sheffield, or between a skin coracle and the *Queen Mary!* (241)

Clearly there is a broad range of social alternatives that could restrain technological advance. And even within the harshest of options, there can be no true 'going back' short of a near-total degeneration of the human race and a virtually complete amnesia. Consider the effect, for example, that a simple knowledge of germ theory and disinfectants could have had, for nearly all of human existence. Such basic knowledge would not disappear even in the face of the most complete technological collapse, and thus the 'Stone Age' gambit fails on both practical and theoretic terms.

Then there is that other common rebuttal: that technological advance is inevitable. "[The socialist] will tell you that it is impossible to 'go back'—as though the hand of progress hadn't been pretty violently put back several times in human history!" (240–241). In fact this is true; there have been a number of setbacks in the progression of technology, but usually only under extreme conditions of social or cultural collapse. When the Roman Empire disintegrated in 395, several Roman techniques collapsed with it, including road-building technologies, aqueducts, architectural construction, and a variety of social organization techniques. In general, when militarily or technologically superior cultures attack and destroy weaker ones, many indigenous technologies are lost—typically those related to agriculture, hunting, and medical treatment. But on the whole, it must be admitted that the long-term trend is toward consistently higher and more sophisticated technologies, greater access to and manipulation of energy, and more intense forms of social organization.

Knowing all this, Orwell is no airy idealist. He is a realist, but his realism borders on fatalism. Even under the most ideal of circumstances, chances for success are small: "Even if the whole of humanity suddenly revolted against the machine and decided to escape to a simpler way of life, the escape would still be immensely difficult" (239)—we are just too dependent, too soft, and the technological mindset too entrenched. In reality, however, it is even worse than this. For all practical purposes, we have already lost:

> We are all dependent upon the machine, and if the machines stopped working most of us would die. You may hate the machine-civilization, probably you are right to hate it, but for the present there can be no question of accepting or rejecting it. The machine-civilization *is here.* And it can only be criticized from the inside, because all of us are inside it. . . . And almost certainly the machine-civilization will continue to triumph. There is no reason to think that it will destroy itself or stop functioning of its own accord. For some time past it has been fashionable to say that war is presently going to "wreck civilization" altogether; but . . . it is immensely unlikely that it will put a stop to mechanical progress. . . . [N]o war is at present thinkable which could wipe out industrialism in all countries simultaneously. (250–251)

Even so, Orwell is not completely nihilistic in his outlook. Revolt, as he sees it, consists of an ongoing resistance to the machine: "those who can see through the swindle of 'progress' will probably find themselves resisting. In fact, it is their special function to do so. In the machine-world they have got to be a sort of permanent opposition" (252). Resistance in the face of inevitability is not futile. We routinely fight against disease, illness, and death, even though we know these are inevitable. As Orwell says, the struggle itself is worthwhile and provides a source of meaning to life; struggle in itself is a kind of victory.

Within two years of Orwell's book, Juenger completed the first draft of his extended essay "Illusions of Technology"—the piece that would eventually be published as *The Failure of Technology.*[6] At that same time, Jose Ortega y Gasset published his own attempt at a metaphysics of technology, in his book *Toward a Philosophy of History* (1939). The chapter on "Man the technician" reiterates several of the concerns cited by his predecessors, and there is little in the way of new insight. Adopting a highly anthropocentric outlook ("man begins where technology begins"), he declares that the meaning of technology "lie[s] outside itself, namely in the use man makes of the unoccupied energies it sets free. The mission of technology consists in releasing man for the task of being himself" ([1939] 1941: 118). But paradoxically, technology makes it much harder for man to realize this goal. Our technological omnipotence points toward an unlimited potential in man, suggesting a kind of bottomless pit, a humanity which is all 'potential' and no 'actual':

> Technology . . . will irretrievably empty the lives of those who are resolved to stake everything on their faith in it and it alone. Just because of its promise of unlimited possibilities, technology is an empty form. . . . That is why our time, being the most intensely technical, is also the emptiest in all human history. (151)

If so, the even more technologically intense era of the 21st century must be emptier still. Human essence is further drained away the more technology advances. Every new development promises more than the one before but succeeds only in further desiccating human nature. Ortega's insight rings true.

POSTWAR CRITIQUES

World War II and its aftermath took a toll on philosophical speculation; apart from Juenger, little of import appeared throughout the 1940s. It was as if the extreme destructive power of technology that was unleashed during the war years took time to be absorbed and processed by the philosophical mind.

The destructive force of technology that emerged during wartime bears further discussion. Technology, as always, is about energy. But this energy can be deployed in either of two basic modes: *information* or *power.* During the war, energy as information was manifest in the increasingly sophisticated communications systems and in the early computer applications. Radar and sonar became vitally important tools, and radio wave communication was critical to the war effort. Computers attained their first practical use as code breakers. On the other hand, warfare is fundamentally about using force to dictate the action of your opponent. Weapons technology is basically a study in delivering concentrated units of energy to specific locations and releasing it rapidly. The technical definition of 'power' is change in energy per unit of time; high quanta of energy, released very rapidly, equate to high-power weaponry. World War II saw the development of a wide range of chemical explosives, primarily new combinations of the older explosives TNT and PETN, though in conjunction with a new material, RDX.

But it was the advent of nuclear weapons that was clearly the dominant milestone in the evolving power dynamic and in humanity's ability to kill;

witness the estimated 140,000 Japanese civilians mostly women, children, and the elderly—incinerated by the United States at Hiroshima and Nagasaki in 1945. Once the atomic bomb exists, "one is under an obligation to use it," as Orwell suggested. It was perhaps inevitable that the first application of a major new energy source would be destructive.

In the longer term, however, peaceful use of nuclear power is more significant. Just as the use of fossil fuels indirectly unlocks the (past) power of the sun—solar energy concentrated in decayed plant matter—so too nuclear fission releases ancient solar energy. In this case, though, it is the energy of long-dead stars. Eons ago, heavy atomic nuclei were formed in the energetic interior of distant stars or in the explosion that accompanied their demise. Uranium was one such element, and countless tons were discharged into interstellar space, eventually becoming incorporated in the dust cloud that became Earth. Uranium and all the heavier elements are dense bundles of energy bequeathed to us from those ancient stars. When we seek it out, concentrate it, and crack open the nuclei, energy flows out—quickly in the case of a bomb, slowly for a nuclear reactor.

On the present thesis, access to new forms of energy by human beings—ourselves complex forms of energy—constitutes the modus operandi of the universe: a superabundance of energy giving rise to yet more complex structures. *We are energy accessing energy.* This process builds new structures that in turn access even more energy. Present nuclear fission seems to have practical limitations, but if these are overcome, or if practical fusion systems materialize, and humanity thus releases vast new quanta of usable energy, we can be sure that new, higher-order structures will appear—structures that further exceed the scale of the individual and thereby thrust him deeper into a condition of irrelevance.

With the 1950s we arrive at the two central works of the mid-20th century: Heidegger's "Question Concerning Technology" and Ellul's *The Technological Society*—both issued in 1954. I examined Heidegger in some detail in Chapter 5, and Ellul is the main subject of the chapter to follow. Hence I pass them over here.

As important as they were, neither Heidegger's nor Ellul's work had immediate effect in the English-speaking world—primarily due to lack of translation. The English version of Ellul's book did not appear until 1964, and Heidegger's essay, surprisingly, not until 1977. For their part, Anglo-American philosophers were busy promoting the scientific mindset, with a decided shift toward analytical approaches. From this perspective, technology poses no problem at all, or at worse, only minor and incidental ones. Thus for most of the 1950s and 1960s it was left to the nonphilosophers to critique the technological society.

One example, the psychologist Erich Fromm, published *The Sane Society* in 1955. Its focus was a contemporary reading of Marxian alienation. Capitalism was still the ultimate cause of this, but now it was a *technological* capitalism that held sway, making the situation much worse. "Alienation as we find it in modern society is almost total," he wrote (1955: 124). "Man has created a world of man-made things as it never existed before. He has constructed a complicated social machine to administer the technical machine he built." In those early days of the Cold War, everyone was concerned about the conflict between capitalism and communism. But for Fromm, these were more or less similar forms of the same ideology: "They both are thoroughly materialistic in their outlook. . . . They organize man in a centralized system, in large factories, political mass parties. Everybody is a cog in the machine" (359). From this he drew an important conclusion: Technology, and technological imperatives, drive all social systems to a common technological base, regardless of their initial states. "Both systems tend to converge," he said. Technology is thus the driving force behind social evolution—not politics, not economics, and not ideology. We can therefore expect similar future effects in both societies, said Fromm. "In the development of both Capitalism and Communism . . . the process of automatization and alienation will proceed" (359). This in turn will impose ever-harsher psychological stresses upon humanity, causing a growth in mental illness and insanity; Fromm has been largely proven correct, as I will show in Chapter 11. Ultimately the modern life of "schizoid self-alienation" is basically a condition of "robotism"—a dehumanized, mindless, joyless existence.

Interestingly, for Fromm, technology per se was not the problem; like Marx, he saw it as a social organizational issue and even argued that "we must retain the industrial method" (360)—though putting it to work in a dramatically decentralized social setting. Only something like a "communitarian socialism" could adequately address our problems, he thought.

Philosophically speaking, though, scant progress occurred between the mid-1950s and mid-1960s. Even among the continental thinkers, who had traditionally been at the forefront of the movement, there was little development. Gilbert Simondon's *Du mode d'existence des objets techniques* (*On the Mode of Existence of Technical Objects*) was published in 1958 but contained minimal insight. Simondon's writing apparently set the stage for subsequent French philosophizing on technology—which, unfortunately, combines the worst aspects of Heidegger's obscurantism and Merleau-Ponty's pseudo-intellectualism. Recent writers, like Baudrillard and Virilio, are cases in point.[7] It is almost as if French rationalism has devolved into a pastiche of images, a kind of philosophical impressionism, devoid of substance. This is regrettable. One can only hope that the French will someday recover their deeper and more lucid philosophical tradition.

INTO THE 1960s

By the late 1950s, technology was rapidly moving into the electronic era. The transistor, invented in 1947, was miniaturized in the first integrated circuits. Computer technology advanced, heading toward the first microprocessor in 1971. Everyday devices, from autos to home appliances, were quickly becoming more complex and more widespread. The social upheaval of the 1960s inaugurated a period of revolutionary thinking. Marcuse's *One Dimensional Man* was released in 1964, becoming something of a counterculture sensation. But the book has not held up well over time and, from the perspective of philosophy of technology, has surprisingly little new to say. Marcuse's main theme is a vaguely Marxian critique of industrial capitalism and of the alienation of the individual. Technology is important as the means of alienation and functional enslavement—but only as a means.

For example, he writes,

> [T]he scope of society's domination over the individual is immeasurably greater than ever before. Our society distinguishes itself by conquering the centrifugal social forces with Technology rather than Terror, on the dual basis of an overwhelming efficiency and an increasing standard of living. . . . Technology serves to institute new, more effective, and more pleasant forms of social control and social cohesion. (1964: xlii, xlvii)

This seems to be Marcuse's key insight into technology, and he repeats it sporadically throughout the book, but with little accompanying analysis.

Of special note, though, is his revolutionary tenor. Faced with the political (but not ontological) determinism of industrial technology, only a political revolt can hope to restore a just order: "Advanced industrial society is approaching the stage where continued progress would demand the radical subversion of the prevailing direction and organization of progress" (16). What this "radical subversion" would consist of is unclear, other than vaguely "refusing to play the game" (257). This "Great Refusal," as Marcuse calls it, "hits the system from without," and thus cannot be internally deflected and defused, as happens with most nominally resistive actions.

But Marcuse is not optimistic about the process: "nothing indicates that it will be a good end." In fact, he says, something like a violent revolution is both necessary and morally justified to restore human freedom and dignity. In a follow-up essay, "Ethics and Revolution" (1966), he argues that the system imposes an oppressive psychological slavery upon the people and, when needed, is more than willing to violently suppress resistance. Peaceful resistance accomplishes nothing because it can be safely ignored by those in power; in fact, they *welcome* it, because it gives the false impression that the system is democratic and responsive to human needs. "Peaceful revolutions . . . do not present any problem" (1966: 135). True change, by contrast, is revolutionary change, and this is not peaceful: "radical and qualitative change implies violence." As the system becomes ever stronger, it increasingly stifles human development and freedom. It does not change willingly, and thus "violence becomes the necessary and essential element of progress" (137). And this is morally justified, he says, because

> the technical apparatus of advanced industrial society is in itself authoritarian, requiring service, submission, subordination to the objective mechanism of the machine system, that is to say, submission to those who control the apparatus. Technology has been made into a powerful instrument of streamlined domination. (147)

And as a consequence, inflicting a "triumphant violence" upon us—and, we might add, upon nature. A violent system, with powerful technology at its disposal, can only be effectively countered with violence—a distressing conclusion, to be sure, but consistent with a thesis of technological determinism.

The revolutionary Zeitgeist seems to have been contagious. Mumford gave a Smithsonian address in Washington in September 1965, subsequently published as "Technics and the Nature of Man" (1966). Having spent much of his career describing and documenting the technical phenomenon, Mumford was now himself calling for a kind of revolt. He attacks the ever-expanding network of interconnected technologies—"megatechnology"—that had been emerging from the beginning of recorded history. In progressively detaching humanity from nature, "megatechnics . . . will create a uniform, all-enveloping structure, designed for automatic operation" (1966: 303). Rather than existing as the tool-using animal, "man will become the passive, machine-serving animal whose proper functions . . . will either be fed into a machine or strictly limited and controlled for the benefit of depersonalized collective organizations." Technological society itself is depicted as a Megamachine, dominating all aspects of life. Sadly, he says, we have largely surrendered ourselves to it: "unconditional commitment to the Megamachine is now regarded by many as the main purpose of human existence." As a result, society has become "increasingly coercive, totalitarian, and—subjectively speaking—compulsive and irrational" (314).

Mumford concludes with his own call to revolution. Our only recourse, he says, is to decentralize the Megamachine and restore it to the human scale: "a deliberate, large-scale dismantling of the Megamachine, in all its institutional forms, must surely take place, with a redistribution of power and authority to smaller units, under direct human control" (316). Bucking the determinist trend, Mumford is relatively optimistic that such action is still possible—though he seems unaware of the pessimism of many of his predecessors.

He reiterates this call a few years later, in volume two of *Myth of the Machine* (1970). "A culture where only the machine embodies order and rationality" is, in a deeper sense, irrational; it means "submission to demonic impulses and drives" (370). Culture is in peril, and "there is no technological solution for this perilous state"—obvious, given that technology itself is the source of the problem. "Only if a sufficiently passionate human reaction takes place will it be possible to reverse this process and give back to the depleted human organism the autonomous functions, orderly processes, and cooperative associations it has almost relinquished." Our present technological society "is in a state of nightmarish disintegration," threatening both human and environmental well-being. For our part, this has both physical and psychological consequences: "the basic conditions for mental stability . . . are constantly being undermined; and as a result our whole power-driven civilization is turning into a blank page, torn to shreds from within by psychotic violence" (371).

Drastic action is therefore demanded. "What is involved, if the human race is not to lose its grip on reality entirely, is something like a profound and ultimately planet-wide reorientation of modern culture." "In short," Mumford concludes, "this is nothing less than a revolt against a power-centered 'civilization.' That revolt had long been overdue: something like five thousand years overdue" (372)—referring to the origins of the Megamachine in ancient Egypt. "Beneath this revolt is a deep and . . . a well-justified fear: that the next step in technological progress may bring about the annihilation of man" (372). The technologies of war, and specifically of nuclear weapons, put us all on the brink of extermination. Unfortunately Mumford provides no specifics, and like Marcuse and others, he offers virtually nothing in the way of philosophical or metaphysical analysis regarding how such a situation has come about. And without some such understanding of what technology is, we cannot assess the adequacy of our response.

Against this backdrop of the revolutionary 1960s, the next two decades were a time of broad skepticism about technological culture. "Dystopian interpretations of technology were popular" at this time, according to Ihde (1990: 6), though most were, strictly speaking, nonphilosophical in nature.

Examples of such dystopian thinking are found in the writings of Theodore Roszak and Ivan Illich, among others. Roszak's work is less a systematic analysis of technology than a series of lamentations and rebukes. His putatively antitechnology books—including *The Making of a Counterculture* (1969), *Where the Wasteland Ends* (1972), *Person/Planet* (1978), and *The Cult of Information* (1986)—are neither revolutionary nor particularly probing; they simply explore the human, social, and psychological damage brought on by modern technological society. Roszak seems to have an aversion to serious, fundamental critiques of society, especially those that might entail any sort of social upheaval, and thus his criticism implies only the mildest recourse.

For his part, Illich's books *Tools for Conviviality* (1973) and *Energy and Equity* (1974) describe the urgent need to reestablish a technology of "natural scale," that is, of an appropriate size and energetic intensity to be manageable, safe, and supportive of social justice. The goal is a "convivial society" characterized by "responsibly limited tools" (1973: xxiv). For Illich, as for Heidegger, technology is essentially about energy; hence his motto for the convivial society: "personal energy under personal control" (12). The destructiveness of large-scale technology lies not in the usage; it is inherent in the technology itself. Thus we cannot blame capitalism, nor—contra Marx—look to socialism or communism for salvation:

> Certain tools are destructive no matter who owns them. . . . Networks of multilane highways, long-range, wide-bandwidth transmitters, strip mines, or compulsory school systems are such tools. Destructive tools must inevitably increase regimentation, dependence, exploitation, or impotence and rob not only the rich but also the poor of conviviality (26).

Illich was greatly concerned about the "radical monopoly" that certain classes of technologies have over particular aspects of society. The automobile over transportation is one example, and compulsory institutional schooling over education is another. Eventually these monopolies build frustration and disempowerment. Reforms will then be sought. But reform action, by addressing only surface issues and not the root cause—namely, the monopoly itself—only serves to strengthen the monopoly's grip:

> The more these reforms succeed in correcting superficial abuses, the better they serve to bolster the monopoly I am trying to describe. [As an example, consider] consumer protection. Consumers cannot do without cars. They buy different makes. They discover that most cars are unsafe at any speed. So they organize to get safer, better, and more durable cars and to get more as well as wider and safer roads. Yet when consumers gain confidence in cars, the victory only increases society's dependence on high-powered vehicles—public or private—and frustrates even more those who have to, or would prefer to, walk. (56)

Thus the ultimate irony: radical technological monopolies ultimately become self-serving entities that actually *frustrate* rather than promote the nominal service they are intended to perform. Automobiles were intended to enhance transportation options and increase the overall mobility of society, but instead they (1) effectively exclude or sideline other options (biking, walking, even mass transit); (2) compel usage, in a functional sense; (3) develop into powerful, self-serving corporations (with auto lobbies, government bailouts, 'too big to fail,' etc.); and (4) produce unintended negative side effects (wars for oil, global warming, traffic fatalities, etc.).

Other examples are not hard to find, says Illich. Consider health care. The medical profession exists, first and foremost, to sustain and grow itself; solving health problems is a secondary and expendable objective. In fact there is an incentive to promote illness, given that it is the source of profit. Public schooling is another case study. The institution of mandatory public education does not and cannot meet its nominal goal of universal education; instead it defines a class of people as dropouts and failures, imposes homogenizing and stultifying standards on all children, and ultimately serves the needs of the technological system for well-trained labor rather than diverse individual human needs. The upshot of all this is clear, according to Illich: technology works through large-scale institutions in a self-serving and self-promoting manner.

The inevitable outcome of this process of technological development will be a "gruesome apocalypse" of social and environmental collapse. In an attempt to stave off collapse, society will find it necessary to isolate and even recreate the human being in order to survive in a depleted and technologically manufactured world:

> Mankind would live in a plastic bubble that would protect his survival and make it increasingly worthless. Since man's tolerance would become the most serious limitation to growth, the alchemist's endeavor would be renewed in the attempt to produce a monstrous type of man, fit to live among reason's dreams. A major function of engineering would become the psychogenetic tooling of man himself as a condition for further growth. (101)

But even this is merely a temporary solution. Limitless growth on a finite planet is impossible; at some point the growth of the system will come to a halt as emerging crises outstrip our ability to respond. The result will be a sudden breakdown of the system. "Almost overnight people will lose confidence not only in the major institutions but also in the miracle prescriptions of the would-be crisis managers." In a moment of brutal enlightenment, "people will suddenly find obvious what is now evident to only a few: that the organization of the entire economy toward the 'better' life has become the major enemy of the *good* life." At this point, revolution and collapse are inevitable. This has happened in the past, Illich reminds us—with the Reformation and the French and Russian revolutions: "The unthinkable became obvious overnight: that people could and would behead their rulers." This process of revolution and collapse Illich calls "political inversion"—a kind of overturning of the entire social order. And we can expect such radical change to be "violently painful" (13), he warns.[8]

In the end, Illich appears torn; his gloomy prognostication anticipates disaster, but he is hopeful that some segment of humanity, at least, will weather the storm. There is a kind of underlying optimism in his work, a faith that we can overcome the pressures of industrial technology. But once again, his analysis is hindered by its lack of metaphysical insight. Illich's analysis is more sociological than philosophical and thus necessarily deals only with surface phenomena. He has no hope of getting to the root cause of the present situation.

ECOLOGICAL CRITIQUES

Environmental philosophy came into being in the early 1970s, and there was a natural affinity between ecophilosophers and technology critics. Arne Naess, founder of the deep ecology movement, emphasized the link between technology and environmental destruction—see, for example, his *Ecology, Community, and Lifestyle* ([1976] 1989). Bill McKibben elaborated on this connection in his books *The End of Nature* (1989) and *Enough* (2004).[9]

But perhaps the most intense focus on the metaphysical and even spiritual dangers posed by technology was that of ecophilosopher Henryk Skolimowski. His early thinking (1966, 1967) was relatively sympathetic to technology, but exposure to Ellul initiated a turn against it. Skolimowski moved from a position of "technological realism" (1968) to mild rebuke (1970, 1971) to stinging critique (1973, 1974a, 1974b). Even at the time of his early work on the topic, in the late 1960s, he recognized that, in contrast to the naïve realism of the day, "technology is an extraordinarily complex phenomenon" (1968: 426) and that it therefore calls for incisive philosophical analysis. The lack of dedicated philosophical effort was self-evident to him, hence his observation that "the philosophy of technology is in a sorry state" (433)—a condition hardly improved to this day.

He was soon reflecting on the two-sided nature of technology: that we construct it, but that it, in turn, constructs or modifies us. In an essay on the role of technology in architecture, he observed,

> The more protected we are against the elements, that is, the more elaborate our artificial (i.e. technological) environment becomes, the more thoroughly it conditions our behavior. In the Stone Age, we were at the mercy of the elements; now we are at the mercy of the shield which protects us against these elements. (1969: 82).

In a like manner, the supposed rationality of technology inflicts a contrary psychological effect upon us. "It is paradoxical that technology, while attempting to make our lives totally rational, succeeds in accomplishing the reverse." Its high degree of order and structure "subject[s] us to the unbearable confinement of pre-conditioned behavior," with the net effect of producing disorders ranging from "eccentricity to paranoia." Summing up, he states that "a 'rational' process which results in making our lives inhuman is not rational but anti-rational" (83).

Into the 1970s, Skolimowski was declaring modern technology "a seducer" and calling for a new, humanistic system that served exclusively human ends. It was increasingly clear to him that technology was in control, that a form of determinism held sway, and that certain "inexorable tendencies" (1972: 437) were bearing down upon humanity. And not only humanity: nature, too, was in its sights. In fact, he virtually defines modern technology as "an instrument designed to control nature" (1973: 50)—and nature suffers as a consequence. In 1974, Skolimowski was tearing down the "illusion of progress": the idea that, because technology is advancing, so too is our quality of life. This, for him, was patently absurd: "the overall balance increasingly shows the progressive trivialization of our lives through increasingly sophisticated technologies" (1974a: 255). Indeed, "the metaphysics of progress is based on an exploitive and parasitic form of philosophy" (257), centered on destruction of the natural world. Technology is homogenizing the human sphere by focusing everything on material consumption and universal technical demands—common standards, interchangeability, optimal efficiency, and so on. This is in fundamental contrast to nature, which tends to promote diversity. "As such, [technology] is becoming an anti-life proposition in the evolutionary sense" (258).

By 1975, Skolimowski was speaking of a "technological consciousness": of technology as so integrated into our way of thinking that we are unable to conceive of nontechnical alternatives. "Technology is a state of Western consciousness in which 'control' and 'manipulation' are dominant features." The technical mode of thinking has become ingrained "to such a pervasive and perverse degree that even when we realize that it devastates our natural and human habitat, our immediate reaction is to think about another technology which will mend it all" (1975: 75)—a situation virtually unchanged in the intervening 40 years. Technology thus imposes its own values on humanity and leads us to its own ends. The presumption of neutrality is naïve and deceptive; dehumanization is part and parcel of the technical program and cannot be separated from it. Skolimowski concisely summarized these concerns in 1978:

> Technology, as developed in the West, is not neutral. . . . When woven into the tapestry of the secular worldview, which in its development emptied out man of his inner life—particularly his ethical and spiritual life—technology cannot but be destructive of man's higher values. No way has yet been found to prevent that, which means that the detrimental consequences of technology are not incidental but endemic; they lie in the very nature of the phenomenon, as it is functioning within a certain worldview. (1978: 51)

By the time of his books *Eco-Philosophy: Designing New Tactics for Living* (1981) and *Technology and Human Destiny* (1983), his position was fully articulated: advanced industrial technology is primarily a means for quantifying, controlling, and manipulating nature and man, and for creating an efficient system of material production. It no longer serves to ameliorate the human condition and in fact dramatically undermines it. Technology fosters a harsh conception of secularism, undercutting any broader and potentially healing notion of human spirituality. It corrupts all aspects of human existence, even the political. With its fixation on manipulation and control, technology inevitably tends to concentrate power and wealth into ever fewer hands—thus eroding even the most democratic of societies. It is in this sense, Skolimowski says, that "technology is a slayer of democracy" (2005: 61). Despite the harsh analysis, there are paths forward. As always, it is not a question of eliminating technology, whatever that may mean, but rather of navigating a shift toward an appropriate, *humanistic* technology that serves the larger needs and visions of humanity.

Owing to the efforts of Skolimowski, Mitcham, and others, a true subdiscipline of philosophy of technology had emerged by the early 1980s. It drew in not only philosophers but also social critics of various professions. On the philosophical side, Don Ihde published his early works around this time, and Borgmann (1984) produced his seminal book, *Technology and the Character of Contemporary Life*—examined in detail in Chapter 6. Noteworthy from the nonphilosophical perspectives are the books of Langdon Winner, and Jerry Mander's (1978) *Four Arguments for the Elimination of Television*. Winner, a political scientist, ably updated Ellul's ideas in *Autonomous Technology*, but, unsurprisingly, did not advance the underlying philosophy. In *The Whale and the Reactor* (1986), he opted for a more purely sociological examination. Mander—an economist by training—crafted an insightful, McLuhan-esque critique of television culture that was tightly focused but consequently of limited applicability.

THE 1990S: GROWTH AND DECAY

By the early 1990s, a strange situation had developed. On the one hand, the field of 'philosophy of technology,' loosely conceived, was beginning to blossom. Ihde (1990), Zimmerman (1990), and Feenberg (1991) were making important contributions. Modern approaches, like social constructivism and critical theory, were becoming prominent—to the point where Braun (1995: 21) could bluntly declare that "technology is a social construct"; it was not a deterministic force, not a thing in itself, certainly not a substantive reality, but rather something wholly constructed by and for human beings. Technological determinism was passé. "This approach," wrote Feenberg at the end of the decade, "has now been largely abandoned for a view that admits the possibility of significant 'difference,' that is, cultural variety in the reception and appropriation of modernity" (1999: 183). Of the traditional critics, only Heidegger retained popularity, providing many a commentator the opportunity to offer passing thoughts on his notorious view of technology.

Continuing the trend from the previous decade, historians, scientists, sociologists, political scientists, and journalists published more works on the subject. Segal (1994), Smith and Marx (1994), Turkle (1995), Sclove (1995), and Tenner (1996), among others, got into the act, bringing a true interdisciplinary slant to the field. Technology itself was, as always, charging ahead. The space shuttle *Challenger* disaster of 1986 put a crimp in the U.S. space program for several years, but from a day-to-day perspective, things were clicking along; computer technology was rapidly expanding, the Internet became a reality for many, and the 'tech bubble' was in the making.

On the other hand, and for no obvious reason, *criticism* of technology—philosophical or otherwise—withered away. This was and continues to be a rather stunning phenomenon: never before has technology attained such a pervasive influence on people's lives—their careers, pastimes, hobbies, social lives, environment, even their very self-image—and yet never have the critics so completely ceded their ground. Technology has simply ceased to be subject to the kind of incisive and penetrating criticism that characterized the preceding 100 or more years.

In place of the traditional, harsh criticism, we have been treated to an interesting spectacle: criticism, but of such a mild variety as to be scarcely worthy of the name. The critics are so fainthearted, their complaints so tepid and conditional, and their analyses so superficial that no insight is afforded into the technical phenomenon, and no substantive response of any kind is called for. The most that we get is a plea for "balance" in our tech lives—of the need for "moderation," "self-discipline," "downtime," and for children, "oversight." Howard Segal's 1994 book, *Future Imperfect*, is a good example; his overarching theme is that technology is, sadly, a "mixed blessing"—as if we needed to hear that message.

The following year saw the publication of Sherry Turkle's *Life on the Screen*. This MIT sociologist and psychologist, who today ranks among the most mediaworthy of technology skeptics, offers only the barest, most facile criticisms. She provides scant insight into the nature of technology and displays little grasp of the severity of the issues at hand.

A revealing—and not atypical—discussion took place on National Public Radio (NPR) in early 2013.[10] In an interview with Turkle, the radio host expressed concern that the various communications technologies, such as email, cell phones, text messages, and social media, are driving people apart and thus damaging intimacy and human relationships in the process—all undoubtedly true. Turkle replied that the matter is "complicated": the anonymity of technological communication allows us to be "unburdened" from potentially awkward, face-to-face conversations but at the same time truncates and compresses them. Communication becomes merely 'connection.' This is all true, once again, but it is only the most superficial and visible aspect of the dehumanizing effects of technological communication. For Turkle, the main drawback seems to be that we are "shortchanging ourselves" through the bypassing of real interaction.

Evidently concerned that this discussion might be heading down an anti-technology path, the interviewer then remarks that "the devices and technology we have [today] improve our lives immeasurably in many ways"; "You're not arguing that we wistfully look back on years past and a different time," she says leadingly. "I'm not arguing for any kind of retro position," Turkle replies; "It is not a Luddite thing to say. . . ." Reassuring her listeners, she adds, "We all get to keep our phones and love our phones"—heaven forbid anyone should suggest otherwise.

I do not intend to be dismissive. The problem of technology clearly admits of a wide range of opinions. If Turkle sees no larger threat to modern technology than that it compresses our interactions and causes us to be "alone together" (2011), she certainly has the right to say so. However, when such mild criticism is represented, implicitly or otherwise, as the *standard* of technological critique; when it is cast as the only allowable form of criticism; when it monopolizes the media discourse—*then* we have reason for concern. Her case is not exceptional. Nearly every present-day intellectual who presents even a nominally critical stance toward technology simultaneously offers a cartload of concessions and qualifications: "I am no Luddite"—the most commonly heard refrain among our so-called technology critics today.[11]

Such 'uncritical criticism' serves at least three important roles in a technological society. First, it allows us to convince ourselves that we do, indeed, accept criticism. Problems arise, and opposing voices are heard. The nominal critics are given due airtime and media coverage, and this fulfills the necessary function of providing a contrarian viewpoint. Never let it be said that our modern society is incapable of self-scrutiny! Second, it allows for some release of the general feeling of technological angst that exists throughout any industrial society. Nearly every thinking person, I submit, has at least subtle and unexpressed concerns about the growing power of technology, and it is psychologically useful to hear these concerns articulated occasionally—if only to reassure oneself that 'they' are looking into these things. Third, and most importantly, it monopolizes the field of technology criticism, effectively displacing other, potentially harsher critics. When the same mildly critical stance is repeatedly called up, the public gets the impression that this is the extent of contemporary thinking, that no 'respectable' intellectual would hold a stronger view, and that, should any such view somehow emerge, it can be forthrightly dismissed. This, in fact, is the truly modern way to control dissent: find and promote your own moderate critics, allow them to monopolize the discussion, and then subtly suppress any harsher critique. This is much more effective and more efficient than outright censorship. But control of the dialogue is essential; and control, after all, is the forte of a technological system.

Throughout the 1990s, then, technology criticism ranged from mild to nonexistent. Segal and Turkle issued their critiques, such as they were. Somewhat more substantial but still tepid critical studies came from Stivers (1999) and Braun. Tenner's *Why Things Bite Back* (1996) earned some notice for its emphasis on the 'unanticipated consequences' problem, but as a historian, he was ill suited to assess the philosophical underpinnings of this complex phenomenon.[12]

As mentioned, philosophers more or less abandoned all attempts at a truly philosophical criticism. With no one willing or able to advance the sociological critique of Ellul or the metaphysical analysis of Heidegger, interest in the problems of determinism and technological autonomy receded. Though oversimplified, the statement by Higgs and colleagues (2000: 4) is understandable: "Technological determinism fell from favor because the arguments advanced to support it seemed crude"—crude only because no one felt compelled to conduct a detailed analysis, nor to formulate a more sophisticated version. I attempt to remedy this in the chapter to follow.

TWO RADICALS

Amid the dearth of truly critical voices in the 1990s, there was a notable exception: Ted Kaczynski. Prior to his conviction for the Unabomber crimes, he was able to force publication of his now-infamous manifesto, *Industrial Society and Its Future,* in September of 1995. Less a philosophical analysis than a call for revolution, the manifesto presents a detailed but succinct statement of many of the central arguments against technology. Its uncompromising tone and thoroughly rational demeanor pose a serious challenge to the advocates of the status quo.

Kaczynski's basic argument is clear enough: (1) humans evolved under primitive, low-tech conditions, and it is these conditions to which we are genetically and psychologically best suited; (2) modern society is radically different than our natural state and thus imposes unprecedented stress upon us; (3) the stress will increase in the coming years as technology advances, and therefore we will have to take increasingly drastic measures to adapt to it—measures that will include genetic, physiological, and psychological manipulation; and (4) there is no way to reform the technological system so as to avoid such a dehumanizing outcome. Hence his logical conclusion: We must act now to tear the system down before it gains overwhelming power against us. If the system survives long enough, he says, it will obtain something approaching complete control over humanity. As a by-product, much of wild nature will be destroyed. And if superintelligent computers emerge, we face the very real prospect of total enslavement or, even worse, outright elimination.

Drawing from and simplifying many of the points raised by Ellul and others, the manifesto systematically walks through the central issues. Human existence must be lived through effort and personal action, at least regarding the basics of life. This is part of our evolutionary heritage—to have direct control over the necessities, to understand the challenges and risks, and, by and large, to attain them. Advanced technological society disrupts this process by removing us from the essentials of life; we no longer directly acquire food, clothing, or shelter, but only indirectly, through money, which is earned in the service of the system. Technology relentlessly reduces the sphere of human freedom by drawing us ever deeper into the growing technical web. In fact, it cannot function otherwise; a technological system must grow in size and complexity and therefore must remove us further and further from personal control over our own lives. We become detached from the essentials of life, alienated from nature, and utterly dependent on a vast artificial system.

There is no hope of fixing things so as to avoid this pernicious outcome. Either we prepare to surrender our humanity, or we plan to take extreme action to save it:

> [One can see] how hopelessly difficult it would be to reform the industrial system in such a way as to prevent it from progressively narrowing our sphere of freedom. . . . [A]ny change designed to protect freedom from

> technology would be contrary to a fundamental trend in the development of our society. Consequently, such a change either would be a transitory one . . . or, if large enough to be permanent, would alter the nature of our whole society. . . . Thus, permanent changes in favor of freedom could be brought about only by persons prepared to accept radical, dangerous, and unpredictable alteration of the entire system. In other words, by revolutionaries, not reformers. (Kaczynski 2010: 69–70)

Those who believe technology works for the benefit of mankind are seriously deluded. "The system does not and cannot exist to satisfy human needs. Instead, it is human behavior that has to be modified to fit the needs of the system" (72). That the system does, in fact, meet some human needs is incidental; it does so only because such action works to the benefit of the system itself. Technology could not advance without a supportive, educated, and moderately satisfied human workforce, and thus the system promotes these very things. If the day comes when we are expendable, it will discard us without a moment's hesitation. In fact, it already discards many thousands of people—the poor, the less intelligent, the uneducated, the elderly. Such people draw more resources than they provide and hence are a net loss for the system. This pool can only be expected to grow in the future.

Kaczynski emphasizes an underappreciated fact: that technology advances through individually benign actions that, collectively, work to our detriment. Hence the paradox of progress:

> While technological progress *as a whole* continually narrows our sphere of freedom, each new technical advance *considered by itself* appears to be desirable. Electricity, indoor plumbing, rapid long-distance communications . . . how could one argue against any of these things, or against any other of the innumerable technical advances that have made modern society? It would have been absurd to resist the introduction of the telephone, for example. It offered many advantages and no disadvantages. Yet . . . all these technical advances taken together have created a world in which the average man's fate is no longer in his own hands. (76–77)

This is a truly insidious process: Every small step appears beneficial, but the combined long-term effect is negative and potentially catastrophic. When asked to point to the specific harms of modern technology, we are at a loss—every concrete device improves some aspect of life. In fact, that is precisely why it exists. And yet the emerging system unquestionably restricts our autonomy and places us under growing pressure to conform and to serve it. We are thereby drawn into an inextricable bind.

In the process of its evolution, the system subjects humanity to increasingly severe hardships. Hence we must be continually bribed, cajoled, seduced, and coerced into playing along. Protechnology propaganda, both explicit and implicit, works to convince the masses that the system serves their needs and that it is under human control. Entertainment functions as a release valve and as a temptation; who doesn't look forward to the latest 3-D action thriller film? Various comforts, amusements, and diversions consume the bulk of people's lives. When entertainment fails, pharmaceuticals are mass produced and mass marketed in order to remove the stress of daily life. Kaczynski presses on this very point:

> Imagine a society that subjects people to conditions that make them terribly unhappy, then gives them drugs to take away their unhappiness. Science fiction? It is already happening to some extent in our own society. . . . Instead of removing the conditions that make people depressed, modern society gives them antidepressant drugs. In effect, antidepressants are a means of modifying an individual's internal state in such a way as to enable him to tolerate social conditions that he would otherwise find intolerable. (82–83)

The same is true for video games, iPods, and social media—all designed to encourage us to embrace the blessings of technology or at least endure its oppressions. For those who cannot or refuse to conform, high-tech surveillance systems and a functioning police state will keep them in line. Control thus expands by a slow, creeping, relentless process of imperceptibly small advances. The average man is powerless to defend himself and inevitably submits.

Confronted with such a situation, it is rational to contemplate "dismantling the megamachine," as Mumford stated three decades before the manifesto. Unlike his predecessors, though, Kaczynski provides some details regarding how this might be achieved.[13] Overall he presents a lucid and powerful case, one that is not easily countered.

From the present perspective, two points stand out. First, Kaczynski offers little in the way of a metaphysical analysis. As we have seen, this is not unusual—and in his case, it is not a substantial defect. Kaczynski's is a pragmatic concern; philosophical underpinnings are, for him, purely incidental. On this point I would disagree. I believe that only by grasping the metaphysical nature of technology can we construct an adequate response. If we underestimate or misinterpret the problem, we will likely fall short of our goal—which is, after all, to produce a high-quality life for humanity within a sustainable environmental context. Increasingly, it appears that high-quality existence and high-tech society are mutually incompatible.

Second, Kaczynski has been all but ignored by the intellectuals, philosopher and nonphilosopher alike. Following his capture in 1996, all attention was focused on the man and his life story; the problem of technology received almost no discussion. The 2010 publication of his book *Technological Slavery*—which, as the first by a prominent American "terrorist," should have been a major media event—garnered virtually no attention at all. On the one hand, this is understandable owing to his notorious background.

But on the other, we have to ask, What is there to gain in doing so? What do we fear? Giving the man "too much credibility"? "Dignifying" him? Promoting copycat criminals? While not baseless, I would suggest these concerns are minimal, particularly given the severity of the problems we face. By refusing to engage with such an argument, we only do ourselves a disservice. We bypass an opportunity for a serious and far-reaching critique of our modern worldview, and we shut off discussion on one line of action, extreme though it may be, that is a viable option—and conceivably, our only option.[14] Whether we are wise to do so, the future will tell.

Kaczynski's manifesto was published in late 1995, and the remainder of the decade passed with scarcely another critical voice.[15] But then, fresh into the new millennium, came another radical critique—this time from technology insider Bill Joy. His article "Why the Future Doesn't Need Us" appeared with great fanfare in April 2000. Here was one of the most tech-savvy people on Earth, who not only granted some credit to Kaczynski but also called for the outright termination of future development of some of our most dangerous technologies—in particular, those associated with the intersection of genetics, nanotechnology, and robotics (GNR).

The 'double-edged sword' theme is prominent in Joy's essay:

> Each of these technologies also offers untold promise. . . . Together they could significantly extend our average lifespan and improve the quality of our lives. Yet, with each of these technologies, a sequence of small, individually sensible advances leads to an accumulation of great power and, consequently, great danger. (n.p.)

Promise and peril, side by side, in all that we do—with the 'promises' leading, all too often, to dependency, corruption, and addiction, even as the perils lead, potentially, to our extermination.

By 2030, Joy argues, it will likely be possible to create submicroscopic, self-replicating, lifelike machines—nanobots. If they are able to replicate under ordinary outdoor conditions, there is potentially nothing to stop them from reproducing by the billions, literally swamping the planet. A simple lab accident or a deliberate act of terrorism could quickly destroy the entire planetary ecosystem. And that is only one of many risks. With threats ranging from runaway nanobots to superintelligent machines that squeeze us out in the battlefield of evolution, Joy is not optimistic about our technological future. And not the distant future—the critical technologies will be available within two decades.

In the face of these potentially catastrophic disasters, Joy's action plan, such as it is, revolves around the concept of relinquishment: of limiting the development of the most dangerous forms of GNR technology. In support, he cites Carl Sagan: "[successful] civilizations place limits on what may and what must not be done, [to] safely pass through the time of perils."[16] Such a stance places Joy dangerously close to the dreaded antitech camp. He is anxious to reassure his readers—"I am not a Luddite"—but yet acknowledges that technology has its limits and that any rational society must vigorously enforce these. Joy's plan thus calls for a "strong ethical code" to supplement and fortify our restraint, along with "vigilance," "personal responsibility," "altruism," and a "new ethical basis" founded on Buddhist principles of love and compassion.

But can relinquishment and new ethics really work? Is such action even a feasible option at this point in time? Consider what our contemporary philosophers of technology have to say. Feenberg (2002) is symptomatic of the situation. For him, "the very project of bounding technology appears suspect" (10). Why is this? "If we choose to leave something untouched by technology, is that not a subtler kind of technical control?" We might rather ask—Is he serious? Does he really expect us to believe that action of any kind is a 'technical' act, and therefore that efforts to counter technology are futile? "How can I ever leave the technical sphere if the very act of bounding a reservation instrumentalizes it?" Well then—apparently our only option is to pack our bags and head back to the daily high-tech grind. Feenberg seems to have a low opinion of the intellectual capacity of his readers.

The technophiles are no less dubious in their reasoning. Kurzweil (2003) argues that unilateral relinquishment of powerful new technologies only leaves the field open to one's enemies; we may need an army of 'good' nanobots, he says, to fight off the 'evil' ones created by some mad scientist. But this is obviously a catastrophically risky strategy. Kurzweil's point is that we are effectively in a state of mutually assured destruction (MAD) and thus cannot afford to come up shorthanded. But as in nuclear MAD, this fact simply raises the level of absurdity. If we are so in thrall to modern technology that we cannot give it up, and that the only response to dangerous technology is yet *more* advanced and *more* dangerous technology, then this not only strengthens the case for determinism, but it argues for immediate and drastic action—something far beyond mere relinquishment.

Joy's article, which remains one of the most-referenced essays on technology in history, could have been yet another ideal lead-in for a renewed philosophical discussion and a reenergized critical debate. But as with Kaczynski, this did not occur. Among more than a dozen important academic works on technology published in the decade following Joy's piece, none gave his concerns more than a passing comment.[17] Most assuredly, none mention Kaczynski at all.[18] Reference to Ellul comes in scattered passages at best. Technological determinism is not addressed in any detail. Only Heidegger merits extended discussion, but even then primarily in the guise of arcane interdisciplinary disputes about interpretation—and nothing on extending or applying his insights. And of these several works, only one—Agar's (2010) *Humanity's End*—is unambiguously critical.[19]

From the mid-1980s, then, through the present day—a period of some 30 years—we find serious technology critiques to be almost nonexistent. Only Kaczynski and Joy argued for substantial action; for all the others, "balance" and "moderation" seem to suffice. And as I have emphasized, these two men are not philosophers. To find the most recent dedicated philosophical critiques, one must go all the way back to Borgmann (1984) and Skolimowski (1983).[20] It is clear that philosophers of technology have almost completely abandoned the critical stance.

What accounts for this? Presuming that they are neither lying nor being coerced into submission, there would seem to be only one explanation: a rare and near-total consensus on the philosophical problem of technology—namely, *that it is not a problem at all.* Contemporary philosophers, they say to us, have studied the matter in detail and concluded that "technology" is neither autonomous nor substantive, that it imposes no determinative pressures on society, that it presents no serious threats to humanity or nature, and that the whole topic is scarcely worthy of the philosophical dime. "Technology" can only be studied as so many specific "technologies" that are socially and contextual dependent, each posing only unique and isolated problems that admit to a variety of localized and modest solutions.

Don Ihde is an illustrative example of this putative consensus. Long considered a leading figure in philosophy of technology, one of his more recent books on the topic is *Bodies in Technology* (2002). The final chapter addresses the interaction between "phil-tech" and "eco-phil"—that is, between philosophers of technology and those of environment. Specifically, he is concerned about the "congenital dystopianism" (114) that dominates both groups. This in itself is a striking statement: environmentalism, to be sure, but for philosophy of technology, one must go back to Heidegger and Ellul, or at least to Marcuse and Skolimowski, to find any such thinking.[21] But to cite the "dominance of dystopianism" in the technology field is either a mistake or a delusion; perhaps the *fear* of dystopianism, but nothing more.

Seeking the proverbial "balance," Ihde attempts to navigate between the optimistic and pessimistic extremes. First he dismisses "Malthusian extrapolations" in which past tendencies—population growth, resource depletion, spread of pollutants, and so forth—are extended indefinitely into the future, with catastrophic implications. Population problems are depicted as insignificant by pointing to the fact that some nations had fallen below replacement levels by the late 1990s.[22] Projections of 9 or 10 billion humans are disparaged as "negative hype" (119). He then writes off "microsolutions"—including the banning of certain harmful technologies, as well as calls to recreate aspects of simpler eras—as mere "nostalgic (or retrospective) romanticism." Retroactive romantics remind him of a "Bob Dole form of environmentalism."

To assuage our concerns, Ihde then cites some environmental success stories: a reduced ozone hole, lower atmospheric lead content, and fish in the river Thames "as far upstream as London." Of course, these are not really success stories as much as they are 'less-damage-than-before' stories. For millennia, atmospheric lead was zero, ozone was plentiful, and fish reproduced freely in the Thames—until humans came along with advanced technology.

Having severely damaged things, we do not now get credit for causing less harm than in years past. But Ihde's conclusion is different: environmental problems "do not entail either abandoning technologies or pulling plugs [i.e. banning]. Instead, they often point to improved and higher technologies" (121). This, of course, is the classic technophile line; any given technological problem always has a (higher) technological solution.

In sum, says Ihde, techno-environmental problems "are almost all complex, interrelated, ambiguous problems that rarely yield to quick-fix or easily grasped solutions" (123). This is difficult to interpret except as an apologist stance. Technology is not the problem, and more likely it is the solution. The problems being complex and ambiguous, we must avoid all simpleminded remedies—such as those proffered by the technological dystopians. In any case, technology per se calls for no special philosophical analysis and certainly poses no unique threats.

Such is the consensus view. The question, then, is this: To what degree is it correct? If it is substantially right, society can rest easy and simply wait for the future blessings of technology to unfold. The situation is so unworthy of investigation that philosophers are justified in directing their efforts elsewhere. Of the isolated technological problems that appear from time to time, we can rely on the good work of our scientists and engineers to resolve them. Technology critics can be rightly dismissed as nearsighted Cassandras who become so frightened by the technological hobgoblin that they conjure up visions of a monstrous autonomous force that is better suited to science fiction than to reality. The media's role is to consign such dystopianism to the purely fictional realm where, in the guise of novels and high-tech motion pictures, it produces large profits and, ironically enough, promotes the further advance of technology itself.

But what if it is wrong? Or perhaps this question: How *could it* be wrong? How could so many philosophers miss the mark? There are several possible explanations. It could arise from intellectual timidity. It could be lack of vision. Perhaps they are too enamored of the benefits of technology to objectively assess the risks. Some might see it as not the kind of thing that can be 'responsible' for problems at all. For others, the issue may be fear—fear of offending the status quo, fear of criticism from colleagues, fear of being labeled a Luddite, fear of institutional (e.g., university) fallout. In Ihde's case, he simply finds no evidence of more fundamental problems and is generally unbothered by the possibility of a technological Armageddon. Perhaps this allows him to sleep better at night.

Of course, the consensus view could be wrong by degrees. It need not be a simple matter of technological bliss or catastrophe. But even if the future plays out somewhere in the middle, it is still a future with substantial, persistent, chronic, disabling, and likely deadly problems. It will still be a future of major environmental loss, of mass indignities and dehumanization, of unhappiness, tedium, stress, addiction, and psychosis. And it could easily be much, much worse.

As I see it, the philosophical consensus has a very small chance of being correct. It is highly likely to be substantially wrong. And the risk of being wrong, even to a small degree, is serious enough for us to rethink our basic metaphysics and reassess the situation at large. As I have said, if we philosophers fail to do so, if we get this one wrong, future generations—or perhaps even the *next* generation—will not forgive us.

* * * * *

To summarize this and the preceding chapter: From virtually the beginning of Western civilization, noted critics have made negative evaluations of the technological phenomenon. This is striking in that, for most of this period, technology was self-evidently good and posed little in the way of immediate threat. With the emergence of an increasingly autonomous system near the turn of the 20th century, the dangers became more apparent, and we saw a corresponding elevation in the expressions of concern. The technological system had come to be seen either as unstoppable—with grave risks for humanity and nature—or so powerful that only radical action could derail its progress.

Philosophers of technology in the past three decades, at least, have largely abdicated their role as critics, finding little cause for alarm. But the public still apparently harbors concerns, even as we gobble up the latest tech toys. There seems to be an ongoing appetite for dystopian motion pictures—*Terminator*, *The Matrix*, *Avatar*—which likely reflects a deep-seated ambivalence. As one concrete indication, I can cite a 2005 Forester Research survey of 69,000 adults in North America that found 51% to be "technological pessimists," meaning, at best, indifference to technology and at worst, outright hostility.[23] There is a strong undercurrent of apprehension that is sure to increase in coming years.

In fact, there is substantial empirical evidence to support such worries. As I will show in Chapter 11, the list of technology-related health concerns is long and growing. Disasters like the 2010 Gulf oil spill and the ongoing debate over climate change all contribute to our sense of unease. Major periodicals publish articles asking if "Google is making us stupid" and if the "technological onslaught is making us crazy."[24] And yet philosophy has completely overlooked the metaphysical challenge. Or worse—looked at the issue and decided there is no challenge at all.

From my viewpoint, this consensus is clearly wrong. The philosophical apprehensions of the past 2,000 years were on the right track. There is a deep and fundamental basis for concern about technology. The situation will undoubtedly come to a head in next few decades. Prudence would dictate preparation—both metaphysical and pragmatic.

NOTES

1. [1900] 1990: 481–483.
2. This countermorality—which ultimately became documented in the New Testament—sought to sway the Arab, Greek, and Egyptian masses to the

Jewish side. Beginning with the monotheistic Judaic God, Paul and friends concocted the story of a Jewish teacher, Jesus, who was God incarnate and who came to earth promising eternal life. For those who did not accept this story, Hell awaited. This carrot-and-stick approach was apparently effective in swaying the superstitious masses to the monotheistic view and away from the values of Rome. The Empire was thus eaten away from the inside and ultimately collapsed. For a powerful elaboration, see Nietzsche's *On the Genealogy of Morals* and *Antichrist*; for a concise summary and analysis, see Dalton (2010).

3. "[N]ature is a structure of evolving processes. The reality is the process" (72).
4. Panpsychism is an obvious corollary to this outlook. See Skrbina (2010).
5. I note here that, for Orwell as for present-day Europeans, socialism involved a moderately capitalist economic system. Thus the central conflict was a *political* one—socialist-capitalist democracy versus autocratic fascism—and not an economic one.
6. See discussion in Chapter 5.
7. See, for example, Baudrillard ([2001] 2012) and Virilio ([2010] 2012).
8. His prescription for survival is one of knowledge and preparation. Illich urges us to understand the causes of the pending crisis and to be prepared for it. He refers to subterranean revolutionaries, those "previously submerged social groups" (105), who will be in a position to act when the crisis approaches. The time of crisis will loosen the grip of the system, allowing revolutionaries their chance.
9. Of tangential relevance here are the antitechnology writings of the "anticivilization" and green-anarchist movements. These groups are primarily activist, not academic. They agitate for change and use theory and history only as needed and as convenient. There is an interesting crossover between anticivilization and antitechnology writing; civilization today *is* technological, and one cannot critique the one without addressing the other. 'Anticiv' inevitably becomes 'antitech' if it is taken seriously.

   Two of the more visible anticiv writers are Derrick Jensen and John Zerzan, both of whom call for the outright termination of technological civilization. Zerzan began his writing career with the book *Elements of Refusal* (1988) and with the coedited volume *Questioning Technology* (1988), but his outlook grew to encompass the whole of industrial society. His *Running on Emptiness* (2002) offers some insightful reflections but nothing significantly new. His recently revised *Against Civilization* (2005) is a disappointment; it consists of a series of very short excerpts from a wide range of writers, loosely organized around the title theme—but almost no discussion of the problem of technology.

   Jensen's prolific output—something like 20 books written, cowritten, or edited since 2000—makes him a great cheerleader for revolutionary change, but unfortunately he too falls notably short in the areas of analysis and insight. His writing consistently weaves together a common set of themes, including environmentalism, racism, anti-industrialism, and education. His two most relevant works, *Welcome to the Machine* and the two-volume *Endgame*, tackle the problem of technological society, but only indirectly. The former addresses specific technological concerns, namely surveillance, security, and nanotechnology; but it has no mention of the deeper philosophical issues. The latter is an extended plea to terminate industrial society, but strictly on pragmatic considerations. In the place of intelligent analysis, Jensen prefers to dish up cutesy slang; thus we get remarks like this: "We're fucked. We're so fucked. Not in the good sense of the word" (2006: 231). Such is the state of contemporary critical thinking about technology, at least within the anticiv movement. The lone exception would be Ted Kaczynski (to be discussed in what follows).

10. National Public Radio, "We need to talk"—interview with Rachel Martin (February 10, 2013).
11. For the record: I am a Luddite.
12. Notwithstanding his pronouncement as "the philosopher of everyday technology."
13. Cf. his essays "The Coming Revolution," "The Road to Revolution," "Morality and Revolution," and "Hit Where It Hurts" (2010: 206–253).
14. If we reply that large-scale dismantling is *not* a viable option, then it seems that the argument is over—we have created a system that is beyond our ability to control and hence will almost certainly crush us.
15. Some misleading works did appear. Feenberg's *Questioning Technology* (1999), for example, does very little questioning of technology.
16. Sagan (1994: 371).
17. Harman (2002), Feenberg (2002, 2010), Williams (2002), Ihde and Selinger (2003), Sturken et al. (2004), Verbeek (2005), Schroeder (2007), Nye (2006), Nusselder (2009), Arthur (2009), Olsen et al. (2009), Ihde (2010), Raunig (2010), Agar (2010), and Kelly (2010).
18. There were a few minor exceptions, of course. Graham (1999: 7–8) allows him a few pages. Kurzweil (1999, 2005) finds some value in his writings, as does Joy. Finnegan (1998) wrote a notably sympathetic piece.
19. During the same period, nonphilosophical critiques were also nearly nonexistent; Fox (2002) and Stivers (2004) were the rare exceptions. A few further works have appeared in the past few years, including Powers (2010), Huesemann and Huesemann (2011), and Morozov (2011, 2013)—but as before, the criticisms are extremely mild.
20. Both did, however, address the topic of technology in later works, though as ancillary to other subjects. See Borgmann (1992, 1999) and Skolimowski (1991, 2005), for example.
21. Ihde cites Jonas as his single example.
22. In the 12 years since Ihde's book, global population has risen from just over 6 billion to just over 7 billion.
23. "The State of Consumers and Technology: Benchmark 2005."
24. See Carr (2008) and Dokoupil (2012).

9 The Case for Technological Determinism

Do we control technology, or does technology control us? To what extent are we controlled? Do only some technologies exert control and others not? If so, which ones and why? And in what specific ways? These questions, which lie at the heart of the discussion on technological determinism—and the related issue of technological autonomy—are surprisingly complex. The wide variety of definitions, descriptions, and analyses suggests that there is more to it than one might initially suspect. In the present context, the metaphysics of the Pantechnikon casts a new light on this situation and allows us to resolve a number of standing concerns. It points to a form of determinism, but of a unique kind: a determinism that grows in strength over time. Correspondingly, there is a sphere of free human action, but it constantly diminishes. I prefer to think of my own view, then, as a kind of vanishing compatibilism.

Determinism, broadly speaking, is the view that future events are fixed or predetermined, given past conditions and certain specific laws of nature.[1] The universe unfolds in a unique and necessary way, determined by its history and operational laws. On many interpretations, modern science suggests a deterministic universe, one that encompasses individual and collective human action. Others, however, point to the subatomic realm and quantum uncertainty as evidence of a kind of indeterminism at the foundation of physics—though it is unclear whether this could have any effect in the larger macroworld of human beings.

Technological determinism is a less formal concept. As generally understood, it refers to the idea that technology advances autonomously, of its own accord and by its own laws, and that therefore the structure of society is ultimately determined by technological imperatives or demands. The literature contains a number of ambiguous, conflicting, and self-contradictory ideas, some of which I will attempt to illuminate. In part, the problems arise from the fact that both 'technology' and 'determinism' are poorly understood, and the conjoining of the two concepts does no service to either.

Still, there are profoundly important ideas underlying the discussion, and a metaphysical analysis can provide some clarity. The pantechnical worldview points to technology, or rather to the technological system, as imposing an ever-increasing pressure on humans and society, thereby exerting a progressively greater influence on our individual and collective lives. It develops autonomously, and as it does so, it presses upon us more forcefully. We are still free agents, but we act against an increasing pressure to conform to technological demands. In practical terms, our sphere of freedom is constantly being eroded. But it exists all the same, and thus we retain the power to act, despite the presence of a progressively determinative technological force.

Before detailing my case, though, I need to present two things. First, something different: a short story that outlines some of the main concepts by placing them in the context of a concrete situation. Second is a look at the evolution of the traditional arguments for technological determinism. With these items in place, I proceed to offer a pantechnical analysis.

A TALE OF AUTONOMY AND DETERMINISM

Consider a more innocent time, not so long ago. The setting is a prominent American university. The story goes as follows:

> One day, out of the blue, a bright young engineering student named Tom Johnson hits upon a brilliant idea—a wonderful new widget that promises to be very profitable. Upon graduation he borrows some money, starts up a small business, and sets to work designing his new product. He builds a prototype, pitches it to a few prospective clients, and soon enough, advance orders begin to flow in. Tom hires a small staff to help out. He rents some manufacturing space and is quickly able to start up a simple assembly line. Sales grow rapidly, and a new business flourishes.
>
> Within a couple years, orders for Tom's widget are so strong that he is able to hire a large, full-time staff—engineers, designers, assembly line workers, shippers, salesmen, and all the support people that go along with them. For practical, prudential, and legal reasons, Tom files for incorporation, and his business is now a formal legal entity: Johnson Widgets, Inc. He is the sole owner, he runs the show, but he is increasingly reliant on his large and growing staff of experts and specialists to meet demand.
>
> A few years later, business remains strong and the corporation hits a milestone: 1,000 employees. Now, a question arises: Is the corporation 'determined' by the people in it? We must keep in mind that the corporation consists of Tom, his 1,000 employees, the new factory building they purchased, various material assets, and several million dollars cash that they hold in reserve. On the one hand, it is obvious that the corporation is fully and completely dependent on its people; should they all suddenly disappear, the corporation collapses. On the other hand, every individual person—except perhaps Tom himself (more on this momentarily)—is expendable. Line workers, secretaries, shipping clerks—they come and go on a regular basis, just as in all corporations.

The same holds for Tom's more educated employees; his individual engineers, designers, financial planners, and marketers can be replaced with nary a bump. Even his top people—his vice presidents, key strategists, management team, and technical wizards—are replaceable, though not without some temporary pain. Even the loss of several individuals—if, say, an executive business plane crashes, killing his top people—is not fatal; the corporation will continue.

The reason for this is clear: the corporation is a technical device. Its charter, rules, and procedures are all documented and formalized. Job assignments, reporting relationships, roles, and responsibilities for every position are written down; they determine how the specific job holder will function. The corporation is a machine, designed and operated for a particular purpose—namely, to produce widgets at a profit. The individual person has minimal autonomy; he or she must function primarily in the role assigned by the corporation. If they did otherwise, the business would function less efficiently and thus less profitably. Furthermore, the larger and more formalized the corporation becomes, the less autonomy each person has. In fact, there is an inverse relationship between the two; organizational autonomy increases as individual autonomy decreases. Effectively it is a transfer from the one to the other.

But we notice something else as well: the corporation is a dynamic, lifelike entity. It has a legal standing, even 'personhood,' in courts of law. 'It' owns things. 'It' draws in resources, metabolizes them, maintains internal order, gives off waste products, and produces outputs.[2] 'It' executes its business plan and strives to maximize profits for its owner, Tom. It does all these things independently of the individuals who work for it, even as it could not function without them. Any one person, constituting a mere 0.1% of the (human) organization, is virtually insignificant. Any ten or twenty people are only slightly less insignificant. Only at the level of, say, 100 people or more do they 'count'—but then *not* as individuals, only as a large group.

The corporation is thus simultaneously dependent on and independent of human beings. It depends on them collectively but is independent of them individually. Individually, the employees—except perhaps Tom—exert little or no control over the corporate system. Once established, it is a self-directing, self-sustaining, and functionally automatic process.

Thus far, Johnson Widgets has been a local enterprise, drawing from and supplying to the local economy. But now Tom, being a typical expansionist capitalist, looks to grow the business. First he expands his product line—no longer just his original widget but now new products that his engineering team has designed. These are products that Tom does not really understand, but he sees their profit potential and thus presses ahead. Needing a significant infusion of cash to finance this expansion, Tom takes the corporation public by listing it on a major stock exchange. This has the benefit of raising a huge amount of cash; but the downside is that Tom is no longer sole owner. Now he is just one of thousands of stockholders. Ownership and control are thus further diffused.

Flush with cash, Tom develops a plan to expand—first nationally and then internationally. He establishes sales offices, supply chains, and customer bases around the country and then globally. At first, the new business model suffers. There is criticism and talk of Tom "overreaching." His handpicked board of directors starts floating rumors about the need for a new president; to his dismay, Tom realizes that even he himself is now expendable. "A Harvard-trained MBA could do a better job," they say.

But Tom weathers this storm, adjusts his plan, and proceeds to grow the business. Thanks to the Internet, ubiquitous computers, free-trade markets, and secure email, Tom is in a good position to go global. Unfortunately, so too are his competitors. In fact, he quickly realizes that he *must* go global or else the overseas competition, exploiting the latest technology, will put *him* out of business. Technology dictates the environment in which the corporation functions.

Having no choice, Tom boldly presses ahead. And thanks to his perseverance, hard work, and a bit of luck, the corporation prospers. Business is good. His employee base grows correspondingly: to 5,000, 10,000, then 50,000. After a few more good years, Tom's corporation has reached a new milestone: 100,000 employees. Tom is now a corporate titan, well paid, and a financial success. His story is trumpeted as a boon to the economy, a provider of jobs, and a great benefactor of the people, for whom his business provides a lucrative income.

With 100,000 employees, and every one of them—even Tom himself—expendable, individual people, or any small group of people, are of virtually no consequence to the corporate body. Each person has a highly structured role and must follow strict procedures if the corporation is to be efficient and profitable. Even Tom himself, as president, has little flexibility; he is a slave to his shareholders and board of directors, who demand ever-greater return on their assets. He can take no action except that which increases profits. The corporation, still wholly dependent on human beings, is functionally autonomous.

And where, incidentally, is 'the corporation'? Formerly it was all in one place, but now, as a global enterprise, it is spread across the planet. Assets, debts, employees, material goods, inventory, ideas, intellectual capital, utilities, contracts . . . all diffuse aspects of this single, vast, interconnected entity. In fact, the real-world technological system is rather like this: a sprawling, multidimensional, diverse collection of things, people, procedures, and processes, all interlinked and functioning in an integrated and coherent manner. But just as surely as the corporation exists as a real and substantial entity, so too does the technological system. Both cannot function without people, but both are completely independent of any one person or any small group.

Tom understands all this but can do nothing about it. He must maintain and grow the corporation; after all, 100,000 people depend on him—not to mention his own lofty salary. The major shareholders clamor for yet greater profits, and pressure grows on him to "cut costs." Seeing the benefits and potential of modern technology, Tom embarks on a new path: *radical automation*. Beginning with the so-called low-skill assembly line workers and moving quickly to the packers, shippers, clerks, and other service staff, Tom systematically replaces his workers with machines. He is so successful that he is even able to repatriate foreign manufacturing plants—but not, of course, the jobs. Production increases, costs decline, profits grow, and shareholders roundly praise Tom's brilliance. Simultaneously, his employee base is now shrinking for the first time in corporate history. Today 100,000, now 90,000, now 75,000, and soon less than 50,000—the trend is inexorable.

Dazzled by his own success, Tom presses even harder; his goal is to become "the model 21st-century technological corporation." He deploys a fleet of "delivery drones" to ship directly from plant to customer. He has an army of robots to tend his inventory and supply parts to his increasingly automated assembly line. Robots soon do all the manufacturing, start to finish. For a time Tom needs a group of servicemen to keep things up and running, but soon enough he procures robots *to repair the robots*. Some of his specialized equipment even has a self-repairing capability due to its intelligent nanostructured material. Most impressively, a master "artificial general intelligence," or AGI, controller regularly supplies Tom with optimization and efficiency ideas. In fact he increasingly pre-authorizes its suggestions: "It knows better than I do," he murmurs.

Employee rolls continue to collapse; the shareholders and board are ecstatic. With technology rapidly advancing, no job is safe. The staff drops to 10,000, to 5,000, to 1,000. Tom is hailed as a visionary corporate leader; he garners accolades from both government and media—ironic, since he is contributing to dramatic unemployment. But never mind that; he is a wealth builder, an economic powerhouse, and a technological success story—and these are what count the most.

After some years like this, the process reaches its climax. Tom succeeds in eliminating—*everyone*. Production is higher than ever, and it is fully 100% automated. It is now as it was in the beginning—a one-man operation. Tom then stops and reflects: Johnson Widgets is now, literally and completely, an autonomous system. *When and how did this happen?* he asks himself. He realizes that it occurred slowly, step by step, over several years. The business began fully and completely dependent upon him, its founder. As he added people, and Johnson Widgets became a true corporation, a machine, it gained a new state of existence: *functional autonomy with ontological dependence* (on humans). This represented a first phase of determinism.

> When machines started to replace the employees, a new phase of determinism began to materialize, one of *true autonomy*. It was not fully realized until just now, Tom sees, but it was underway for many years. 'Autonomy,' he now understands, is a *process*—one of progression along a continuum, in conjunction with a proportionate reduction in ontological dependence. His humble little business made the full transition from a one-man, zero-autonomy organization to a one-man, fully autonomous organization—or, in truth, a *zero*-man operation. Tom realizes that he is utterly irrelevant. He clears out his office and heads home.
>
> That night, Tom dies in his sleep. The next morning, a delivery drone brings a bouquet of flowers to his widow. A note is attached: "We are sorry for your loss." And the assembly line never misses a beat.

Technology is like this—with one important difference. When Tom's employees were displaced, they simply became unemployed or found other work. In the real world, as technology gains autonomy, there is no being unemployed. We can't leave. We are stuck within the planetary system as the technosphere envelops us. We suffer unending loss of autonomy and progressive disempowerment, even as the system increasingly dictates our living conditions. The price we pay increases, bit by bit, every day—until we die in our sleep.

AN EVOLVING ARGUMENT: THE ORIGINS OF DETERMINISM

At this point, it is worthwhile briefly assessing the traditional arguments for technological determinism. There are two important parallels here with the situation surrounding the historical criticisms, as addressed in previous chapters. First, in both cases the arguments are generally as valid today as when they were written—given that the conditions and pressures of technological society have only increased with time. Second, and as before, I am compelled to provide a fairly detailed history. The problem of technological determinism is at once too important, too poorly understood, and too ill documented to let its development pass with only a few cursory remarks.

The case for some form of technological determinism is ancient, being found, implicitly, in Plato's story of the Ring of Gyges. I examined the main plot line in Chapter 7, where I cited it as evidence of his critique of technology. In a supplementary paragraph, however, Plato adds a twist: Give identical rings to two different men: the most just and the most corrupt. The outcome is unsurprising: "both would follow the same path" (360c4). That it, both would misuse the power of the ring to dominate and plunder. This occurs because technology is unconcerned with such human qualities as moral integrity or personal values. It drives all men, good and bad, to the same end. The technology is determinative.

Regarding more recent arguments, it is often claimed that the thesis began with Marx; but in fact we find a related precedent in Hegel. In his Jena Lectures of 1805/6, he argued that the mechanization of labor distances the worker from his product and that this has negative consequences for both him and society. Not only that, but the worker is compelled to participate in just such a dehumanizing and alienating system; he has no choice. "In the machine, man terminates his own formal activity and lets it do all the work for him. . . . The more mechanical the labor becomes, the less value it has and the more he has to work in this manner."[3] Mechanized work is "dead" work, according to Hegel, and it progressively displaces living work of the human being.

When the system of industrial production rises to a position of prominence within society, it becomes an autonomous and quasi-deterministic force in its own right: "If need and labor are elevated to universality, there emerges . . . a life of the dead with its own momentum; this system moves hither and thither blindly and primitively in its agitation and, like a wild animal, demands constant strict control and restraint." For the worker, such a condition is devastating; his life becomes "totally stupefying, unhealthy, insecure, and faculty-stunting." It was precisely this analysis that was succinctly summarized by a post-Hegelian commentator in 1857: "The most basic underpinnings of our physical and our spiritual life are being ripped apart and transformed by the triumph of technology (*Triumphe der Technik*)."[4]

It was Marx, though, that formulated the first rough outline of a technological determinism. Even in his early manuscripts of 1844, he was already aware of the debilitating effect of mechanized labor processes. In modern industry, the worker is "depressed spiritually and physically to the condition of a machine." Furthermore, "since the worker has sunk to the level of a machine, he can be confronted by the machine as a competitor" ([1844] 1961: 24–25). Collectively, this capitalist system of mass production operates with an uncanny force: "all is under the sway of *inhuman* power" (126).

Just three years later, in *The Poverty of Philosophy*, Marx gives his classic statement on the subject.

> Social relations are closely bound up with productive forces. In acquiring new productive forces men change their mode of production; and in changing their mode of production, in changing the way of earning their living, they change all their social relations. The hand-mill gives you society with the feudal lord; the steam-mill society with the industrial capitalist.[5]

Society is driven by conditions of production, which in turn are driven by technological forces. Technology is thus the ultimate determinate of society.

A decade later, in *The Grundrisse* (1857), Marx extended his thesis. The industrial system, he argued, has become something in its own right, a thing autonomous and organic. As the means of labor are absorbed by this organic being, it becomes transformed into "an automatic system of machinery." This system "is set in motion by an automaton, a motive force that moves of its own accord." The workers become "no more than the conscious limbs of the automaton." As such, the system is virtually alive and even ensouled: "The machine, which possess skill and force in the worker's place, is itself the virtuoso, with a spirit (*Seele*) of its own in the mechanical laws that take effect in it" ([1857] 1971: 132–133).

This idea is stated most concisely and explicitly in 1859, in *A Contribution to the Critique of Political Economy*:

> In the social production of their existence, men inevitably enter into definite relations, which are independent of their will, namely relations of production appropriate to a given stage in the development of their material forces of production. The totality of these relations of production constitutes the economic structure of society, the real foundation, on which arises a legal and political superstructure and to which correspond definite forms of social consciousness. . . . At a certain stage of development, the material productive forces of society come into conflict with the existing relations of production. . . . From forms of development of the productive forces, these relations turn into their fetters. Then begins an era of social revolution. The changes in the economic foundation lead sooner or later to the transformation of the whole immense superstructure.[6]

As before: economic production is the foundation, or structure, of a society, and this is determined primarily by technological "forces of production." Above this lies the superstructure of politics and civic life, which is determined by the structure below. As technology changes, so follow changes in both economic structure and political superstructure.

We do not know if Marx intended to advance a truly technological form of determinism or something based in a broader conception of the notion of productive forces. Suffice to say that technology, for him, plays a unique and dominant role in the life of society. We get a sense of this from one other brief comment, found in his last major work, *Capital* ([1867] 1976: 493):

> Technology reveals the active relation of man to nature, the direct process of the production of his life, and thereby it also lays bare the process of the production of the social relations of his life, and of the mental conceptions that flow from those relations.

There is, to be sure, dispute about the intention and implications of these short passages, which together constitute nearly all that Marx has to say on the topic. It is unclear what he means by his form of determinism, nor even to what degree we should take him as seriously holding the view.[7] Ideologically, it would seem, Marx cannot truly be a technological determinist at all.

In his larger scheme, social evolution is driven primarily by class struggle and only secondarily, at best, by technology. Furthermore, in his vision of the proletarian revolution, the workers would take command of the means of production, including the technological means; thus it is they who are in control, not technology. And he is utterly lacking a metaphysical analysis, something that severely limits his thinking. It is as if Marx could sense the fundamental technological determinism at work but willfully blinded himself to that fact—because the implications would dramatically undermine his entire project.

Marx, of course, was not alone in his struggles with determinism. As I explained in Chapter 7, Butler also observed this phenomenon and wrote about it in his two essays of 1863 and 1865—though, as with Marx, offering little more than passing remarks. Weber's "iron cage" and Veblen's notion of technology as the "chief force" in shaping our lives were also suggestive of determinism. So too with Scheler's insight that "the machine has come to dominate life." Even the likes of Whitehead and Spengler, as explicit as they were, did little more than state a claim of strong determinism; they did not analyze it and did not attempt to explain it. This is striking: even as vital an issue as technology determining our future brought forth little more than incidental observations.

Into the late 1940s, it fell to the nonphilosophers to continue the analysis. Anthropologist Leslie White published an extraordinary work, *The Science of Culture*, in 1949. On his view, human culture is composed of three layers, in something like a pyramidal structure. The base layer is technology. On top of this lies the sociological layer and at the peak, ideology. The technological base consists not only of tools and machines but is a truly integrated system comprising "material, mechanical, physical, and chemical instruments together with the techniques of their use" (364)—a surprisingly farsighted definition, one that anticipates Ellul. The sociological layer includes "interpersonal relations expressed in patterns of behavior." At the highest level we have ideology, consisting of knowledge, beliefs, philosophy, religion, science, and literature, among other aspects.

As the foundation of the cultural system, technology plays the primary role. "This is," says White, "as we would expect it to be; it could not be otherwise" (365). Technology connects humanity to nature and serves as the means by which society accesses and manipulates its vital sources of energy. The result is a true technological determinism: "A social system is a function of its technological system." But the other two layers are not powerless; they in turn affect and "condition" the technology at the foundation of the system. He thus arrives at a more sophisticated form of determinism, one in which technology drives the system as a whole but is affected in various ways by sociology and ideology.

> The technological system is basic and primary. Social systems [the sociological layer] are functions of technologies; and philosophies [the

> ideological layer] express technological forces and reflect social systems. The technological factor is therefore the determinant of a cultural system as a whole. It determines the form of social systems, and technology and society together determine the content and orientation of philosophy. This is not to say, of course, that social systems do not condition the operation of technologies, or that social and technological systems are not affected by philosophies. They do and are. But to condition is one thing; to determine, quite another. (White 1949: 366)

Energy is the key, and technology is the gateway to that energy. Everything depends on society's ability to access and manipulate ever-growing sources of energy, and progress occurs only through advancing technology. All in all, White's approach provides the first significant advance in theorizing about determinism in nearly a century.[8]

At about the same time, sociologist Roderick Seidenberg wrote his own milestone work, *Posthistoric Man* (1950). Examining the vast sweep of time, he describes a process of human society undergoing periodic upheavals and revolutions, each time becoming ultimately reorganized on a higher and more complex social plane. "[M]ankind is being ushered relentlessly through a series of unstable equilibria . . . toward an ever more stringently organized state" (25)—a trend he says is irrevocable. This "underlying drift" of social evolution "is perceptible in ever sharper outline as a kind of morphological determinism embracing every aspect of the human scene."

Driving this process forward is—the machine. As realized in modern society, the machine, technology, is "a unique and perfect paradigm of organization," which functions as "a series of interrelated means contrived to achieve an explicit end." Itself the product of an orderly design and construction process, the machine can only exist and operate in an orderly environment; and indeed it creates order simply by functioning. Therefore, via a self-sustaining feedback process, "order demands order." When the machine, as an instance of innovative technology, is set loose in its social setting, it triggers a wave of organization that ripples out into the world, altering the whole system. Seidenberg has an evocative metaphor at hand: Just as when a seed crystal is dropped into a supersaturated solution and catalyzes a complete phase transition of the fluid, the machine "crystallizes" new forms of order: "the machine, like a crystal, has brought about a unique, sudden, and profound change of structure in the fabric of society—the crystallization of society in the patterns of a collectivized and ever more intensively organized world" (28).

The result, again, is a form of technological determinism. The "necessitous character" of this crystallization causes "a universal transmutation of values in every aspect of existence." At present we seem to be experiencing a rapid phase transition, says Seidenberg, but this will not last. Just as the crystallized fluid quickly reaches a state of complete transformation, so too will we, conceivably, attain a social state of "perfected adjustment," a kind of "end state of growth," in which we all function as atomic elements in a vast, fully realized social machine. This "impending collectivization of man" seems inevitable; and, unfortunately, "we have no reason to believe . . . that man is exempt from a fate that appears universal." Seidenberg's determinism holds little prospect either for human freedom or well-being.

JACQUES ELLUL AND *THE TECHNOLOGICAL SOCIETY*

Thus we arrive at 1954, the year of Heidegger's and Ellul's most important statements. Heidegger was covered in some detail in Chapter 5, but it is worth recalling here his view on the issue at hand. Technology, he says, "holds complete sway over man" as a kind of fate, but one that has at least a few theoretical escape routes. If we see technology as neutral or see ourselves as masters of it or of nature, this is a sign that we have looked right past its essence and thus are captive to its determinism. On the other hand, if we see it as the ultimate danger and hold this danger always before our eyes, we are open to a "freeing claim" that may yet be our salvation. So too the fine arts, as the epitome of the benign form of revealing (Bringing-forth), are, in some ill-defined way, a possible way forward. Such, in any case, was his view in 1954.

Twelve years later, though, his outlook had changed. In the *Der Spiegel* interview, he admitted that technology is something "that humanity . . . cannot master." If it now takes a god to save us, mankind itself has no hope. The saving power of art and the freeing claim have apparently collapsed into nothingness. Technological determinism is complete, and our fate is effectively sealed.

It is difficult to reconcile this shift in outlook with Heidegger's larger metaphysical picture. Did we have still have a chance to save ourselves in 1954 but lost it by 1966? Perhaps, but unlikely. The probability that there was a tiny metaphysical opening that closed up during just those few years is small; we can safely exclude this option. Therefore, in stating that we require gods for salvation, Heidegger seems to be implicitly admitting that he was previously in error—that we *never* had a freeing claim and that there was *never* any saving power in Enframing. Had he been honest with himself, he would have been forthright about this change in outlook. To err is no sin, but to paper it over with later contrary pronouncements is needlessly deceptive. And this is no minor terminological point; the future of humanity is at stake. In the end, Heidegger's inability to craft a consistent metaphysics reveals itself both as a confusion about determinism and as the absence of a forward-looking plan of action. "I cannot help you": his lasting message to humanity.

Turning to Ellul's great work, *The Technological Society* can scarcely be assessed or even summarized here. But it is necessary, in the present context, to outline at least his "characterology" of technology, with an emphasis on the question of determinism. And this is indeed a key issue, not the least because Ellul is often presented as the paragon of technological determinists.

This is understandable but evidently incorrect—as he himself informs us in the very same book.

The work was written in 1954, but with the publication of the 1964 English translation, Ellul revised the text at several points and added a short author's foreword. The latter is particularly relevant and revealing. He opens by denying his alleged pessimism regarding the human condition in the face of an overpowering "technique."[9] Offering neither speculation nor judgments nor a plan of action, his self-stated objective is "concerned only with knowing whether things are so or not" (xxvii). "I do not seek to show," he says, "that man is determined, or that technique is bad, or anything else of the kind." Still, he is compelled to respond to the charge that he lays out a "rigorous determinism" that invalidates all possibility of free action.

Flatly denying the charge, Ellul insists that he "do[es] not deny the existence of individual action or of some inner sphere of freedom." Instead, he simply holds that individual action can do nothing to affect the larger social trends that are now dictated by technology. "I do not maintain that the individual is more determined today than he has been in the past; rather, that he is differently determined" (xxix). Previously, social customs and taboos dictated our actions; today, it is technology. "We are conditioned by something new: technological civilization." Technological determinants "operate in increasingly wide areas and penetrate more and more deeply into human existence." Denying that this represents a kind of fatalism, Ellul continues by arguing that if—*only* if—mankind abdicates its responsibility for deliberate action, then, and only then, will the determinism play out. If we do not "mak[e] a stand against these determinants, then everything *will* happen as I have described it, and the determinants *will* be transformed into inevitabilities."

The foreword ends with a reiteration: "Reality is itself a combination of determinisms, and freedom consists in overcoming and transcending these determinisms" (xxxii)—but then they were never true determinisms in the first place. We need to see things dialectically, says Ellul; we must accept that "man is indeed determined, but that it is open to him to overcome necessity, and that this *act* is freedom"—but then it never was truly a necessity.

It seems that Ellul is awkwardly attempting a reconception of the central term. As normally understood, determinism implies a truly determinate situation, one of inescapable and unalterable conditions. Obviously he sees things differently. For him, it refers simply to a system of strong pressures or forces that, lacking countervailing forces, will hold sway. Technology presses upon us strongly to act in certain ways; and yet it is still apparently within the domain of human free will to oppose it—and to do so successfully.

I think his intuition is correct, but the matter needs to be put into sharper focus. To 'determine' is, literally, to set a limit. The word derives from the Latin *de-terminare*—that which is 'of a terminus,' something bounded or limited in some way. This concept can be useful here. As Ellul seems to intend, and I concur, human freedom operates only within certain circumscribed limits. Consider it an instance of *constrained freedom*. It is rather like a child who is let out to play in a fenced-in yard; he has complete freedom to do as he likes, within the confines of the established boundary. So too with humanity: We have freedom to act, but only within limits defined by the technological system in which we live. Unfortunately, as the system grows in size, strength, and complexity, the limits become ever more restrictive. We never lose our freedom completely, and it may even increase in certain narrowly defined domains. But overall, and considered in total, the scope of our freedom diminishes.

Technophiles often reply that technology does not eliminate our freedom but expands it. There is a fragment of truth here, but it is highly misleading. For example, when we make the shift to a high-power, high-speed transportation system consisting of motorized vehicles, we obviously travel faster and farther than before. We gain freedom in these specific modes of movement. But we must simultaneously accept the many constraints that go along with such a system: a network of paved roads, mass auto production, rules and regulations, speed limits, driver's licenses, auto insurance, mass oil production and refining, traffic injuries and fatalities, climate-changing greenhouse gasses, a military capable of enforcing access to petroleum reserves in foreign lands—the list goes on. One gains freedom to move, but only where the roads allow. If one replies that we can drive 'everywhere,' this only surrenders the point; the world is now little more than our roadway.

Even at the most basic level of human movement, we are constrained. Prior to the automobile society, a person could walk freely, with virtually no constraints, in any direction he chose and for as long as he liked. Civic life was organized around the walker or the man on horseback, and one could maintain a high quality of life with few limitations. Today, civic life is organized around the automobile, and the walker has highly restricted spaces and options. His sphere of freedom has been constrained and distorted by automotive technology.

The technophile responds: "But people prefer high-tech transportation; if they did not, they would enact laws changing it." Again, a misleading reply. No one "chose" to introduce a high-speed transportation system in the first place. No king, no president, no legislature sat down and said: Now we shall have such a system. Rather, the system emerged and evolved in a highly diffused manner, over a long period of time, as a function of countless individual actions that had little or nothing to do with creating a high-speed transportation system per se. "The wishes of the people" had nothing to do with creating the system, and "the wishes of the people" can have nothing to do with changing or eliminating it. Rather, it is as Ellul described—a self-evolving and autonomous system that has grown up organically in our midst. It circumscribes and progressively restricts our overall sphere of freedom, even as it expands our technological options in certain narrowly defined ways.

The discrepancies of the foreword vis-à-vis the main text are attributable, I believe, to two main factors. The first is Ellul's lifelong commitment to Catholicism.[10] As a believer in God and an immortal soul, he is necessarily bound to a view of humanity in which we have ultimate free will, even over a totalitarian force like technology. Second, the near complete absence of (secular) metaphysics in his work suggests that he, like so many others, did not penetrate to the core of the technical phenomenon. And lacking such a metaphysical framework, one can easily slip into inconsistencies. Even his prescriptions for action fall short. He closes the foreword by noting that we can overcome the power of technology only "by an act of freedom, of transcending it." Fine—but how is this to be done? "I do not yet know." In the end, he has no more to offer than Heidegger.

Ellul, then, was by his own admission a nondeterminist. And yet everything about his discussion of technique suggests a strong, even totalitarian perspective. The contrast is striking. Perhaps we should take him at his word: He offers us only descriptions, an account simply of "whether things are so or not." Let us then consider strictly his assessment of technology in light of the pantechnical framework that I propose.

THE CHARACTEROLOGY OF TECHNIQUE

Ellul's book begins with an ominous statement: "No social, human, or spiritual fact is so important as the fact of technique in the modern world. And yet no subject is so little understood" (3). Technique has penetrated every aspect of human existence, resulting in an increasingly mechanized and dehumanized life. Like a modern-day curse of Midas, "technique transforms everything it touches into a machine." At this point, its character changes:

> When technique enters into every area of life, including the human, it ceases to be external to man and becomes his very substance. It is no longer face to face with man but is integrated with him, and it progressively absorbs him. In this respect, technique is radically different from the machine. This transformation, so obvious in modern society, is the result of the fact that technique has become autonomous. (6)

As an autonomous force, it can be neither controlled nor stopped. It has attached itself to us and now threatens to consume us.

The second chapter, on the "characterology of technique," is central to the whole work. Here Ellul describes in great detail the seven primary characteristics of modern technology: rationality, artificiality, automatism, self-augmentation, monism, universalism, and autonomy. Let me take these in turn.

The first two are straightforward, and Ellul spends little time on them. That technology is rational is self-evident: "Every intervention of technique is, in effect, a reduction of facts, forces, phenomena, means, and instruments to the schema of logic" (79). Technology is the physical embodiment of rationality, a manifestation of logos—consistent with my own thesis, though hardly controversial.

The second characteristic, artificiality, highlights the fact that technology is fundamentally opposed to nature. Technique

> destroys, eliminates, or subordinates the natural world, and does not allow this world to restore itself or even to enter into a symbiotic relation with it. The two worlds obey different imperatives, different directives, and different laws which have nothing in common. (79)

This is in sharp contrast to my own proposal and is, in fact, unjustified. I claim that both worlds are the *same* world, that both obey precisely the *same* imperatives and directives, and that they have *everything* in common. Perhaps Ellul, as a theologian, is comfortable with incongruous dualisms; I am not. The only valid distinction here is that man-made technology is second-order nature, 'nature-made nature' as I have called it, and that this fact distinguishes technique from first-order nature, the Technê-Logos as manifest by universal processes. There is a distinction, but it is one of degree and not kind. There is no ontological divide here, and it is crucially misleading to suggest otherwise. That Ellul gives no more than two paragraphs to this aspect is indicative of an analytic deficiency.

He is on firmer ground with the other five characteristics. Automatism refers to technique's tendency to be self-directing and to operate in the single-minded pursuit of efficiency. Given a specific technological problem or challenge, the most efficient solution will be sought. All other factors are largely disregarded: human needs and desires, aesthetics, religious dictates, environmental concerns, political demands—all these are of little consequence. Even capitalism, seen by many as the ruling power of our age, is swept away by the advancing technological tide:

> It is the automatism of technique . . . that endangers capitalism and heralds its final disappearance. [This] is so because technical progress acts automatically. The choice between methods is no longer made according to human measure, but occurs as a mechanical process which nothing can prevent. Capitalism, in spite of all its power, will be crushed by this automatism. . . . There is nothing left to do but wonder at a mechanism that functions so well and, apparently, so tirelessly. But, above all else, no finger must be laid upon it, nor its automatism interfered with. . . . Man is stripped of his faculty of choice and he is satisfied. He accepts the situation when he sides with technique. (82)

As a consequence, and contrary to the views laid out in the foreword, any meaningful notion of human freedom is destroyed:

> Nothing can compete with the technical means. The choice is made *a priori.* It is not in the power of the individual or of the group to decide to follow some method other than the technical. The individual is in a

> dilemma: either he decides to safeguard his freedom of choice . . . thereby entering into competition with a power against which there is no efficacious defense and before which he must suffer defeat; or he decides to accept technical necessity, in which case he will himself be the victor, but only by submitting irreparably to technical slavery. In effect he has no freedom of choice. (84)

The fourth characteristic is self-augmentation. Technology not only acts by itself (automatism), it also *grows* by itself—it "engenders itself." Just as most living things become larger with age, technology inevitably advances and increases its reach. It furthermore grows at an accelerating rate—as Kurzweil and many others have since emphasized. Ellul sums up these points in the form of two laws of self-augmentation: (1) "In a given civilization, technical progress is irreversible." (2) "Technical progress tends to act, not according to an *arithmetic*, but according to a *geometric* progression." This second law is usually expressed as the contrast between linear and exponential growth.[11]

Ellul is quick to note that it is *not* true to say that technology *never* retreats. Technology is bound to a given civilization, and when that civilization declines—as all civilizations must—the corresponding technology falls with it: "Arrest and retreat only occur when an entire society collapses" (89). This is an important counterpoint to those who would argue that technological progress is inevitable. It is true only within the context of a thriving and growing civilization. Civilizational collapse means technological regress. Of course, in the very long run, on the pantechnical thesis, technology does inevitably advance, but only where the cosmic conditions allow—namely, abundant free energy and an amenable material foundation. But nothing in my thesis suggests that technology must always advance, at all times, everywhere in the universe. As with all natural processes, there will be countermovements, retrograde actions, and opposing forces in operation, all of which could mark a slowing or even temporary reversal of the larger universal bias toward order, complexity, and transcendence. In both human and cosmic terms, then, there are many opportunities for technological retreat—even as the long-term trend is in the other direction.

The question of regress is interesting on yet another account. Many technophiles today argue *in favor* of technological determinism precisely because they see it as an argument to promote yet further growth. Since technology is bound to progress, they say, we ought to "go with the flow" and ride the technological wave to some glorious future. Resistance, they tell us, is futile. This is in striking contrast to the traditional determinists, like Orwell and Ellul, who insist that such determinism is fundamentally *opposed* to our interests and thus ought to be resisted. In a notable metaphysical irony, determinism is thus promoted by members of both the pro- and antitech camps. And of course we also have the more orthodox academics, including most historians and conventional philosophers, who argue that technology is not determinative at all. A strange situation indeed.

Fifth: Monism, or holism, expresses the interconnectedness of all aspects of technology. All parts are linked, and none can be fully understood apart from the whole. Tools are used to make tools, and systems function only within technical contexts. This fact has, for Ellul, three important consequences. First, the use of any one machine implies, in some sense, a use of them all. When we purchase a car, fly on an airplane, or log on to a computer network, a vast array of technologies is utilized. A single automobile has 30 or 40 microchips in it, miles of wiring, and a host of toxic substances—all of which are precisely engineered products in themselves, each requiring, in turn, an entire complex of other technologies. Simply buying gasoline for that car contributes to a globally interconnected network of mining, processing, shipping, and information technologies. And yet we never see these effects; our individual actions are too small and the global technology system too large and complex. This, I might add, is yet another reason technologies can never be considered neutral.

Second, and as a consequence of the foregoing, we are not free to eliminate the so-called bad parts of technology and keep only the good—another common but unfounded myth. As Ellul explains, one is tempted to ask the following question:

> Could not atomic engines and atomic power have been discovered without creating the bomb? To reason thus is to separate technical elements with no justification. . . . If atomic research is encouraged, it is obligatory to pass through the stage of the atomic bomb; the bomb represents by far the simplest utilization of atomic energy. . . . [Once discovered, the bomb must be used] because everything which is technique is necessarily used as soon as it is available, without distinction of good or evil. This is the principal law of our age. (98–99)

The third consequence is that the monism of technology yields such a high degree of complexity that the consequences of any particular action can never be anticipated—consequences that are almost inevitably for the worse.

> Man can never foresee the totality of consequences of a given technical action. History shows that every technical application from its beginnings presents certain unforeseeable secondary effects which are much more disastrous than the lack of the technique would have been. (105)

An exaggeration, perhaps; surely not every technology is so constituted, but it is certainly true that many modern technologies have produced stunningly destructive side effects—sufficiently so to outweigh the benefits of the technology itself. This situation grows much worse as technologies become more powerful and as their effects take longer to play out. To take the obvious example, fossil fuels have been accessed and used in large amounts for some 300 years, but it is only now that we are seeing the impact on the global climate. If climate change leads ultimately to the collapse of industrial society, as some are warning, then we—those of use who survive, that is—will have to reassess the value of ever using fossil fuels in the first place. That decision may yet prove to be the biggest mistake in human history. But it was a mistake that could never have been anticipated. It was not a case of poor planning or inadequate study; it was an intrinsically unforeseeable situation. But this is the nature of advanced technology.

This problem of unanticipated consequences is greatly exacerbated, says Ellul, by the fact that technology demands to be used quickly. It is expensive and time consuming to develop, and large profits await. Time is rarely taken to fully understand even the first-order effects, let alone secondary or other consequences. Rapid implementation of powerful new technologies significantly magnifies the problem.

Some of our most catastrophic examples involve the environment, as Ellul realized by 1964.[12] Climate change was not yet well understood, but he cited cases of industrial, chemical-intensive agriculture producing eroded soils and toxic foods, of DDT working its way into cows' milk, and of petroleum extraction damaging fragile desert ecosystems. Today we can add such industrial legacies as acid rain, depletion of atmospheric ozone, antibiotic-resistant bacteria from both human misuse and industrial animal agriculture, and the growing impact of greenhouse gases.

The monistic ensemble that constitutes technology extends even to the human sphere, to the techniques of humanity itself. Technical monism places tremendous demands—physical, psychological, and social—upon ordinary people in their daily lives. People reflexively resist, and the system responds with temptations, coercions, and bribes. Ultimately a sophisticated form of propaganda must be applied, and even a subtle brainwashing, to convince the masses that an increasingly pervasive technological system is in their best interests:

> A great effort was required of the individual, and this effort he could not make unless he was genuinely convinced, not merely constrained. He must be made to yield his heart and will, as he had yielded his body and brain. And so the techniques of propaganda, education, and psychic manipulation came to reinforce the others. (115)

In the face of the adverse pressures of technological monism, there is a simple-minded reply: We can fix the system, reform it, in order to mitigate the negative consequences and promote the beneficial ones. But the monistic and integrated nature of the system does not allow this. As Ellul says, "It is useless to hope for a modification of a system like this—so complex and precisely adjusted that no single part can be modified by itself. Moreover, the system perfects and completes itself unremittingly" (116). It repairs itself, patches holes, bounces back—just as a living organism heals from a wound or illness. The system resists efforts to be redirected in any way, to the point that the very possibility of meaningful reform is called into question.

The sixth characteristic, universalism, is a recognition that technology crosses all geographic, cultural, economic, religious, and political boundaries. It is everywhere functionally the same, though perhaps at different stages of development, and is thus the driving force in the homogenization of world culture. Universalism explains why cities such as Bangkok, Nairobi, and London are becoming more alike by the day. It effects a meeting of East and West, North and South. In recent decades, it has become commonplace to state that American culture pervades the planet; in fact it is a *technological* culture that has taken hold everywhere, America included. The U.S. is simply at the leading edge of much of technological evolution and thus serves as its emissary. When American business or government establishes a foothold in a foreign land, a technological invasion is sure to follow.

Technology is tearing down and remaking civilization worldwide. Indigenous and traditional cultures are no match; they have been, or will soon be, obliterated. The moment they are touched by the blessings of modern technology, their fate is sealed. The apologist for the system has a ready response: Cultural destruction is unnecessary; it is the result of corruption, poor planning, or simple errors of judgment. Ellul replies:

> I do not know if it is necessary, but nevertheless it is so. A civilization which is collapsing cannot be recreated abstractly. It is too late to turn back and enable these worlds to live. What has been given them is not simply well-being. This well-being presupposes a transformation of all life. . . Technique cannot be otherwise than totalitarian. (124–125)

Finally, we have the autonomy of technique. Technology is a whole, complete phenomenon. It is self-sufficient. It operates independently of other human spheres; or more correctly, it is the driving force in these realms:

> Technique elicits and conditions social, political, and economic change. It is the prime mover of all the rest, in spite of any appearance to the contrary and in spite of human pride, which pretends that man's philosophical theories are still determining influences. . . External necessities no longer determine technique. Technique's own internal necessities are determinative. (133–134)

Technological autonomy is now evident even in those areas previously believed to be exclusively human: morality and spirituality.

> Since it has put itself beyond good and evil, it need fear no limitation whatever. It was long claimed that technique was neutral. Today this is no longer a useful distinction. The power and autonomy of technique are so well secured that it, in its turn, has become the judge of what is

> moral, the creator of a new morality. Thus, it plays the role of creator of a new civilization as well. . . . Technique in itself is neither [good nor evil], and can therefore do what it will. It is truly autonomous.

Technological autonomy exists only in an inverse relationship to human autonomy; every gain it makes comes at the direct expense of our own freedom. As the system's autonomy accelerates to unbounded heights, our freedom collapses to zero. Thus its victory over humanity is assured:

> No technique is possible when men are free. . . . It is necessary, then, that technique prevail over the human being. For technique, this is a matter of life or death. Technique must reduce man to a technical animal, the king of slaves of technique. Human caprice crumbles before this necessity; there can be no human autonomy in the face of technical autonomy. (138)

Once morality and freedom have been remade in the technological mold, the final step can be taken: domination of the spiritual realm. Traditional human spirituality is grounded in mystery and a sense of the sacred. These things, says Ellul, are necessities of human life. Unfortunately, the power and autonomy of technology are such that they completely destroy such traditional notions, which are shown to be meaningless in the technological context. However,

> man cannot live without the sacred. He therefore transfers his sense of the sacred to the very thing which has destroyed its former object: to technique itself. In the world in which we live, technique has become the essential mystery. . . . Technique is the god which brings salvation. [By self-definition,] it is good in its essence. (143–144)

The outlook is grim, to say the least. Technology will triumph, and all we can do now is document its progress. "The victory of technique has already been secured. It is too late to set limits to it, or to put it into doubt" (130). Despite Ellul's initial disclaimer, one can be excused for seeing here the hardest of hard determinisms.

Is technology, then, truly unstoppable? Is complete domination by technology inevitable? His text argues in the affirmative, but if we return to Ellul's foreword, we find three "external factors" that could yet change the course of history and circumvent the looming determinism.[13] These factors are: (1) something approaching a global, presumably nuclear, war; (2) intervention by God; or (3) a reassertion of our native freedom and autonomy by a sufficiently large mass of people. The first would require a "general war" that eliminates most of humanity and functionally destroys technological society throughout the world; certainly this would have a dramatic impact on the future of our species, but no one can plan for such a thing.

The second is mere wishful thinking; one would sooner hope for intervention from benign aliens than wait for God to save us. Only the third factor is conceivable and actionable. As Ellul sees it, only if a growing number "become fully aware of the threat the technological world poses to man's personal and spiritual life," and if they exert their freedom "by upsetting the course of this [technological] evolution," then the pending determinism may yet be avoided. But this single line is essentially the sum total of his recommendation to humanity—at least, in the subject work.

FURTHER THOUGHTS

Critics of Ellul, and even some of his supporters, typically act as if the man wrote only a single book. Quite the contrary; Ellul was a prolific writer on many subjects, and he explicitly addressed technology several times after *Technological Society.* In fact, he wrote half a dozen articles on the topic and two further lengthy books: *The Technological System* ([1977] 1980) and *The Technological Bluff* ([1988] 1990). To review the important ideas in all these works would take me too far afield, but a few comments are warranted.

The Technological System includes one chapter dedicated to "technology as a determining factor." Here he acknowledges that there are multiple causal factors at work in society, but even in such a situation, "it is not impossible to discern one that is more effective, more constraining." He continues:

> If [in our analysis] we keep finding that same element, we have to accept it as determinant. . . . If this factor allows us to take into account a large number of data in the society being examined, and if it permits us to understand the correlations and differential structures of those data, then we have to admit that this factor has a "strategic" place and an exceptional role. (53)

The situation is as I outlined earlier—technology is *a* determinant, and even the *prime* determinant, but not the sole determinant. Human action, social impulses, and ethical concerns can alter the path of development; they can contribute to the form of technical realization; but they cannot negate the fact of realization.

Something of a more positive action plan appears in Ellul's essay "The Search for Ethics in a Technicist Society" ([1983] 1989). Reiterating his earlier view, he states that "Technique has become the dominant factor in the Western world. . . . It is on technique that all other factors depend" (23). To his critics who charge that human beings are in fact in control, he responds that they are "always neglecting to specify which human beings. Is the 'who' not important? . . . Who is supposed to impose ends or get hold of the technical apparatus? No one knows" (26). A valid point: it is always some anonymous "they" who allegedly override technical imperatives. "They" are the nameless and faceless humanity who are allegedly consciously steering our technological society toward human ends—ends that never materialize.

Of greater interest, though, is Ellul's contrast between "technicist ethics" that serve the needs of the system, and his opposing theory, an "ethics of nonpower," that can serve as a model of resistance. It has four components. The first is the namesake, nonpower, which for Ellul means not doing everything that one has the ability to do. It councils restraint, prudence, and caution. It is a withholding of one's power, not an exertion of it. The second component, freedom, comes about by this very process: of withholding and judiciously exercising one's power, and of not yielding to technology's demands for maximal power consumption. One is then free to act on an ethical rather than a technical basis. Third, those who act thusly will find themselves in conflict with technological imperatives; they must therefore embrace this conflict, accept it, and allow it to be a source of creative change. The only way to avoid tension with the system is to yield to its demands, which ultimately means to adopt technicist ethics. But this is surrender; a society that is conflict averse will ossify. Ellul states the matter concisely: "Conflict is a survival value for the whole of humanity" (34)—recalling Orwell's notion that those concerned about human welfare must stand in "permanent opposition" to the system.[14]

The final component, transgression, refers to a willingness to usurp the bounds established by technology. Here we find a hint—just a hint—of rebellion. Technology lives by establishing limits, says Ellul, and in transgressing those limits we confront it directly; we defy its dominance in our lives, and we actively seek to subvert the system in order to reestablish truly human ends:

> Transgression must deal with reality. Reality is technique itself. Transgression will therefore take the shape either of the demythologization of technique, or a challenge to the imperatives of action based on technique, or a questioning of the conditions imposed on people and on groups so that technique is able to develop. (34)

Such action will involve "desacralizing" technology, criticizing the perpetual illusion of progress, and taking note of the real costs of its advance. It will even involve a kind of technological devolution, a partial dismantling of the system, as a necessary condition for human survival: "Transgression with respect to technique will take the form of destruction of the faith that people place in technique, *and the reduction of technique to a point that it is nothing more than a producer of haphazard and insignificant objects*" (italics added). How precisely this is to be done—and its likelihood of occurrence—are left unspecified.

By the time of his final book on technology, in 1988, Ellul had become fatalistic. The determinisms predicted in his original 1954 work were playing out. "I can say that social, economic, and technical developments have confirmed in its entirety what I said thirty years ago. I have no need to correct or modify anything" (xii). The small window of time in which certain less technicized societies had to effect change was closed. "It is now too late to change the course of technique. We have lost a decisive opportunity in human history" (xiii). And so we are back to Heidegger: only a god can save us.

SUBSEQUENT DEVELOPMENTS

Ellul was the last major thinker to take seriously the problem of determinism. Apart from him, only a few scattered voices were heard—some from surprising quarters. For example, British astrophysicist Fred Hoyle's *Of Men and Galaxies* (1964) included a series of "extrapolations into the future." Among these were some thoughts on the future of technology and on debunking the idea that government or scientists control its advance.

> In principle, you might suppose that scientists could stop the development of technology, or at least direct its development, by agreeing together on some policy. At one time I used to believe this myself. Now I see it just could not work like that, not merely because scientists would never reach agreement, but for a deeper reason. . . . The situation is that scientists produce science, and you must not expect them to produce anything else. (59)

By "anything else," Hoyle means things like practical policy or social tradeoffs, actions that might conceivably exert control over technological development. He continues:

> Who then is in the driver's seat? If not governments, if not scientists, who? Nobody. We are traveling in a vehicle that guides itself. . . It is my belief that . . . we are still in the grip of natural processes, we are not in charge of our own destiny. (59)

. . . a view in notable agreement with my own.

A few years later, economic historian Robert Heilbronner wrote a short but influential essay, "Do Machines Make History?" (1967). Examining the question of whether "there is a fixed sequence to technological development," he answers in the affirmative: "I believe there is such a sequence," and that technology "follows one and only one grand avenue of advance" (336). As evidence he cites simultaneity of invention, the lack of dramatic technological leaps, and the general predictability of advance.

But as the essay develops, we find that his position is highly qualified. In acknowledging the role of social influences, Heilbronner reverts to a position of "soft determinism." To emphasize this fact, he argues that "technological progress is itself a social activity," "the course of technological advance is responsive to social direction," and "technological change must be compatible with existing social conditions" (342–343). More problematically, he concludes the piece by identifying determinism with the unique social and economic circumstances of the present day:

> Technological determinism is thus peculiarly a problem of a certain historical epoch—specifically that of high capitalism and low socialism—in which the forces of technical change have been unleashed, but when the agencies for the control or guidance of technology are still rudimentary. (345)

Heilbronner optimistically presumes that we will overcome the problem in due time—though we are now approaching 50 years hence, and the awaited control has yet to materialize.

One might have expected support for determinism from the other technology critics of the 1960s, but generally speaking, it was not forthcoming. Roszak, for one notable example, found Ellul's determinism "outrageously [and] crushingly pessimistic" (1969: 6, 294).[15] For Roszak and many others, determinism meant fatalism, and they were not about to be cast in that light. At worst they might "personify" technology, but this is something far short of a deterministic stance.

Even those critics who did touch on determinism did so superficially. Jonas (1979), for example, gave merely passing comments along these lines. Technological progress is "an inherent drive which acts willy-nilly in the formal automatics of its modus operandi as it interacts with society"; "the juggernaut moves on relentlessly"; and "technology is destiny" (35). But he offered little in the way of elaboration, and almost nothing that would pass for a philosophical explanation.

In the past two or three decades, philosophical support for technological determinism has all but evaporated—not because the issue has been "settled" but evidently because it is no longer fashionable to contemplate a technologically dictated future. The topic is now the province of technophiles—who use it to support the alleged inevitability of their cause—and journalists. Among the latter is Thomas Friedman. His book *The World Is Flat* (2007) includes an analysis of 10 "flatteners," or factors that will drive future events; they are all technological. "I am a technological determinist! Guilty as charged," he declares (536).

DETERMINISM IN THE PANTECHNIKON

As physical beings in a physical world, we are subject to a wide range of forces and pressures. Depending on circumstances, they exert more or less influence upon our lives. These forces—ranging from physical quantities like friction, inertia, and gravity to moral and social pressures—act in conjunction with human agency, circumscribing its range of efficacy. We oppose gravity every time we stand up, but our power of resistance is limited—we cannot jump more than a meter or two into the air. We oppose frictional and inertial forces simply by moving about the world, but this consumes energy and tires us out. We act 'independently,' but we all are subject, more or less, to social norms and pressures to act in certain ways and to think in certain ways. The greater and more relevant these pressures are to a given situation, the less free we are—and the more we are determined. Determinism exists along a continuum; it is not a binary state.

Technology is one of the forces of the world. On the pantechnical thesis, the universal process of Technê-Logos drives evolution forward, creating order, complexity, and intelligence along the way. Like gravity, thermodynamics, and quantum physics, it is a constant of the universe—a natural law that in no sense depends upon human agency. It gave rise, in fact, to humanity and human agency. Technê-Logos supplies a steady pressure toward transcendence of physical and mental forms; it is a universal bias toward value, order, and 'the better.' It operates as a law of nature—perhaps the *fundamental* law of nature—and thus is autonomous and inexorable. It seeks to realize the maximum order possible, given the free energy available to it; this is the true meaning of the so-called quest for efficiency. This fact is no more mysterious than that gravity seeks to pull all masses together or that the second law of thermodynamics seeks to maximize randomness. All laws are teleological in this sense. None of these forces act unopposed, and they all win and lose at various times and under various conditions. Overall, though, the striving for order—the realization of Logos—is the dominant force in nature. In the long run, order prevails in the cosmos. The universe is a Pantechnikon, after all.

To continue the comparison with other natural forces: Gravity, for example, exists throughout the universe, but the form and intensity of the gravitational field varies widely, depending on the physical structure at hand. In interstellar space, gravitational force is almost negligible. On the surface of a planet, moon, or asteroid, it is significant. In the vicinity of a neutron star or black hole, it is overwhelming. To a much lesser degree, it applies to the ordinary, mid-scale things around us. Every individual physical object, from a locomotive to a human being to a grain of sand, possesses its own small gravitational field, one that is uniquely shaped by the structure and dynamics of the object itself. And every object responds uniquely, if often imperceptibly, to the fields of those things surrounding it.[16]

In a similar manner, Technê-Logos exists throughout the universe. As with gravity, the shape and manifestation of this force is a function of the specific structures that realize it. For the subject at hand, the most relevant manifestation of Technê-Logos is the technological system that we, as means, have constructed. In response to the universal pressure toward order and complexity, we created a system that embodies the universal technological force and that in turn exerts a pressure back upon us. Depending on the point in history, we can see that the system has exerted more or less pressure, depending on its intensity at that time. In the earliest days of primitive technology, the system was a low-intensity force, and it had little effect. Throughout history, as technology waxed and waned in the various human cultures and civilizations, the corresponding technological system was a lesser or greater determinant of human behavior. Today, in the Western world, we have constructed a potent, far-reaching technological system, and it has become the dominant influence on our lives.

We need to recall the nature of the technological system. It is composed not only of our various tools, machines, and devices but also of the procedures, rules, and organizational principles of society. This has been true throughout the history of humanity. Even a million years ago, when our tools consisted of little more than sticks, sharpened stones, and perhaps crude clothing, our simple hunter-gatherer societies had their own functional rules: how the band was to be organized, decisions made, mates selected, dangers confronted.

In the early historic era of ancient Egypt, large-scale societies were created, and sophisticated cultural and religious techniques were developed. The pharaohs organized masses of men into the first 'machine'—a megamachine, to use Mumford's memorable term—in order to build the pyramids. The Egyptian system was thus comprised of a range of technologies: organizational but also physical (woodworking, copper and bronze metalworking, ceramics, papyrus, wheeled vehicles), and biological (wheat agriculture, domestication of sheep, cattle, and pigs). Their society was not something that stood over and above their technology; it was a part of their technology. It is therefore misleading to ask if or to what degree technology 'determined' Egyptian society. The social order was one of several technologies that collectively formed their technological system. And it was the system that imposed determinative forces upon the individuals.

The same is true today. The formal structure of our modern society, including its economic and political subsystems, is but one component of the larger technological system. The social structure acts in combination with the myriad physical components—computers, automobiles, airplanes, the Internet, and so on—to form the overall technological system in which we live. This broader system-level conception of technology, which reaches back to the work of White and Ellul, is necessary in order to grasp the full effect of technical influence.

When we speak, then, of technology as a determinant of human lives, there are in reality *two forces* at work. The first is the universal component, the pantechnical field, which exerts itself as a constant pressure toward complexity and order. Like gravity, we are immersed in it. Like gravity, it is invisible but makes itself felt all the same. In the human sphere, it is a subtle, steady pressure toward greater order, advancement, wisdom, and transcendence. By following this subtle pressure, we are realizing the cosmic

Logos—sometimes benignly, sometimes detrimentally. It is the foundational force behind all technological development and in fact all creation.

The second component is that caused by the technological system itself—the very system that we have created. Unlike the pantechnical force, this component varies dramatically, from almost negligible in ancient times to a position of predominance today. Unlike the pantechnical force, it is neither inevitable nor inexorable. It has gone into retreat many times in human history, most typically when the host civilization underwent radical transformation or collapse. The nature of the technological system, and thus the pressure applied to humanity, always changes with societal conditions and therefore, in a very loose sense, may be called a "social construct." But the fact that there *is* a technological system is not a construct, nor is the underlying pantechnical field that gave rise to it. To declare technology per se as a social construct is facile and misleading; I take up this issue again in the following chapter.

From this perspective, we gain a new outlook on both technological autonomy and determinism. The technological system, while dependent on human beings, derives its strength *and its autonomy* from the broader pantechnical force at work in the universe. The system is both and at once autonomous and dependent. It power, its motive force, is beyond human influence, but its specific form and manifestation are wholly dependent on the modes by which we elect to create it.

Take a specific example. Is nuclear power a dependent or autonomous technical force? It is both: the energy of a nuclear reaction is part of the structure of reality, but the form that it takes—a bomb, a breeder reactor, a fusion reaction—is completely dependent on our technical skill and collective decisions. The present-day technological system functions in much the same way: a universal, autonomous force operating within the confines of a specific human implementation.

Regarding the determinism, we now explicitly see the point that was implicit in Ellul. There is a distinction between the *technical phenomenon* and *human action*. The fact that we are confronted with an autonomously advancing technological system does not mean that there is no scope for human action. We retain agency even as the system grows in power. But the system thereby becomes a greater determinant of our action, imposing physical and psychological pressures to act and think in certain ways. Over time, it takes progressively more effort to resist the technological pressure. Over time, we are less and less able to act in a truly free manner. Gradually and inevitably, we submit.

We can also see now the basis for the justified criticisms of both Ellul and Heidegger. Ellul gave us a strikingly detailed and prescient analysis of the technological system but was utterly lacking in a metaphysical foundation. His philosophical system thus seemed empty, devoid of the necessary underpinnings; it became, as Borgmann (1984: 9) rightly noted, "its own unexplained explanation." Thus Ellul fell prey to charges of demonization and even mysticism. But empirically speaking, he was correct: the system does indeed exert a powerful and growing pressure on humanity to conform to technological imperatives.

In Heidegger's case, he produced an insightful analysis of the underlying force, the Enframing aspect, of technology, but neglected the specific manifestation—the technological system and its various individual techniques—by which the Enframing power became tangible. Heidegger realized that there was an unstoppable, ontological force at work, one that offered insight into the very nature of being. Here he was absolutely correct. But in neglecting the tangible aspect, his theory too, in its own way, became empty, arbitrary, and mystical.

Each man had one piece of the complete picture: Heidegger, the background force of technology, that which I have called the pantechnical field; Ellul, the nature of the specific implementation of that field, in the technological system. Both men's views were incomplete, and thus both were subject to justifiable philosophical rebuke. The pantechnical framework unifies these outlooks in a broader and more precise metaphysical matrix. It proposes a more complete and grounded ontological structure to the technical phenomenon.

A TWO-PHASE PROCESS

The technological system, then, exerts its own deterministic pressures upon humanity, even as we have constructed the system ourselves. In this sense, we are the willing agents of our own repression. But the deterministic pressures have not been uniform throughout history; they correspond, as I have said, to the degree of intensity of the system in a given place and time. A progressively more articulated, more fully realized technical system will exert a proportionately greater influence on human lives, and thus, in the same measure, reduce the overall scope of human freedom. In order to better understand this situation, we need to analyze the system's role in further detail.

My claim is that technological determinism is, overall, a two-phase process. It is obvious—and tautological—that all of (human) technology to date has been human constructed. Technological history so far thus constitutes an initial phase in its realization on Earth; I call it the *anthropogenic phase*. It began some 2.5 million years ago with our earliest stone tools and has progressively grown in sophistication, albeit with periodic setbacks, ever since. In the earliest days, our tools exerted virtually no pressure on us. We used them at our pleasure, and their adverse effects were minimal. Among the forces of the world that dictated our actions, such primitive technology was nearly inconsequential.

The most significant of the early technologies was surely the controlled use of fire, which emerged around 1 million years ago. The campfire, as a tool of socialization, protection, and warmth, must have imposed certain behavioral expectations on early mankind. No doubt it was a welcome tool, and it clearly granted new powers to humans; but at the same time, it created *structure*: new skills were required to produce and maintain it, new cooking techniques had to be mastered, and new social techniques emerged. Evidently it was successful: we continued our small-band, foraging ways for another million years.

Hence the first mode or manifestation of the anthropogenic phase. Let us call it the Foraging Era. It began with the advent of the genus *Homo* more than 2 million years ago and carried on throughout our hunting-gathering existence, terminating some 10,000 years ago.

As we developed agriculture, permanent settlements, and larger societies, our various technologies came to exert more pressure on us—if only in the form of more stringent social customs, norms, and taboos. With large, stable societies came social hierarchies, and with these came laws, rules, regulations, and enforcement, all of which further circumscribed individual autonomy. Importantly, though, the restrictions imposed by the new technologies were outweighed by the benefits of social order: primarily security, stability, and culture. The technological system gave more than it took. It was a net gain for humanity. It allowed us to maximize our individual potential with a minimum of technological imposition. It marks the beginning of what I call the Human Era—the middle period of anthropogenic determination.

A number of specific peaks occurred within the Human Era. These are well known: ancient Egypt (circa 3500–1000 BC), the Minoan civilization of Crete (3500–1100 BC), the Harappans of the Indus Valley (2600–1900 BC), Mesopotamia (3100–500 BC), Athens (600–300 BC), and Rome (500 BC–400 AD), to name but a few in the Western and Middle Eastern worlds. Into the Christian age we could cite, for example, the achievements of the Abbasid Caliphate (750–1250) centered in Baghdad, and the growing Western European culture beginning with early Renaissance, circa 1200 AD. Each civilization developed its own unique technological system, even as successful innovations spread. In each case, I claim, the *human* achievements were considerable and outweighed the disadvantages of steadily growing technological oppression. Needless to say, things were not perfect; women and minority rights were minimal, brutal and corrupt tyrants had their day, and slave labor was not uncommon. But equally obvious is the fact that no social order is perfect, all suffer from flaws and injustices, and none can eliminate human suffering. The best societies in human history aimed not for perfection but for greatness. And many of them succeeded.

The Human Era came to a close by the early years of the second millennium. With the development of such things as metal horse gear, gunpowder, and the compass, the technological system began to outstrip our limited ability to adapt. Tools became more complex, and our machines became more energy intensive. Perhaps most decisive was the mechanical clock, developed in the 1200s. It marked a profound shift in the anthropogenic phase of technology. From that time on, humanity was ruled by the clock; it established the heartbeat of the modern technological system. As Mumford said, "the clock, not the steam engine, is the key-machine of the modern industrial age" (1934: 14).

We may mark this development as the beginning of a new period, a Technological Era. Our new tools and machines gave us unheard-of power, but they imposed equally powerful constraints—and, importantly, failed to increase our quality of life. Indeed, not long into the Technological Era, things would get markedly worse; witness the Black Death of the mid-1300s. It was almost as if the cosmos were warning us that we had overstepped our bounds and that danger lie ahead.

The story of the evolution of the Technological Era need not be repeated here. Suffice to say that, as more energy-intensive techniques came to dominate human existence, we found our sphere of free action further modified—expanded in some ways, restricted in others, but thoroughly denaturalized in every case. Through the industrial era and into today's information and computer age, we have more power than ever, and yet we suffer proportionately greater losses. A high-speed transportation system whisks us around town or across the globe in a blink. But this comes at a price: highway deaths, infrastructure costs, fuel shortages, oil wars, and global climate change. Computers and the Internet deliver vast amounts of information to us instantaneously, but they stupefy us with data, accelerate our lives, and give yet greater power to those anonymous, depersonalized, large-scale institutions of the world. In a similar manner, all the benefits of technology, in every sphere, grant us, as individuals, certain powers and abilities, but they all come with myriad strings attached—strings that we often cannot see until years later. This is the hallmark of the Technological Era: a sequence of apparently beneficial advances contributes to a technological system that is, in the end, *used against us.* With every advance, the costs further exceed the gains, and thus we fall farther behind. This is the chief characteristic of the third wave of anthropogenism. It dominates us to the present day.

THE ROAD AHEAD

We seem to be approaching the end of the anthropogenic phase of technological determinism. But this is not good news. Computer speed, memory, and power continue on their path of exponential, upward growth, just as Ellul foresaw in the 1950s. Within a few decades they will exceed human intelligence. Scientists are also working on self-replicating machines and organisms, which paves the way for an era of literally autonomous technology. If we continue to develop such things—whether intelligent computers, nanomachines, or biotech creations—we then open the future to a truly autonomous technological system, one that can literally advance of its own accord. A superintelligent, self-constructing, and self-reproducing system will constitute a dramatic second and *autogenic* phase of technological determinism.

The transition to an autogenic system will undoubtedly be a profoundly dangerous and potentially catastrophic event. Upon achieving full independence from humanity, the system will no longer be directed, even nominally, toward human ends. Even now, at the end of the anthropogenic phase, we can see that technology increasingly develops in opposition to human well-being—as I will demonstrate in Chapter 11. In the second phase of determinism, humans and nature will be little more than mere resources to be deployed or disposed of at will. In even the most benign realization, autogenic technology will repeatedly spin out of control, producing an endless stream of unanticipated and disastrous side effects. As Drexler, Joy, and others have warned, even the simplest mistake with such powerful technology could mean complete catastrophe.

This is not science fiction, and it is not a problem for the distant future. Kurzweil (2005) famously predicted that the advent of superintelligent computers—which he calls "the singularity" and others call "artificial general intelligence" (AGI)—would occur by the year 2045. His more recent writings reiterate this prediction and supply data to confirm it.[17] But far from fearing this event, Kurzweil welcomes it—with open arms. Phase two ultra-technologies will allow us "to address the grand challenges of humanity, such as maintaining a healthy environment, providing the resources for a growing population, overcoming disease, vastly extending human longevity, and eliminating poverty" (2012: 278–279). "We will merge with the intelligent technology," he says confidently, including having "intelligent nanobots in our bloodstream" to keep us healthy, destroy cancer cells, and extend the capacity of our brains.

As to the worries mentioned, Kurzweil will have none of them. "These technologies are not the vanguard of an intelligent invasion that will compete with and ultimately displace us"—though he gives neither defense nor justification for this claim. His argument, such as it is, is based on a kind of continuity thesis: Technology has always produced problematic side effects, and we have always handled them in due course; the future will be no different. Yet he seems to overlook the implications of his own prediction. A postsingularity world will be profoundly different than the present, and its dangers correspondingly greater. It is the height of irresponsibility to suggest otherwise.

To summarize: On the pantechnical view, technology is a manifestation of order in the universe anywhere the conditions of matter and energy allow—specifically, fluid matter and abundant energy. This will yield more complex and ordered systems that thrive on those conditions. This is a law of nature and must be dealt with accordingly. Throughout the history of the Earth and into the foreseeable future, the conditions are right for continued growth of technological order. Humanity is thus subject to two components of technological pressure: a subtle but inescapable force due to the universal pantechnical field and an explicit, dominant, but variable force from the technological system. Both impel us to embrace and promote technical advance, regardless of our own best interests or the welfare of the planet.

To this point in history, humanity has been an integral part of technological complexity; but this will not always be the case. Very soon, relatively speaking, technology will be fully autonomous and thus truly self-augmenting and self-evolving. The vital question for humanity now is how we will respond in the remaining few decades of human dependence.

With very few exceptions, philosophers of technology have either ignored these metaphysical issues or underestimated their importance. They view technology simply as a social construct and consequently misinterpret the meaning of determinism—either dismissing it as the result of a groundless and inexplicably mysterious force or denying its existence outright. Both views miss the mark, but the latter is the more problematic because of its predominance among philosophers and other intellectuals; I will elaborate in the following chapter.

When we focus on the human role in technology, we miss the larger metaphysical context. There is a permanent and immutable pressure toward complexity and order in the universe. This is the true meaning of technological determinism. But even if we cannot stop it, we can still alter the direction of this advance, we can shape it, and we can accelerate or delay it. It is in this fact that our true saving power resides.

NOTES

1. This is equivalent to causal determinism.
2. Every corporation therefore has a corresponding energy flux density, as described in Chapter 4.
3. *Jenaer Realphilosophie* (I, 237), cited in Cullen (1979: 67).
4. Rudolf Haym, cited in Rockmore (1992: 215).
5. *Collected Works*, vol. 6: 166.
6. *Collected Works*, vol. 29: 263.
7. See Shaw (1978, 1979) for a more detailed discussion.
8. White held to the same basic scheme in his *Evolution of Culture* of a decade later, though with the addition of a fourth layer at the top of the pyramid: sentiment or attitude. But technology was still "the basis and determinant of cultural systems."
9. Ellul prefers the word 'technique' to 'technology' because the former suggests something broader and all encompassing, a point essential to his outlook. Technique is "nothing less than the organized ensemble of all individual techniques which have been used to secure any end whatsoever" (in the words of the translator, John Wilkinson). It comprises any means to any end. As such, it includes not only tools and machines but also procedures, organizations, laws, processes, living organisms, even human beings—anything organized or designed to achieve a certain goal.
10. Albeit of a nontraditional kind.

11. Kurzweil (2005) has a useful elaboration of this point.
12. The sections on environmental impact were updated for the 1964 English edition.
13. Foreword, p. xxx.
14. See Chapter 8.
15. Whether it is pessimistic is, of course, beside the point. The relevant question is, Is it true?
16. Technically, we know that every object exerts a calculable gravitational force on every other object in the universe, no matter the size or distance.
17. See Kurzweil (2012), for example.

10 Replies and Refutations

The question of determinism is of central importance to any metaphysical study of technology. In the previous chapter I outlined several arguments in favor of various forms of determinism, both from traditional sources and from a pantechnical outlook. Here I will examine several antideterminist views in order to allow an assessment of the full picture. At the same time, I consider other related refutations of the technology critics. If it hopes to have any force, the technophile viewpoint must be supported by solid responses to the various critiques and give independent arguments in favor of technology; flat assertions of its beneficial nature will not suffice.

The English publication of Ellul's book in 1964 seems to have stimulated something of a philosophical response by those eager to deny his strong claims. Habermas, for example, explicitly rejected determinism in the late 1960s: "It is true that social interests still determine the direction, functions, and pace of technical progress" ([1968] 1970: 105).[1] The dominant capitalist interests, he says, coincide with those of the system as a whole, thus creating the appearance that the system is independent and "quasi-autonomous." But if social interests *always* coincide with technological interests, this becomes an unprovable and unfalsifiable assertion. To defend his claim, Habermas must show us that, on some occasions at least, social interests exert definitive control *contrary* to the interests of the technological system—and this he does not do.

Most philosophers, however, defended their critiques of substantivism and determinism by simply denying their existence. When pressed for a more specific defense, many cite a single man: Samuel Florman. An engineer by training, Florman took it upon himself publish a nominally philosophical book titled *The Existential Pleasures of Engineering* (1976), extolling the virtues of science and technology.[2] The book is notable as one of very few efforts, in history, to give a defense of technology by attempting a systematic critique of the "doctrine of antitechnology." Since the release of his book, Florman is considered by many to have offered a definitive refutation of the substantivist and determinist perspectives, and of their accompanying "demonization" of technology. Borgmann (1984) cites him precisely as such an authoritative source.[3] Mitcham (1994: 142) calls him a "withering critic of humanities philosophers of technology." And the recent anthology by Katz and colleagues (2003) includes an excerpt from his book. It is thus worthwhile examining Florman's arguments, along with those of some more recent apologists.

Florman targets five individuals as representative of the antitech club: Ellul, Mumford, Roszak, biologist René Dubos, and lawyer Charles Reich. Reich is now largely forgotten, but his *Greening of America* was a popular book in the early 1970s, lauding the youth counterculture that was inaugurated the previous decade. Dubos is best known as an early scientific environmentalist and social critic, but he also challenged conventional views on technology; his book *So Human an Animal* won the Pulitzer Prize in 1968. Though coming from significantly different backgrounds, all five men "are united in their hatred and fear of technology," according to Florman (1975: 54).[4]

Rather than examining the actual arguments of the condemned five, Florman runs through a list of contentious quotations and then summarizes the overall antitechnology position in six points:

> (1) Technology is a "thing" or a force that has escaped from human control and is spoiling our lives. (2) Technology forces man to do work that is tedious and degrading. (3) Technology forces man to consume things that he does not really desire. (4) Technology creates an elite class of technocrats, and so disenfranchises the masses. (5) Technology cripples man by cutting him off from the natural world in which he evolved. (6) Technology provides man with technical diversions which destroy his existential sense of his own being. (56)

Such summaries typically threaten to turn into straw men, which are then dispatched with relative ease. Points three and four, for example, are relatively insignificant in the larger debate. And certainly the five writers make many more points than these in their wide-ranging critiques. But I will let the summary stand as representative of at least some of their common concerns.

Upon declaring these ideas to be "obviously false," Florman then proceeds to refute each in turn. Of the independence or autonomy of technology and its substantive nature, Florman declares that "sober thought" shows that this is simply untrue; that technology "is not an independent force, much less a thing, but merely one of the types of activities in which people engage" (57). It is something that people freely and willfully do, at least under normal circumstances. Where they are compelled, he says, it is due to other people's demands and not technology's own doing. There is no technological determinism, only people making individual choices—which, in the aggregate, *appear* as an autonomous force. It is a "logical absurdity," he says, to depict technology as "a demon, a force, a thing-in-itself."

I addressed this issue in detail in the previous chapter, but in brief, the situation is this: It is true that "people control technology" only in a highly restricted sense. Virtually everyone assumes an anthropocentric definition of technology, that is, as manmade artifacts, devices, and procedures. Under this assumption, we are obviously needed to drive technology forward. If all human beings vanished overnight, technological development would halt instantly. Effects of our technology would linger for centuries, primarily in the form of pollution, climate change, and nuclear waste,[5] but progress itself would stop. This could conceivably change within a few decades if self-replicating systems are developed or a technological singularity is reached. For the time being, though, dependence holds, and this gives us a small and diminishing window of action. But to make the unconditional claim that "people control technology" is misleading in the extreme.

Given that we are still in the anthropogenic phase, then, there is a small degree of nominal control. What this means in a practical sense is less clear. One important factor is the issue of scale. Individually or in small groups, it is obvious that we can take specific action to affect the role of technology in our personal lives. Each of us can minimize or even eliminate the use of automobiles, computers, cell phones, the Internet, and so on—though not without consequences for our social lives and careers. At the level of society as whole, and at the present time, it is equally clear that we can do virtually nothing to restrain technological progress. Governmental funding priorities, legislative action, and corporate investment can modestly impact development, but only to a small degree. Advanced technology now pervades the realm of society as a whole; it is a collective effort, demanding the involvement of thousands or millions of people. No one person, as a *public* person, can change this—not our political leaders, not our corporate leaders, not our religious leaders. At most they can make minor, temporary alterations in the course of events.

Florman thinks otherwise. As he sees it, technology is simply human activity, and human activity is self-evidently under human control; therefore technology is under human control and thus is a completely nondeterministic phenomenon—QED. The logic is valid, but it yields an irrelevant conclusion. My claim is this: If society cannot fundamentally alter the rate or direction of progress, and if we cannot willfully halt or reverse it, then we do not control it. I further claim that we are proceeding irrationally in the face of myriad severe risks—involving environment, accident, terrorism, military warfare, morality, psychology, even human nature itself. We proceed because "we must" or "we want to" or "we need to"—but proceed we do. And beyond the level of the individual or small group, we apparently lack even the option to do otherwise. I believe the evidence is in my favor.

Let us put this demand to Florman and his followers: *Conduct for us an experiment, one that will prove beyond reasonable doubt that human beings have meaningful and decisive control over technological progress.* This is not too much to ask. What experiment could they perform? That is, *actually* perform, in the real world of modern industrial society. I submit that there is no such action that could be done. Actions of a few individuals would be meaningless. To be convincing, it would take a large group—say, a whole nation—exerting definitive control over a specific and widespread technology for an extended period of time. Laws regulating technology or prescribing ethical guidelines are clearly insufficient. It would require halting development, banning future actions or products, or otherwise relinquishing some substantial technological component of their society.

Compare: If I were to ask a pack-a-day smoker to demonstrate that he is not controlled by his cigarettes, what would I require? Surely an hour or two without smoking is insufficient. Is one day enough? A week? A month? Clearly this is a question of statistical confidence. If a pack-a-day smoker, who has done so for years, can go one week without any cigarettes, I have some confidence that he can quit at will and thus is in control. If he goes one month without, I have very high confidence. In any case, there is a clear and relatively decisive experiment that can be performed.

But we must be careful: What if he switches to a pipe, and then says, "See, I'm not controlled by those cigarettes!"—are we to be convinced? Obviously not. What if he cuts down from, say, 20 cigarettes per day to one per day—has he proven his point? Perhaps. It would require, I think, a dramatic reduction or outright abstention over a significant period of time in order to have confidence that our friend is truly in control.

The same holds for technology. A given society would have to willfully conduct a large-scale reduction or abstention of a pervasive technological product over a significant period of time to prove that people are in meaningful control. Is it theoretically possible? Yes. Is it *practically* possible? Evidently not. Can we imagine a modern, industrial nation banning all cell phone use for one month, simply to prove that it is in control? Or shutting off the Internet? Or banning all automobile use? Or grounding all airplanes? Or shutting off all electricity? Such things are out of the question. Theoretical control by society at large is meaningless. It would be like a smoker or alcoholic saying, "*Theoretically*, I can stop tomorrow, and this alone proves that I am in control"—a bad joke, at best. Only the actual deed can prove the fact, precisely because it is the actual deed that is in question.

Florman, then, has no hope of proving his case. All he can do is make assertions and proclamations, pointing to the trivial obviousness of human dependency. In the meantime, technology advances at an accelerating rate in a vast array of fields—because we *want* it that way, claim Florman and his compatriots.

A variation on the antideterminist stance is that society and social/cultural values play a significant role in technology development, and that therefore technology is not driven solely by its own intrinsic imperatives. This is the approach followed by Feenberg and other social constructivists.[6] But this fails as well, because society and social values are *themselves* technologically determined. It is beyond dispute that modern society is a *technological* society; to then argue that "society" is an intervening factor in technological development—and that therefore technology alone does not dictate its own progress—is logically defective. There is no "society," independent of technology, that can exert a directing power on technology. Social values and norms are already conditioned by technology; it is meaningless to treat them as if they were independent entities, untainted by the technical sphere. Our values are *technological* values, and our norms are *technological* norms. If these things are said to control technology, then we have, at best, *technology* controlling technology—but this is precisely the thesis of determinism.

On the other hand, there are at least three compelling empirical facts that support a deterministic stance. First, as I cited in the previous chapter, is the convergence of all societies around the world to a common technological structure. Two centuries ago there were dramatic and fundamental differences among the world's major cities, but today even Third World urban centers look increasingly like New York or Los Angeles. Neither geography, religion, culture, nor economy matter. Capitalist or communist, developed or developing, Christian or Hindu, East or West, North or South—all societies converge to the same technological standard over time. This is because technology is the determining—but not sole—factor driving human society worldwide.

Second, the fact that we persist in promoting technological advance, both socially and individually, in the face of known and growing direct risks to our health and well-being is itself evidence that we are controlled by the technological system rather than vice versa. This is the subject of the chapter to follow.

Third, and relatedly, our negative environmental impact again indicates that we cannot stop and cannot help ourselves. Even if we disregard all value in nature itself, the fact that we may well be destroying our own basis for life is comparable to the drug addict or alcoholic who continues in his self-destructive behavior, precisely because he is not in control. Our very devaluing of nature in favor of our own technological lifestyle is potent evidence that technological values surpass all others—just as the heroin addict devalues all other people and things around him in order to acquire his next fix. And if technological values reign supreme—even though they are not our *only* values—then we are predominantly controlled by the technological phenomenon.

We see then that Florman's first point, the thesis of technological autonomy, is not undermined. His simplistic argument is based on the trivial dependence on human agency, but this, as I have shown, does not defeat the determinist claim. Our present condition of anthropogenic determinism presumes human dependence, as well as a small, and shrinking, degree of freedom. These things are not incompatible.

WORK, CONSUMERISM, AND THE TECHNOCRATIC ELITE

Of Florman's six-point list, items two and three are given only the briefest treatment. On the question of "tedious and degrading work," he blithely claims that work has *always* been tedious and degrading. Premodern agriculture and stone temple construction were alike products of drudgery and pain, no less than the modern assembly line. Physical labor has always been hard, and even administrative or bureaucratic work has, in its own way, always been stressful and demeaning. For Florman, the fact that people seem to prefer the drudgery of the factory to that of agriculture, or city stress to country stress, demonstrates that technological society is preferable to premodern ways of life.

The first reply here is that preferences, of course, are irrelevant. An addict prefers his drug over abstention, but this says nothing about the virtues or vices of drug abuse. But more to the point, we may reply that premodern labor at least had the virtue of being direct and immediate, in the sense that the worker was intimately connected to the results of his efforts. Apart from prison or slave labor, past workers generally produced goods and services in line with their own skills and interests. Their work benefited both themselves and society in a way that was largely transparent to all. It fit directly and intimately into the larger social order. And they had direct or near-direct control over the immediate flow of events. Perhaps their work was tedious and hard, but at least they had *a place*—one of meaning and value. With the industrial era, the phenomenon of worker alienation arises, as Marx has ably shown. The worker gets paid, but his effort serves primarily to benefit others, especially the capitalist owner. And as the scale of work increases, each individual's share of effort comes to a smaller and smaller portion of the end product. He is progressively controlled by larger organizational forces and has less personal autonomy. The psychic effects of such alienation are difficult to estimate, especially when occurring throughout a large population. But they are assuredly not positive, and they likely go far in explaining the growing psychological maladies that we see in modern society.

As to the consumption issue (point #3), it is certainly difficult to try to disentangle 'authentic' from 'artificial' wants. If advertising generates new wants for new products—as it clearly must, given the tremendous expenditures on it—are these necessarily *illegitimate* desires? As to the specific claim that technology forces us to consume things that we don't really want—something that Florman calls a "canard"—we are best to simply state the facts. A modern technological system produces more and newer products for consumption than in the past. The more advanced the technology, the greater the number and diversity of products. Where there are more products, there are, of necessity, more desires; things are produced, by definition, to meet a need. The products are also more highly engineered over time, more technologically advanced, and are the result of more sophisticated production processes. Thus we consume ever more 'technology,' whether we realize it or not. It is therefore more a case of 'knowing not what we consume' rather than 'consuming what we do not want.'

There is also the phenomenon of enforced consumption. When we are given options to choose from, but all of them are *technological* options, in this sense we have no choice. If you want to function in modern American society, you need to drive or ride in a high-tech automobile. There are plenty to choose from, but choosing to do *without* is generally not a viable option. Traditional alternatives, such as walking, horseback riding, or biking, are difficult or impossible. The private car—once a luxury toy and object of curiosity—is now an item of effectively mandatory consumption. There are more recent examples. No one really "wants" to purchase antivirus software for their computer, but we do, and we must. The same could be said for software upgrades, the latest smart phones, or computer tablets. We must consume the newest products, or we are functionally disenfranchised.

Thus, much of what passes for consumption in a technological society falls under conditions of either ignorance or necessity—and frequently both. These things cannot be changed; they are intrinsic to a technological existence. We may not like it, but we must accept it. It is true that we can (usually) choose not to consume, but often at a high and growing price. When most everyone is consuming a given technological product, we are generally under compulsion to do so as well. If we do not, we are cutting our social ties in ways small and large. We are unable to interact with others in the expected manner. We may become functionally unemployable. The more we exercise our nominal freedom to not consume, the more isolated we become. Most people will not accept this, and so they consume things that they "do not want." Florman seems to grasp little of this.

His fourth concern, that of control by a technocratic elite, leads to a somewhat conciliatory response. He accepts that we are indeed dominated by a relative handful of individuals. But as with the case of tedious work, Florman argues that life has always been this way, and thus technology is not to blame. Every large society, he says, has been hierarchically organized. Of necessity this requires top-down leadership by a relative few.

Regarding the specific claim that technology has increased the power of the few and thus exacerbates the inequity, Florman baldly denies it; precisely the opposite, he says. Technology has given more power to the average man and thus increased his freedom relative to the dominant few. As evidence for this, Florman cites the fact that crime is undiminished—an odd claim from an advocate of the blessings of technological society. That the government is unable to stop or dramatically reduce crime is a measure of the freedom that people have—and if the criminals have greater freedom with technology, so much the better for the average law-abiding citizen.

Furthermore, says Florman, society has made long-term progress toward freedom, commensurate with the growth of technology. "Slaves are now free," "the disenfranchised can now vote," "rigid class structures" are vanishing—though how these are attributable to advancing technology, he does not say. The determinist may respond here that slavery and disenfranchisement have not vanished, they have simply changed forms. We now have a system that promotes technological slavery and technological disempowerment, and these affect far more people than did the older, more primitive forms.

As to the notion that the technologists themselves have more or too much power, Florman responds rightly that they themselves disavow such a thing. Engineers and scientists complain that they are as disempowered as anyone—true enough. This is because power, in a large technological system, is diffused throughout. It resides in no one person or station but is woven into the fabric of the system itself. The many feedback loops, redundancies, and alternate pathways ensure that the system is robust, reliable, dynamic, and self-correcting. Florman does not see things this way. On his view, it is "the wealthy, the clever, and the daring—and their friends" (1975: 60) who hold the real power, as always.

But of course, even he must agree that their power is substantially magnified by the technology that they wield. And it is only via obeisance to the dictates of technological society that the wealthy become wealthy and that the rulers come to rule—random cases of inheritance and nepotism notwithstanding. In a sense, the system acts as a filter, ensuring that only those favorable to its interests succeed in obtaining power. In this way, technology grants power to those of its own choosing. Recall that, from the pantechnical perspective, the technological system is a functioning collective being, with an intelligence, will, and value system in place. These qualities are an outgrowth of a universal Pantechnikon that is teleologically oriented. Hence our manmade system acts with purpose, toward specific ends, and in its own perceived best interests—interests that are increasingly in conflict with those of individuals and nature.

ALIENATION FROM NATURE

Nature, then, is Florman's fifth point: specifically, that we are separated from it via our daily high-tech existence. Once again, he does not deny the actual fact but disputes that it is caused by technology. Rather, he says, technology is as likely to *increase* our contact with nature if we so wish. It puts new means of travel and living at our disposal, provides new safety and security alternatives, and, via increased wealth, supplies us with the monetary means. But in reality few people choose to do this. The wealthiest, in fact, seem to spend the least amount of time in nature—indicating to Florman that people don't really love it all that much. "There is tedium in the countryside," he says.

Furthermore, he asks, where is the evidence of harm? Have city dwellers really suffered so much over the centuries? "I can see no evidence that frequent contact with nature is essential to human well-being," he says. "Millions of families" have spent "happy years" in generic city apartment complexes; "millions of people" have spent "pleasant working lifetimes" in modern office buildings. Urban people "are no more dehumanized" than peasants performing their daily chores, and there is nothing to the claim that city life is "emotionally crippling." And even if it is admitted that we need some element of nature in our lives, why isn't a city park, a small garden, or even a balcony flowerbox sufficient to meet our needs?

We have two distinct issues here, and both are important. Let me address the second one first. It is true that we lack a good definition of nature, at least with respect to conditions of human well-being. Total land area, accessibility, degree of manicure, context, and variation all affect perceived naturalness. Often just simple exposure to grass, trees, and sunshine is sufficient to improve mental states, and physiological health generally. Ulrich (1984) showed that hospital patients heal more quickly when they have a view of trees rather than of brick walls or a windowless room. Benefits have also been shown for mental fatigue (Kuo 2001), social interaction (Sullivan et al. 2004), and general health (deVries et al. 2003). Less clear is how much and what kind of nature is beneficial and to what degree. For example, Fuller and colleagues (2007) found that increased plant and animal biodiversity in city parks correlates with increased psychological benefit. Berman and colleagues (2012) showed cognitive benefits for depressed individuals based on a 50-minute walk in a park-like setting—obviously requiring a reasonably extensive area of greenery. In general, the conclusion seems to be this: the more extensive and diverse the nature, the better. This should come as no surprise.

Such a conclusion finds support in our evolved human condition. For most of human existence we lived in hunter-gatherer bands of a few dozen people. The bands moved across open wilderness, having little interaction with other humans—perhaps a few times per year, and in some cases with an annual gathering of a larger tribe or clan. Even under temperate climatic conditions, our population density was extremely low: typically between 5 and 100 people per hundred square miles.[7] By comparison, the state of Montana today, considered very sparse, is about 700 people per hundred square miles. At the high end of modern society we have Los Angeles County, at about 230,000, and Manhattan, at nearly 6,900,000. Manhattan is thus about 100,000 times as dense as our evolutionary norm. This cannot but have adverse consequences.

The point here, of course, is that the genetic foundation of our bodies and minds is centered on small groups of people whom we intimately know, roaming amid vast openness. Our success as a species is adapted to this precise mode of existence. Our abilities to solve problems and to anticipate dangers all evolved under such conditions. We are most at home in open and pristine wilderness, and therefore the further removed we are from this experience, the greater the likelihood of psychological disorder. And conversely, the greater the degree to which we are able to return to such a condition, the better for our psyches. The true 'wilderness requirement' for the human species is so vast that we cannot hope to attain it in the present age. But obviously we can move in that direction. The policy implication is clear: We should work to restore vast wilderness in every nation and every bioregion, strictly for the sake of our mental well-being. Every step in that direction will be an improvement, and no amount of wilderness is too much. Thus it is obvious that Florman's "parks" and "gardens" are deeply inadequate; at best they can temporarily alleviate the more pernicious symptoms of physical and mental stress.

But the other issue—the question of harm caused by city life—is particularly interesting. Urban living is the quintessential technological existence. It is in the modern city that we are confronted most directly and most persistently with an artificial, man-made environment. For North America, Latin America, and Europe, urban areas account for more than 80% of total population. Thus it is vital that we examine this lifestyle for evidence of the impact of technology on human well-being. Here, if anywhere, we ought to see prominent adverse effects of the technological society.

Consider the basic conditions of urban living. The city landscape is almost completely artificial,[8] plant and animal diversity is minimal, and air quality is degraded. Open water and wetlands are typically nonexistent. People spend a large portion of time indoors. They are constantly in the presence of other people, at unnaturally high densities. And they live amid a high-energy, high-speed pace of life. Urban living unquestionably carries physical risks, including higher rates of violent crime and greater prevalence of conditions such as alcoholism and sexually transmitted diseases. But Florman ignores these. Instead, he presses the case for *psychological* well-being. So let us examine that aspect.

If only anecdotally, it has long been known that city life seems to promote various neuroses and psychological disorders. Florman saw "no evidence" of systematic problems—but how hard did he look? A bit of research might have drawn him to analyses from previous decades. Simmel, for example, wrote a famous and insightful critique, "The Metropolis and Mental Life," in 1903.[9] Or he might have tracked down the 1939 book by Farris and Dunham, *Mental Disorders in Urban Areas*. Or perhaps he would have found the seminal British studies by Hare in the mid-1950s.[10] But still, it must be granted that widespread, detailed research on the psychological effects of city life was not widely known in the 1960s and early 1970s.

Today, however, this is no longer the case. We now have extensive research that allows us to reject Florman's claim that city life causes little to no psychological harm—and not just Florman, but all those who would argue that a technological existence has minimal impact on our mental well-being. The evidence is damning. Researchers have known for decades that, statistically speaking, rates of mental illness are unquestionably higher in cities than in rural areas. In the early 1990s, Lewis and colleagues (1992: 138) found that nonschizophrenic mental disorders were 34% more prevalent in large cities.[11] Focusing specifically on eating disorders, van Son and colleagues (2006) found that bulimia was roughly five times more likely in urban areas. Preen and colleagues (2010) performed a meta-analysis of recent, high-quality studies, determining that generalized anxiety disorders were 21% higher in cities; generalized mood disorders, 39% higher.

Of special interest is the most severe mental illness, schizophrenia. Again, it has long been known that schizophrenia rates are higher in urban areas; see Eaton (1974) or Torrey and Bowler (1991) for supporting data. However, it was commonly believed that this was caused by the movement of previously afflicted individuals into the cities—what is known as the "geographical drift" theory. Lewis and colleagues were the first to debunk this idea, in their 1992 study. They eliminated the drift hypothesis by examining where individuals were raised—a city upbringing, medium or small town, or a rural upbringing. They found a 65% higher rate of schizophrenia for those raised in large cities versus countryside. Small and medium towns had 28% and 39% higher rates, respectively.

These results have since been corroborated by others. Zammit and colleagues (2010) examined more than 200,000 Swedes born in the 1970s, finding a 32% higher rate in Swedish cities. Even more troubling was the work of Krabbendam and van Os (2005), who performed a meta-analysis of 10 studies conducted between 1992 and 2004. Their work demonstrated that "the rate of schizophrenia in urban areas is around double the rate of that in rural areas" (795).

The data are clear and uncontroversial: the more urbanized the setting, the greater the likelihood of mental illness. What are not well understood are the specific causal factors and the specific neural mechanisms. On the one hand, city life necessarily reduces contact with nature, and this in itself has detrimental effects. Berman and colleagues (2008, 2012) have shown that interaction with nature improves cognitive functioning, both for normal healthy adults and for those with depression. But it remains to be proven that removal from nature is a causal factor in mental illness. Traditional causal factors—such as air pollution, noise, childhood social standing, obstetric complications, infectious diseases, and social isolation—seem to have been largely ruled out.[12] Some researchers have therefore suggested that population density and social fragmentation, or "social stress" generically, are significant factors. Meyer-Lindenberg simply states that "certain aspects of metropolitan life appear to incline the brain toward mental illness" (2013: 59). In other words, the technological environment itself, as a whole, seems to collectively work on the human psyche in adverse ways.

In fact, researchers have now been able to measure changes in the brains of people raised and living in urban centers. Using an fMRI scan, Lederbogen and colleagues (2011) found dramatically increased activity in the amygdalas of people who had lived for a number of years in an urban environment over those from rural or small-town settings. Of those raised in the city, they found a linear relationship in the activity of a related brain area, the perigenual anterior cingulated cortex (pACC)—that is, the least active pACCs were found in those with entirely rural upbringings and the highest activities in those that were entirely brought up in large cities. This is important because the amygdala is the brain's danger sensor, and overactivity is implicated in depression, anxiety, and violent behavior. Perhaps more significantly, changes in the pACC have been connected to schizophrenia.[13] This is unsurprising, given that the pACC is a control center for amygdala activity. But if city life is damaging both a key sense organ in the brain *and* its control region, the potential for psychological disorder is magnified significantly.

The mere fact that the brain is visibly and objectively altered for the worse by city life is a profound fact of modern existence. Meyer-Lindenberg even notes that, simply from the fMRI data, he can determine the extent of the subject's urban life: "In each experiment, we could readily identify city residents by brain scan alone: urban life had marked all of them with telltale hyperactivity in the amygdala" (2013: 60). High-tech existence is scarring our brains and causing untold human suffering. Florman's naïve claim about the harmlessness of urban, technological life is shown to be groundless.

A few final points can be made. First, the cited data say little about other important psychological disorders, including ADHD, autism, and depression—though we would not be surprised to find comparable increases in urban environments.[14] Second, we might actually expect to see *decreasing* differences between urban and rural conditions in recent years, given that rural life is becoming increasingly technologized. Even some of the most remote locations now have wireless Internet access and cell phone connections, and the maladies associated with these products can thus be expected to appear almost anywhere. Third, the fact that schizophrenia, for example, has a low overall rate of occurrence does not imply that all those without schizophrenia are suffering no ill effects. The pressures and stresses that exist, exist for all. It is the genetic component that varies, and this explains why only certain individuals succumb. But all are feeling the effects of technological pressure, and all are likely manifesting these stresses in some way or another. The lack of a positive diagnosis does not mean all is well. Fourth, we do not know how permanent the damage is. How long does a hyperactive amygdala stay in that condition? Can it be restored to a normal state? We seem to have no answers at present. And finally, and perhaps most troublingly, we do not know the effects on children—except to say that, in general, developing brains are much more susceptible to these kinds of influences. It bodes ill for our future generations.

This all points to a further question: What should be done? Some would say: carry on as usual. The technophilic advocate of city life can explain, "Yes, we know there are stresses, and yes, we know the risks increase substantially with urban living. But we willingly accept these risks, as do millions of others. We know that a modern city is an evolutionarily unnatural situation, but obviously we can handle it, at least reasonably well, so what's the problem? We'll take our pills, visit our psychiatrists, and deal with it the best that we can. That's what we've always done." To which I give the following reply: In fact many people do *not* know the truth of the risks, because they do not follow the latest scientific research, and the popular media generally underplay such risks. Second, children are thrust into the stresses of city life unwillingly, and they unfortunately suffer the severest consequences. Third, the cost of such disorders is certainly large but widely dispersed via large-scale health care systems and insurance corporations; hence those who promote and profit from a high-tech existence never bear the full cost of their actions.

But the main point is this: Technology critics have long claimed that a high-tech urban life is demeaning, dehumanizing, and psychologically damaging—and they have largely been proven correct. The fact that many city dwellers avoid the worst of the disorders is incidental. The stresses are there, they are real, and they seem to be increasing. Technology is the root cause, and the prospects for the future are not good.

DIVERSION AND INDIGNITY

Coming to Florman's sixth and final point, he rejects the claims that technology ties us up with trivial diversions and meaningless pastimes, and that this situation is ultimately eroding our very sense of being. On Florman's version of the antitech stance, the average man merely thinks he is satisfied with his trivial amusements; his satisfaction is illusory, and true happiness eludes him. He cites Mumford's "total scorn" for the average man's own belief that he really does enjoy these things.

This is a difficult issue for either side to press. Rather like the "inauthentic desires" problem mentioned earlier, we have a significant concern here: How we can we tell authentic happiness from a trivial, demeaning, technologically induced happiness? *Is there* such a thing as demeaning or undignified or unjust happiness? I believe there is—the chemical-induced happiness of a drug addict, for example, or the sadistic happiness of bullies, brutalizers, and warmongers. There is the happiness of ill-gotten gains and the happiness that is an avoidance of some well-deserved pain. Happiness is not an intrinsic good, as Kant recognized long ago.[15] Simply because someone says they are happy is insufficient reason to declare it a moral good. Hence the antitech critics are justified in condemning certain pleasures as undignified, unhealthy, or unworthy of our best human nature. Florman has no awareness of the subtleties involved here.

But let us take happiness at face value for a moment. Then consider a basic issue: trends in happiness over time. Are people generally happier now than in past decades? They *should* be happier, shouldn't they? Technology is progressing, getting better all the time, and it is presumably designed with the intention of satisfying our needs and thus with making us happy. That is the point of technology, after all.

Or is it? For that matter, what *is* the point of technology? On my view, there is no "point" at all; as the enactment of order and complexity via human beings, it moves like a slow-motion wildfire, spreading here and there, jumping fire lines, sparking new outbreaks, dying out in one quarter even as it rises up again in another. We influence the process but do not control it. We are the means but not the end. Our happiness is incidental to the larger process underway; it matters only to the extent that we fan the technological flames ever harder in the directions that we prefer. For the time being, our quest for happiness promotes technological advance. When that quest is no longer an aid to its progress, our happiness will become irrelevant. Or worse: If our quest becomes an obstacle, then it will be removed, ruthlessly.

Even a brief look at the data is revealing. Based on self-reporting, the Western industrial nations show a remarkable consistency in happiness, at least back to the mid-1950s in the U.S. and the mid-1970s in Europe.[16] The trend lines are virtually flat. But technology has advanced significantly over this same time. Given that there is no increase in happiness, the share of happiness per quantum of technology is *declining*. That is, we adopt new technological products, features, and capabilities all the time, and yet people are no happier than they were decades ago. We incur the costs—monetary, physiological, psychological, and environmental—but gain nothing in return. We live with the disadvantages without earning an offset. We are proceeding—downhill.

RECENT APOLOGETICS

In the years following Florman's defense, new variations on the antideterminist and protechnology themes have emerged. Borgmann (1984) says almost nothing explicit about determinism, though he does in principle equate it with substantivism. Referencing Florman, he brushes off both views with a wave of the hand: "Talk of such an obscure and pernicious power is easily dismissed as a demonizing of technology" (10).

Ihde is nearly as dismissive. Adopting a phenomenological stance, he argues that technology does not warrant a hard determinist conclusion. It does, however, support "latent telic inclinations" (1979: 42), something that supplies a mild sort of pressure toward certain ends. In *Technology and Lifeworld*, he adds that phenomenology suggests a "double ambiguity" within technology—that any given object is amenable to multiple uses, and that any given end has multiple technological means—and that consequently there exists an "indeterminacy to all human-technological directions" (1990: 139); in other words, antideterminism holds. Summarizing a brief case study on various writing technologies, Ihde states that "in none of these variants does the technology 'determine' the style or the type of composition—but it does 'incline' toward some possibilities" (142). Interestingly, he adds that when such inclinations are projected onto the larger scale of society, "the result is closer to predictable"—suggesting that a kind of determinism emerges at higher levels of organization.

Ten years later, though, he was anxious to reassure his readers of his antideterminist stance. Briefly addressing the topic in his 2002 book *Bodies in*

Technology, Ihde cites the "convergence" of uses of advanced technologies among various user communities, as when astronomers, television producers, and medical technicians all use digital image enhancement techniques. Such a reading of convergence "flirts with old notions of *technological determinism*," he says—immediately cautioning us that "one must be careful not to revive this simplistic explanation" (2002: 131).

The mid-1990s was a time of critique from outside the discipline of philosophy as well. Historians Leo Marx and Merritt Smith issued a compilation of pieces opposing determinism in 1994, in their anthology *Does Technology Drive History?* Historians, it seems, have an intrinsic predisposition toward seeing history as the product of individual or collective action and thus are almost congenitally blind to other possible factors. Smith defines the central concept in varying ways—initially as "the belief in technology as a key governing force in society" (1994: 2), and later in the following passage:

> 'Technological determinism' is a curious phrase. The gist of it is heartbreaking in its simplicity: the belief that social progress is driven by technological innovation, which in turn follows an 'inevitable' course. (38)

Such a view, he adds, is "subversive" because it "upends the Second Law of Thermodynamics, conjuring up a nonentropic universe in which dust never settles . . . and everything is always new and improved, forever young." Clearly Smith has a highly restricted, if not to say erroneous, conception of the Second Law. His statement has been amply rebutted in Chapter 4 and in the work of Eric Chaisson. But Smith can barely contain his contempt:

> We scholars of technology and culture lament the stubborn tenacity of technological determinism, but we rarely try to identify the needs it identifies and attempts to address. On the face of it, our brief against this variety of superstition resembles the academy's response to creationism: How can something so demonstrably wrong-headed continue to sway adherents? (38–39)

In my terminology, Smith is viewing technological history in its anthropogenic phase, seeing it as implemented strictly via human beings, and thus concluding that humans have decisive control over the course of events. He does not even consider the possibility that larger forces may be at work, acting through humanity, and that these may effectively override human authority. Instead he prefers to issue arrogant and unjustified assertions.

At the same time, Schmookler (1995: 253–254) offered a short but interesting analysis of the situation.[17] Upon initially stating that "a thorough critique [of] technological determinism . . . is a complex undertaking"—perhaps explaining in part why it has never been done—he then provides a short recap of its two "essential points." First is the claim that "societies with substantially the same levels of technological development can be culturally very different." His examples are ancient Rome and China, which, he claims, had a "common agrarian approach to the problem of subsistence" yet had clearly different social orders. But this is an invalid claim. To cite "agriculture" as the common technological basis for those two societies is meaningless. Obviously every society in human history, from 10,000 years ago through the Industrial Revolution, had such a basis. Do we therefore demand that they all must have a common sociological structure, if we are to defend determinism? Certainly not. Rome and China had vastly different tools and construction technologies at their disposal, along with obviously different environmental conditions and socio-cultural histories. Even if, by chance, they had independently and at the same time hit upon the same key technologies, we would never expect instantaneous transformation into common social orders. Technological determinism becomes manifest only over time. It is a convergence process. This is going on everywhere in the world today, as cell phones and the Internet invade the farthest reaches of humanity. Once these technologies become available, societies converge under technological pressure to common modes of existence.

Schmookler's second point is that determinism "fails as an explanation of change." There are two subproblems here: (A) It is simply empirically disproven, he says, that changes in technology lead to or cause changes in culture. He cites two other sources arguing that the opposite has occurred, namely that culture drives technology. But this claim presumes that culture is a preexisting entity, independent of technology, which then influences technological progress—which is untrue, or at least begs the question. All societies are effectively technological societies; their influence on technological progress is then simply a matter of technology steering itself. And this is the thesis of determinism.

The other issue is (B) that determinism provides no mechanism "to account for the movement in the supposed engine of the whole process" (254). Why, asks Schmookler, did agrarian societies suddenly give way to industrial societies? And why in different places at different times? Standard defenses—such as Leslie White's—are merely descriptive rather than explanatory, he says. In fact, one could charge Ellul with the same failing. But as I have argued throughout, a pantechnical analysis provides just such an "engine" for change—it is the teleological thrust toward order in the cosmos. This in turn is a manifestation (following Chaisson) of the expansion of the universe, which is ultimately driven by something we barely understand, namely, dark energy. The overall process is a realization of Logos—of order, intelligence, and mind—in the various material forms that we see in the universe.

There is yet one other issue behind Schmookler's critique that is particularly relevant. A number of times he states that if we expand the concept of technology to include society and the social order, we then produce a tautological argument, and the concept of technological determinism thus becomes meaningless. *Prima facie*, this is a concern, as it is precisely what I and others—most notably Ellul—have done: generalize the concept of technology to include, at least, the entire human realm. In this case, there is nothing 'apart' from technology, and therefore all causation must be of a technological nature. This is true, but it is not a tautology—anymore than the claims that all material reality has mass, is subject to gravitational force, is equivalent to energy, and so on. That there are common, underlying, metaphysical principles to the world does not make the whole system logically incoherent. And ruling them out *a priori*, as Schmookler appears to do, is to make a wholly unwarranted metaphysical assumption and thus to fall into error.

Another voice from that time, physicist Ernest Braun, also largely accepted the popular view, that "technology is a social construct" (1995: 21) and that technological progress does not lead to an "inevitable and immutable outcome." And yet there arises an "apparent contradiction": social progress—movement toward a better life—waxes and wanes, but technology "advances monotonously despite the fact that it is a social construct" (28). Braun offers three reasons for this. First, there is an "inner logic" to technology that drives it relentlessly forward, toward the better, the faster, the more efficient. Second, society generally endorses this progression, which is wholeheartedly embraced by those business leaders who profit from its advance. Third is the unconditional support provided by the military and defense industries. The combined effect is such that technology "appears largely autonomous"—but it is only an appearance, Braun assures us. He thus marks out an unusual position: technological progress ought to be "decoupled" from social progress, but at the same time technology itself is a social construct, and hence nondeterministic. Suffice to say that Braun's case is unconvincing.

Philosophers were not completely absent from this discussion. Joseph Pitt, for example, presents something of an argument against determinism in his book *Thinking about Technology* (2000). Technological autonomy, he claims, exists only in a trivial sense—as when an inventor loses control of his invention, or when unforeseen consequences of some device occur. Such things create merely the illusion of autonomy, not the real thing. He continues his brief inquiry, though providing almost nothing in the way of a generalized rebuttal. His strongest claim is that science and technology are interlinked, and therefore that neither can be seen as independent of the other. But this is a meaningless assertion, because (A) he does not define what he means by 'science,' and (B) it is clear that technology has a far more fundamental ontological standing than does science—if only because its historical origin vastly predates that of science and because of the obvious spread of technology into the animal kingdom, at a minimum.

In the end, Pitt resorts to a kind of *ad hominem* attack; determinists offer little more than "outrageous examples" (2000: 88) to prove their point. Only "misdirected philosophical analysis" could support such a view, something that casts us "into the world of science fiction." Ultimately, determinists "fall victim to a kind of intellectual hysteria that makes successful dealings with the real world impossible." Right—as if that definitively resolves the issue for us. One suspects that Pitt has been reading a bit too much of Merritt Smith.

A few years later, historian David Nye presented a sort of weak compromise position that recalls the stance developed by Braun. Like virtually all of his colleagues, Nye disavows determinism. Noting, with understatement, that "few historians argued that machines determine history" (2006: 19), he observes that, nonetheless, "a surprising number of people [today] speak and write about technologies as though they were deterministic." To demonstrate this fact, he presents a (very) brief overview of its advocates—though covering an odd array of characters, including Marx, McLuhan, Toffler, Marcuse, Roszak, and Foucault. Ellul enters the discussion but merits all of one paragraph. From this sporadic and haphazard recap, Nye concludes that "technological determinism lacks a coherent philosophical tradition" (31)—when in fact it is his own account that is incoherent. Near the end of the book, he leaves no doubt where he stands: "this view is incorrect, no matter how plausible it may seem" (210).

Be that as it may, Nye does concede that technology possesses a kind of force or power to it after all. Taking a phrase from fellow historian Thomas Hughes, Nye endorses the concept of "technological momentum"—the idea that technologies, once in place, have a "rigidity and direction that can seem deterministic to those who use them" (211). But this momentum is entirely our own doing, he says, and in no sense reflects an intrinsic quality of technology. "Technological systems are socially constructed" (219), he emphasizes. We have complete control over their development and evolution—not to worry.

Attempting to close his book on a positive note, Nye states that "we are not *necessarily* evolving toward a single world culture, nor *must* we become subservient to (or extinct in favor of) intelligent machines" (226—italics added). But we *are,* nonetheless, evolving toward a single *technological* world culture, and we *do* face the very real threat, in the near future, posed by superintelligent machines. Nye's statement comes off as mere wishful thinking, with little theoretical substantiation and despite empirical evidence to the contrary. The point must be emphasized: *If he and the other antideterminists are correct, we have nothing to fear.* Since technology is a "social construct," we will simply steer it away from any such disastrous outcomes. On the other hand, if the determinists and I are correct, many of the worst outcomes will likely materialize. This is not an idle debate; time will tell.

FEENBERG'S ACRITICAL THEORY

The most persistent foe of determinism in recent years is surely the philosopher Andrew Feenberg. Adopting a "critical theory" approach that derives from Marcuse, Foucault, and ultimately the Frankfurt School, he objects to both instrumentalist (neutrality) and substantivist theories. For Feenberg, technology is under human control and is thoroughly political in nature. Critical theory emphasizes the particular over the universal; there is no "Technology," only various technologies, each of which must be assessed in its unique circumstances. Consequently there is no technological essence—both Heidegger and Ellul are fundamentally mistaken, on his view. Most importantly, technology poses no fundamental threat to humanity. In fact, just the reverse: it presents the opportunity for freeing humankind from various forms of social domination. Technology causes harm *only* via the "antidemocratic values that govern technological development" (Feenberg 2002: 3). Fix this, and all will be well. "Radical democratization can thus be rooted in the very nature of technology" (34).

Feenberg's terminology calls for a brief aside. The word 'critical' is problematic here because of its multiple meanings. First, it can refer simply to a process of assessing or analyzing, and then passing judgment. This in fact is the original context—deriving from the Greek verb *krinein* ('to decide'), which led to *krisis* ('judgment')[18] and *krites* ('judge'). In principle, the judge is a neutral party and may decide either for or against the issue at hand. A critic is then "one who expresses a reasoned opinion" on some matter. However, in common usage, 'critical' has taken on a more negative intonation—as in the critic who is "given to harsh or captious judgment." In the ordinary vernacular, then, a critic is one who issues a *negative* assessment, and thus to be critical is to judge "severely and unfavorably." In fact this is now the primary definition of the term.[19] To further complicate things, we often use 'critical' in the sense of 'vital,' 'crucial,' or 'decisive'—as when an action of some sort is particularly important. This counts as a third usage.

To add yet one more layer of confusion, within philosophy and social theory we have had, for decades, the term 'critical theory,' which derives from the work of Horkheimer, Adorno, and others of the Frankfurt School. For them, a theory is critical if it is aimed at human freedom or, in a slightly broader sense, as an approach that is simultaneously explanatory, practical, and normative.[20] This is apparently connected to the third definition above, in that critical theoretic approaches have a particular moral urgency that exceeds that of ordinary inquiry.

'Critical' in the first, original sense is what is meant, for example, by a 'film critic.' The film critic's job is to evaluate, assess, and pass reasoned judgment on motion pictures. He is presumably objective; the film at hand may be given a positive, negative, or mixed assessment. In his objectivity, the film critic is, in one sense, *neutral*—he assesses without bias or predisposition. On a larger scale, however, the critic is self-evidently *favorably disposed* toward motion pictures as a whole; he necessarily sees them as valuable and important contributions to modern culture. It is his self-chosen profession, after all. To be a critic in the first sense is to presuppose the value and importance of the subject at hand.

Throughout the present work, I have used 'critical' exclusively in the second and more common sense—as a negative evaluation. Technology critics are, for me, those who assess it as harmful, dangerous, or otherwise detrimental. Feenberg, however, evidently draws inspiration from the critical theory of the Frankfurt School. For him, critical theory of technology addresses human emancipation and thus is intrinsically political. This is certainly an admirable goal, but unfortunately it introduces a subtle deception: Feenberg's "critical" is in fact *entirely uncritical in the common sense.* From his perspective, technology is an essential tool for the restoration of democracy and human autonomy. The only problem, he says, is that modern technology reflects "the values of a specific [anti-democratic] social system and the interests of the ruling classes" (2002: 14). Technology is a vital means for human liberation; it is only being misapplied by the ruling capitalists. In this sense Feenberg is a classic Marxist.

Others in the philosophy of technology, not of Feenberg's persuasion, use the term 'critical' in the original 'film critic' sense. For them, to be critical of technology is simply to assess and evaluate it—or rather, to evaluate specific technologies within specific contexts. This allows them to call themselves technology critics without even the slightest inclination toward a negative, condemnatory, or harsh assessment toward technology as a whole. Instead they are like film critics: predisposed to a generally positive view of the subject matter. Whereas I, to reiterate, use the term exclusively in the negative and now primary sense.

The result, sadly, is a significant confusion over the concept of a technology critic. It now has three distinct meanings: as a negative judgment, as a nominally objective assessment, and in Feenberg's sense, as a normative process of human emancipation. In the second and third usages, there is an implicit and unexamined assumption that technology as a whole is essentially good and beneficial. Criticism can therefore only be focused on narrow and specific applications, never on the phenomenon collectively—to the point where they even deny that there is such a thing as a technical phenomenon. This cooptation of the concept of criticism is both fascinating and revealing; it would seem to be a subtle way of ensuring that negative criticism—in the common sense usage of the term—is diluted and emasculated. "Of course we allow technology criticism," they can now say. "In fact, that's all we do."

But back to the subject at hand. Feenberg (2002) explicitly states that we need not fear technological determinism. "The choice of civilization is not decided by autonomous technology, but can be affected by human action. There is no one single 'technical phenomenon' that can be rejected as a whole in the manner of Ellul" (14). In Feenberg's hands, technology becomes a murky, relativistic entity; it is an "ambivalent process of development suspended between different possibilities," a "scene of struggle," a "social battlefield" (15). Consequently, it is not a "thing" to be feared but rather a process to be mastered for the welfare of humanity. But that, of course, is precisely the issue at hand—whether it is even possible to do so. One cannot presume what one is trying to prove.

Feenberg's hostility toward determinism has been a consistent marker of his work. His 1999 book *Questioning Technology*—which, as I noted previously, does precious little 'questioning'—opens with this very topic. Determinism, he says, was historically linked both with technological neutrality and with the idea of progress. Technology existed, traditionally speaking, to achieve human ends and to advance the species. As the aura of inevitability came to surround progress, so too did it attach to technology. In time the notion of progress came to be questioned, but not the inevitability of technology. Thus emerged a corollary to determinism—that technology advances without regard to human well-being.

Since both the positive and negative views of determinism accept, if only implicitly, an underlying technological essence, Feenberg labels such views as "essentialist." He rejects this idea, and indeed states that the purpose of his book is to offer "a critique of essentialism" (1999: 3). He then explains the hallmarks of social constructivism, including, most importantly, its debunking of determinism. In his view, technology begins in an initial state of unformed potentiality. Social forces then act upon these broad potentialities to lock in and fix a given implementation or a few specific applications. This process of "closure" gives rise to a specific technology that is divorced from its origins; it becomes a so-called black box in which the early potentialities and social choices are effectively lost to history. From our present perspective, we see only a finished and "inevitable" end product, one that appears to be the foregone result of a deterministic process. "This," says Feenberg, "is the source of the deterministic illusion" (11).

It's a nice story, but it suffers from the same problems I cited earlier: namely, that it presumes that 'society' is an independent actor, unaffected by technology itself. The social choices that shape a given device are, in reality, technological choices, ones that, on the whole, promote the refinement and advancement of the technological system. They do *not* act on behalf of human interests, as Feenberg and the constructivists assume. They are social choices, but always choices in favor of technical advance.

Feenberg's follow-up work, *Transforming Technology* (2002), emphasizes from the start his ultimately instrumentalist and optimistic view: "I argue that the degradation of labor, education, and the environment is rooted not in technology per se but in the antidemocratic values that govern technological development" (3). Only an "innovative redesign of technology" (18) can remedy these problems—presuming, of course, that technology is fully under human control. He rightly observes that "the goal of a good society should be to enable human beings to realize their potentialities to the fullest" and that the present system imposes "limitations on human development" (19)—understated but true.

Later he provides a two-part definition: Determinism is based on (1) the idea that technical progress follows a fixed pattern in all societies and (2) that the social order must adjust and adapt itself to this progress (138–139). In light of the preceding chapter, we can see that this is a restricted definition; the second point is generally valid, but the first one has not been claimed by any adherent. Of particular concern to Feenberg is the idea that capitalist imperatives, especially the profit motive, have no role in such a deterministic outlook, and this notion is clearly anathema to him. But his refutation of determinism is limited to a simple denial of each of the two points: technological development is overdetermined by technical *and social* factors, and both society and technology must adapt to each other (143).

Feenberg's most recent attempt to refute determinism, *Between Reason and Experience* (2010),[21] opens with a striking introductory comment by sociologist Brian Wynne. The constraints and limitations imposed by modern technology, Wynne says, "should not be allowed to lead to the logical non sequitur, namely the long-discredited but perverse and persistent theology of technological determinism" (2010: x). Quite a statement!—and one that smacks of desperation. This alone should cause the reader to suspect that the case for determinism remains strong and that ulterior motives are at work in those who oppose it.

As before, Feenberg accepts that modern technology is "increasingly alienated from everyday experience," but once again states that "this is an effect of capitalism" (xvii)—setting the stage for his neo-Marxist analysis. His techno-optimism remains intact; the Internet, for example, advances the agenda of democracy because it supports "interaction" and "participation" and thus "democratic rationalization" (3). But this end is frustrated because of distortion introduced by "the masters of technical systems"—by which he means "corporate and military leaders" and "professional associations" of doctors and engineers (5).

He then reiterates the two-part definition of determinism cited earlier, adding that, in this view, "technology would thus resemble science and mathematics by its intrinsic independence of the social world" (8). Though I would not put it quite this way, that is in fact the sentiment behind my pantechnical thesis. Feenberg then adds that, although the two pillars of determinism appear "obvious," it is "surprising" to realize that neither holds up to scrutiny. The first pillar has been refuted, he claims once again, by the social constructivists, who have shown that technology is significantly influenced by social factors.

The canonical example of the bicycle is then invoked. As we recall from old photographs, early development was divided between 'speed' (large front wheel) and 'safety' (smaller, equal wheels) designs. For various social reasons, the latter won out. But looking back retrospectively, it appears that the standard design was inevitable and thus a foregone conclusion—the deterministic view. For Feenberg and the constructivists, this is proof that social factors influence technology and therefore that determinism is false. As I see it, the outcome—a realization of safety and efficiency—was rather the most conducive to technological advance. In the end, the balanced two-wheel design provided both speed *and* safety; thus it was, in fact, a 'foregone' conclusion.

In his attempt to undermine the second pillar, Feenberg calls to mind the example of child labor. Technology, it would seem, wanted to put all hands to work in the early factories, and thus the technological imperative would press even women and children into hazardous work conditions. But society intervened, invoked humanistic concerns about justice, fairness, and human rights, and put an end to these evil practices. Such, anyway, is the constructivist account.

But their point goes unproven. What actually happened? Children ceased to be factory workers and became "redefined socially as learners and consumers" (12)—in other words, one could say, as future servants of the technological system. It was found, purely for technical reasons, to be inefficient to have children spend their lives in hard labor. It was more conducive to technological growth for them to go to school and learn how to be scientists, engineers, mathematicians, and otherwise good little promoters of technical development. And as more educated consumers, they would thereby promote wealth production and surplus profits that in turn could be used to fund technological progress. Pulling children out of factories was *the* technological solution—precisely as determinism would have it.

Thus Feenberg's attempted refutation fails. His claim that "technological development is constrained by cultural norms originating in economics, ideology, religion, and tradition" (16) cannot be maintained. His bald assertion that the situation with modern technology "is not a metaphysical condition" (26) is patently false—obviously so on the pantechnical view. Lacking a deeper philosophical understanding, his plan to "democratize technology" will certainly fail. Once again, this is an empirical question: Let us see if truly democratic values flourish in the future or if they decay. Then we will be better able to assess Feenberg's view.

Ultimately, for Feenberg, determinism would mean capitalist domination and political and social injustice. It would require accepting environmental trade-offs. It would imply a "transhumanist" future of cyborgs, drugs, and genetic modification (52–53). Since these things are bad, he says, we must jettison all talk of determinism. But in fact he has it exactly backward: It is *because of* technological determinism that we are facing these dreadful outcomes. And only by facing up to this determinism can we give ourselves a chance to avoid them. I share his concerns; but embracing modern technology will only make them worse, not better.

The situation in present-day philosophy of technology is truly remarkable. There simply are no opposing views to Feenberg's 'uncritical critical theory'—no attempts at a deeper, metaphysical analysis, no truly *critical* arguments expressed, no challenges to the dominant technophilic orthodoxy. Serious contrarian views are not allowed to penetrate the broader discourse. There is no real debate, period. Philosophers have abdicated their role as insightful critics of the human predicament, and at least with respect to technology, have fallen in line with the dominant power structures of society. Without doubt, technology is winning—and everyone likes to be on the winning side. But it is particularly tragic when philosophers, as the embodiment of the wisdom and conscience of the human race, supplicate themselves to forces detrimental to human well-being. This cannot but have adverse consequences for our collective future.

If Feenberg and the constructivists are correct in assuming that humanity has decisive control over technology, then we ought to see, very soon, definitive resolution to some of our most urgent problems. Surely it is not in the long-term interests of the ruling capitalists and other "masters of technical systems" to destroy the global environment, as is happening at present. Surely they do not wish to see their army of workers suffer from the current wide range of mental and physical maladies. Surely they would like to have well-educated, wise, and critically thinking youth—all the better to promote the glories of economic prosperity. But if, in fact, these things are *not* happening, if the health of humanity and nature are *not* increasing, and if our attempts at technological education are *not* succeeding, then perhaps we must abandon Feenberg's naïve optimism and accept the alternative—that technology proceeds more or less autonomously, regardless of human concerns. Case studies and empirical data can be of great use here; these are the subject of the chapter to follow.

NOTES

1. It is clear that Habermas was responding to Ellul, as he cites him in a footnote in this very passage. See p. 127, n. 20.
2. A prerelease excerpt of the book was published in 1975 as the article "In Praise of Technology" (Florman 1975).
3. Borgmann (1984: 252, n. 7, n. 15) and (278, n. 4).
4. Citations will refer to the 1975 article, as this was his first presentation. The quotations are identical to those in the book.
5. For an interesting speculation, see the 2008 documentary *Life After People*.
6. See, for example, Feenberg (2010: 8–11, 39, 146). I elaborate below.
7. Native Americans of pre-Columbian North America were more dense than this, ranging from about 50 to 500 per hundred square miles.
8. Even the parks and trees are typically well manicured or otherwise technologically altered.
9. Which would have been readily available to Florman, in books such as Sennett (1969).
10. See his 1956a, 1956b. Cited in Krabbendam and van Os (2005).
11. Defined as urban areas of greater than 200,000 people.
12. See Krabbendam and van Os (2005: 797).
13. Meyer-Lindenberg (2013: 61) notes that Japanese researchers "have found structural changes in the pACC in patients with schizophrenia."
14. More on these ailments in Chapter 11.
15. See *Groundwork of the Metaphysics of Morals* (Kant 1996: 49).
16. World Database of Happiness (http://www1.eur.nl/fsw/happiness).

17. Schmookler's focus is on the work of Leslie White (1949, 1959), which I examined in the previous chapter, but his argument may be taken as a generalized critique of determinism.
18. A 'crisis' is literally a turning point or a moment of decision.
19. Though, oddly, the verb 'criticize' has as a primary meaning the more neutral 'to evaluate' and as a secondary meaning the more negative one of 'stressing the faults of.'
20. See J. Bohman, 'Critical Theory,' Stanford Encyclopedia of Philosophy (plato.stanford.edu/entries/critical-theory).
21. I use the term 'recent' advisedly; most of this book is a self-anthology of articles that were, at that time, between 2 and 18 years old. In particular, the passages that I quote here (to follow) date from 1992.

11 Three Case Studies

Metaphysics has consequences. In a pantechnical universe, technological development proceeds autonomously and deterministically. In its early years, technological determinism was relatively benign; it was characterized by complete dependence on human beings, and thus it served, or appeared to serve, human interests. As it evolved through its first phase, technology increasingly sacrificed human well-being for the sake of its own development. The latter part of phase one, which we are in today, is marked by little regard for humanity or nature. Technological advance becomes a self-serving end. This is in preparation for the second phase of determinism in which technology achieves full autonomy from humanity. This point—which some are calling the singularity—seems to be within a few decades of attainment.[1]

If the pantechnical thesis is valid, we should find evidence of progressively greater harm inflicted on people and on nature. Such harm does not appear suddenly but rather emerges slowly and subtly over centuries. Previously I defined the anthropogenic phase of determinism as consisting of three modes: an initial and lengthy period of foraging technology, a peak period marked by agricultural and early civilizational technologies, and from the year 1200 AD, a third period, one of decline—the present technological and industrial era.

The initial Foraging Era was one of slow but gradual progress; technological harms were thus minimal. Agriculture inaugurated our peak period and, as such, foreshadowed our decay; where we are no longer ascending, we are declining. It is for this reason that agriculture was, at once, our greatest success and our greatest mistake.[2] A few thousand years into the Human Era, around 6000 BC, we learned how to work metals. This took us another step forward but introduced a range of potential and actual harms. For centuries, the damage was generally localized and limited in severity. Once we crossed the threshold into the Technological Era, technology accelerated, and so therefore did the adverse effects. The rapid advances of the Industrial Revolution brought proportionate decline in human and environmental well-being, a situation that has continued through the present day.

Empirical data are an essential aspect of a synthetic metaphysical approach and are particularly relevant to the study of technology. It is notable, therefore, that few of the historical critics cited in Chapters 7 and 8 supplied supporting data. Of course, for much of history it would have been nonexistent; but even into the late 20th century, when such data did exist, few found it necessary to employ them. Ellul was the notable exception, and his three technology books offer much in the way of empirical data to defend his thesis.[3] But the others were in large part satisfied to perform theoretical or otherwise abstract philosophical analyses, relying ultimately on the intuitive force of their argument to sway the reader. This lack of data is unfortunate because, as I have said, if technology is causing harm, we ought to be able to quantify it, track it, and anticipate future trends. Furthermore, this harm and the underlying technological causes will advance, on my thesis, despite our wishes and efforts to the contrary. This happens because the process of technological progress is ultimately beyond our control; we are the means but not the ends.

CASE ONE: ENVIRONMENTAL DAMAGE

My primary concern in this chapter is human well-being, but the environmental situation warrants an examination—in fact, a far more detailed one than I can conduct here. At present I can only offer some brief comments. Two points hang over this discussion. First, all man-made damage is technological damage. Unaided humans have little power to adversely affect the natural world. With the advent of energetic tools, we began to significantly damage nature, and as the power of those tools increased, so did the damage. Second, everything we know about evolution suggests that nature has been continually developing toward both greater biodiversity and stability as organisms evolve toward increasing sensitivity and complexity. The global ecosystem was on a directionally upward trajectory for some 4 billion years—until the appearance of the technological animal.

At that time, for the first time in history, one species gained the power to detrimentally affect virtually every other on the planet. We like to think that we are making things "better" by draining wetlands, converting "useless" land to agriculture, and otherwise maximizing the use of the Earth. In fact, we are drawing down the biocapacity of the entire planet for the sake of increasing our numbers and our material standard of living. Prior to mankind, the Earth was maximally efficient; it was producing as much diversity, food, and energy as sustainably possible at each point in planetary history. To paraphrase Leibniz, the Earth was the most perfect world possible. Any large-scale intervention by humanity could only reduce this perfection; the larger the intervention, the greater the fall. Through our technology, we intervened in the most energetic manner possible. We are now seeing the consequences.

Consider our impact from prehistoric days, when we numbered less than 10 million globally. Our oldest technologies, crude stone hand tools dating back more than 2 million years, did little more than improve our hunting efficiency; they had a small overall effect.[4] The first dramatic impact came from the controlled use of fire, circa 1 million BC. Hunter-gatherers, for example, set landscape fires for a variety of reasons, including hunting, pest control, and territorial management. Certainly such events had a major effect on the affected ecology, altering species diversity and vegetation growth patterns. But for the most part it was localized and recoverable damage, given that fire is a naturally occurring phenomenon and that human numbers were low everywhere.

The earliest recorded account of environmental damage is in Plato's *Critias*. In a fascinating passage, he describes the situation in Athens "nine thousand years" prior to his time—which, if taken literally, would place events around the earliest shift from hunting and gathering to agriculture.[5] Plato describes the dramatic loss of topsoil, stating that "Attica of today is like the skeleton revealed by a wasting disease" (111b7). The once dense forests are gone, honey bees are declining, and the natural springs have disappeared. This again was a localized effect, but it shows the damage that a small number of people with primitive technology can do over time.

In Plato's day, global population was around 100 million. By 1000 AD, we were at 300 million; by 1500, 500 million. We hit 1 billion around the year 1800 and 2 billion in the mid-1920s. Today we stand at 7.1 billion, on our way to 10 billion by the mid-2040s if the worst projections materialize.

Understandably, most all people wish to live a luxurious Western lifestyle, but the effect on the planet is considerable. Through our numbers and our technology, humans now appropriate as much as 40% of the net primary productivity of the Earth, which is the total value of surplus plant energy at the base of the global ecosystem.[6] As our numbers increase—by some 210,000 every day—this can only get worse. We have claimed 83% of the planet's surface for our use and are cultivating nearly one quarter of it.[7] Our domesticated food animals—the direct products of industrial technology—now comprise about 20% of total terrestrial animal biomass.[8] We clear-cut or otherwise deforest approximately one football field of land every two seconds.[9] Inevitably, such impact is driving other species to extinction; roughly 20% of all vertebrates and 40% of all amphibians are under imminent threat of extermination.[10] By most assessments, this is bad news indeed.

Looming above all are the potentially devastating effects of global climate change. I offered some thoughts on this topic in my discussion of Borgmann in Chapter 6, but a few additional points are worth mention here. The degree and cause of warming have been debated for years, but gradually over the past decade, scientific consensus has concluded that the warming is real and is largely due to human activity. Already in 2005, the head of the U.S. National Academy of Sciences was testifying that "nearly all climate scientists" agree that the cause of warming comes "mostly from the burning of [fossil] fuels."[11] Later that same year came reports that the Siberian permafrost was melting, threatening to cross a "tipping point" of positive feedback due to methane release.[12] Loss of Arctic sea ice was also cited as providing yet another potential mechanism for the same feedback effect. By 2006, prominent scientist James Lovelock—of Gaia fame—was declaring the situation near hopeless: the Earth's climate "is soon to pass into a morbid fever that may last as long as 10,000 years." He added that "before this century is over billions of us will die," and the remainder will be huddled at the poles, struggling to survive.[13]

For at least 10,000 years prior to the industrial age, atmospheric carbon dioxide levels were near 275 parts per million (ppm). From around 1850 they began to rise and by 2008 stood at 385 ppm. The latest data indicate concentrations in the Arctic of 400 ppm, with a global average of 395—levels not seen on this planet for at least 800,000 years. And they are increasing at a faster rate than even the worst-case scenarios of just a few years back.[14] At this pace, global temperatures could rise between 5.2 and 7.0 degrees Celsius by the end of the century, along with a concomitant sea level rise of some 2 meters. The fifth Intergovernmental Panel on Climate Change (IPCC) report of 2013 stated with 95% confidence that human technological activity is the primary cause of warming and that humans have been the sole cause of all warming in the past 60 years.

The consequences from just this one unanticipated side effect of technology will be profound. A UN–funded report from 2009, *The State of the Future*,[15] warned that, with such levels of climate change, "much of civilization will collapse." In a similar vein, Brown (2009) argues that climate-induced food shortages alone could have the same effect:

> For most of us, the idea that civilization itself could disintegrate probably seems preposterous. . . . Our continuing failure to deal with the environmental declines that are undermining the world food economy . . . forces me to conclude that such a collapse is possible.

This would be the ultimate irony: a technological civilization that so damages the global ecosystem that it ends up destroying itself. Perhaps a kind of cosmic justice is at work after all.

It is hubris to think that we can improve on nature. Prior to our intervention, nature was optimally productive. The system functioned as an organic whole, with all parts coordinated and working toward the ongoing evolution of planetary life. With our technology, we have been able to disrupt the organic wholeness of the planet for our own short-term benefit. But this is simply stealing from other life forms and, in the long run, engenders a countereffect that more than offsets the temporary gain.

At this point in time, our options are limited. It is likely too late to circumvent the major effects of climate change; we can only prepare to survive and perhaps to mitigate the damage. Longer term, there seems to be just one viable plan: work to restore the planet, as closely as possible, to its evolutionarily stable state. This would mean returning to the conditions prior to the first energetic intervention by humanity—in other words, the state of the world circa 10,000 BC. It would mean restoring native species to their previous conditions, cleaning up any residue of industrial toxins, reducing carbon emissions to as low as physically possible, and returning vast areas of the Earth to something like a wilderness state. And soon—the longer we wait, the larger the task, and the closer we edge to our own extinction.

CASE TWO: EDUCATIONAL TECHNOLOGY

For several years now, governmental and corporate leaders have impressed upon the public the need to utilize educational technology (ET)[16] in primary and secondary schools to optimize learning outcomes. Nearly two decades ago, President Bill Clinton stated that "every classroom in America must be connected to the information superhighway." His goal at the time was to connect "every classroom and every library in the entire United States by the year 2000."[17] And he largely succeeded: At that time, 65% of schools had access; by 2000, the figure was 98%. The shift in individual classrooms was even more impressive. In 1996, just 14% of classrooms had access; in 2000, it was 77%. By 2005, 100% of schools had access, as did 94% of all classrooms. During the Clinton years, federal funding for technology in schools rose from $23 million in 1993 to $872 million in 2001—an increase of 3,600%.

The trend continues today. President Obama aims to connect every individual child to the Internet via high-speed wireless devices. "I am directing the FCC . . . to begin a process that will connect 99% of America's students to high-speed broadband Internet within five years"—that is, by 2018. He spoke at a "model school" in North Carolina, where, "starting in third grade . . . every student in the district gets a laptop and high-speed Internet in the classroom." Obama's administration will "partner with private companies" to lay the infrastructure, which will mean "a big new market for private innovation."[18] His model school ranked second in its state in terms of student achievement, and the implication is that all schools that supply advanced ET will see similar exemplary results.

All this is consistent with my pantechnical metaphysics. It predicts that technology will be relentlessly promoted in all areas of society, particularly those involving schools and youth—which represent the future of the technological society. It further predicts that such developments are counter to the best interests of the people affected: in this case, school-age children. Let us see what the data show.

To build a rational case for ET, it would be necessary to show three things: first, that it produces clear evidence of improved learning—better retention, increased analytic abilities, improved critical thinking, greater creativity; second, that it does not cause adverse or harmful side effects; and third, that it functions cost effectively as compared with other pedagogical alternatives such as lower class size. All three are necessary conditions. As it happens, ET fails on all three counts.

The focus of this section is the first issue; the second is addressed in the section to follow. I will not say much on the financial aspect, except to note that the United States spends well over $10 billion per year on primary and secondary school technology. A single school district—Los Angeles Unified—alone has a $1 billion technology plan, which works out to about $1,500 per student.[19] Cost studies typically underestimate ongoing expense requirements for such things as hardware and software upgrades, training, maintenance, and infrastructure; these now exceed $5,000 per computer per year. Despite the costs, school bonds for technology upgrades are routinely approved; citizens generally view ET as a prestige item, authorizing such spending without even a basic understanding of the tradeoffs involved. In studying school district plans over several years, I have yet to see a true cost-benefit-harm analysis performed or even discussed.

ET is thus an easier sell to the public than just about any other aspect of school spending, despite the fact that it has strikingly little academic benefit to its credit. It is worthwhile examining some of the data on this matter. For years, the public have been told that ET is a cutting-edge pedagogical tool that will raise test scores and improve American global competitiveness. One state's "Freedom to Learn" program listed several expected benefits, including: "improved student motivation," "improved student learning," high state-wide test scores, and "improved teaching."[20] A follow-up ET plan stated that "educational technology is addressed in this plan as a powerful means of improving student learning."[21] The National Education Technology Plan of 2004 was titled "Toward a New Golden Age" and spoke of "the often dramatic improvements that we are beginning to see in student achievement" (p. 7). The 2010 plan calls for a "revolutionary transformation" in education, stating that ET will be "pivotal in improving student learning" (p. 7). Technology will "enable 24/7 learning" and prepare our young "to be more productive members of a globally competitive workforce."

Based on such statements, one would assume that ET actually has produced a significant improvement in learning outcomes, test scores, and competitiveness. But this is not the case. Data questioning ET's effectiveness were becoming known as far back as the early 1990s. One important early meta-analysis by Kulik (1994) examined 97 studies for effect sizes on learning improvement. At first glance, the data look encouraging: the average effect size from the group was 0.35, indicating a significant positive correlation.[22] But a deeper look reveals some questions. Of the 97 studies, 71 were performed prior to 1984—that is, were more than 10 years old at the time. Also, 16 of these studies showed negative effect sizes, indicating *worse* outcome after the introduction of ET. Of six computer-aided instruction applications, the highest-performing one—Logo—showed a high degree of inconsistency, with a few studies showing very high effects and the majority very low ones. Kulik suggests that "the strong positive outcomes may have been produced by unusual evaluator rapport with Logo students during testing, or even unconscious bias" (1994: 21). Of the other five applications, "the record is also unimpressive." "Computer-managed instruction, for example, seldom produces significant positive gains" in either elementary or high schools. In fact, their effectiveness "seems similar to the record compiled by diagnostic and prescriptive systems that use only paper-and-pencil and printed materials". Finally, Kulik observes that the longer-term studies show lower effects than the short ones—suggesting that any improvements are ephemeral.

Soon-to-follow research supported the view that ET gains are ambiguous or nonexistent. One model school system in Romulus, Michigan, conducted a "$24 million experiment that came up short." The school superintendent said that "the MEAP [standard test] scores had to shoot up and that didn't happen."[23] Even worse, scores at four of five elementary schools declined from the previous year. Other studies confirmed this situation. The *San Jose Mercury News* examined 227 California schools in 1995.[24] Overall, their analysis "showed no strong link between the presence of technology—or the use of technology in teaching—and superior achievement." And a number of schools with more technology underperformed those with less.

Sensing the futility of academic improvement, an executive director of the Computer Learning Foundation stated that "we view [achievement gains] as almost irrelevant." What matters, she said, is preparing children for the workplace. In 1997, the *Los Angeles Times* reported that "technology is still a promise, not a panacea."[25] "Hefty investments in school computers have, thus far, produced few academic gains at most schools," they wrote. One former state education official, William Rukeyser, said, "Basically, we're making some very expensive, long-range decisions in the absence of any evidence that shows this is the solution to the problems of public schools." In a revealing comment, he added that "the nearest thing I can draw a parallel to is a theological discussion." "There's so much an element of faith here that demanding evidence is almost a sign of heresy."

Other skeptical voices soon appeared. Koblitz (1996), for example, argued that computers should not be the focus of mathematics education, citing four areas of concern: drain on resources, bad pedagogy, anti-intellectual appeal (meaning a focus on gadgets rather than on thinking), and "corruption of educators" (due to profit motives). Cuban (1996) wrote that "the persistent dream of technology driving school and classroom changes has continuously foundered in transforming teaching practices." Matthew (1997) found "no statistically significant difference in reading comprehension" when third graders used interactive CD ROM storybooks versus print editions. Other criticisms came from various quarters, including Pepi and Scheurman (1996), Baines (1997), Sava (1997), and Oppenheimer (1997).

ET defenders had their replies. Reeves (1998: 50) claimed that the skeptics overlooked "hundreds of studies that support the impact of media and technology on achievement"—though he neglected to cite any of these. Others blamed the teachers for not being creative enough in their ET usage, or the generally immature state of ET itself; newer products, they assured us, would yield demonstrable gains. This latter claim is a perennial favorite—'if the present technology does not show results, don't worry, the new stuff will.' It is a conveniently unverifiable proposition.

Criticisms continued into the new century. McKenzie (2000) noted that "we have no evidence that widespread introduction of networked computers into all classrooms enhances the learning of students." "To the contrary, we have mounting evidence that the mere presence of equipment accomplishes very little." Regarding learning outcomes, he added that "there is a serious lack of credible studies showing how student learning may change as networked technologies are introduced." The same year, a major report was issued by the nonprofit group Alliance for Childhood (Cordes and Miller, 2000). Their study, *Fool's Gold*, summarized much of current research and added new skepticism regarding possible adverse health effects—to be discussed in the following section.

Despite such warnings, school boards pressed ahead. *Time* magazine reported on the growing trend toward "a laptop for every kid," citing developments in Connecticut, New York, and Maine.[26] And the tech industry responded as well. The Software Information Industry Association (SIIA) issued a press release on its newest report that "clearly demonstrates that ET has increased student achievement, enhanced student self-concept and attitude about learning, and improved interaction involving educators and students."[27] But the report supplied neither data nor the reports to back up the claims. Tellingly, SIIA president Ken Wasch added that "the simple addition of computers to schools does not directly translate to higher test scores, and never will."

Soon thereafter, newer and more detailed studies emerged, showing further adverse results. Angrist and Lavy (2002) examined the extensive rollout of computers in Israeli schools. They wrote that "there is no evidence . . . that increased educational use of computers actually raised pupil test scores" (737). The results were generally close to zero, except for *negative* outcomes in fourth- and eighth-grade math students. They concluded that "significant and ongoing expenditure on ET does not appear to be justified by pupil performance results to date" (761), adding that, by comparison, "traditional inputs—reductions in class size and increased teacher training—have had substantial achievement benefits."

At about the same time, a new meta-analysis by Waxman and colleagues (2002) was released, showing a "small, positive" effect on learning. But there were problems here as well. They found a "small, negative effect" on "behavioral outcomes," suggesting adverse psychological consequences. Of 138 effect sizes, fully 30% (42) were either zero or negative. And perhaps most troubling, they found that many of the studies themselves were methodologically suspect. For example, only 20% used the most rigorous approach: a randomized pretest–posttest control group design. Often the positive results were "merely descriptive in nature" or "anecdotal." They noted, unsurprisingly, that unpublished Web-based studies "have dramatically higher effect sizes than published reports." And they found evidence that "many Web-based reports are sponsored by agencies that have obvious conflicts of interest associated with the results." Finally, they cited another study showing that, in more than half the cases, "the researcher was hired to do the research by either the vendor or school district involved in the study"—again, more conflicts of interest.

Then in 2004 the bottom began to fall out for the ET cause. Fuchs and Woessmann (2004) examined nearly 175,000 15-year-old students across 31 Organization for Economic Cooperation and Development (OECD) countries, assessing student skills in math, reading, and science. The study also queried school and home learning environments as related to computers and ET. With respect to student performance, a controlled multivariate analysis of the data shows overall a "negative [correlation] for home computers, and an insignificant [one] for school computers."

The data assessed student proficiency versus both "availability" and "use" of computers, both at home and at school. Availability at home was correlated, universally, with *decreased* performance. Those having one computer saw a reduction of 5.9 achievement points (AP) in math and 6.6 in reading. Those having more than one computer experienced decreases of 13.9 AP in math and 15.6 in reading. Regarding availability at school, there was no statistical difference between schools "strongly lacking" computers and those with "no lack at all," in either math or reading. In other words, "computer availability at school is unrelated to performance."

As for computer use at home: Use of certain applications, including Internet, email, and educational software, does correspond to slightly higher performance (4 AP in math, 5 AP in reading). However, this result "may simply reflect that more able students tend to be more likely to get Internet access and educational software at home." Regarding use at school, those using computers or Internet several times a week showed a *marked decline* in performance, significantly below those who never used a computer in school at all (decrease of 7 AP math, 5–10 AP reading). The authors speculated that this result "might be explained by a *true negative effect* of excessive computer use at school," that is, computers directly causing reduced performance. Their overall conclusion was not encouraging:

> Thus, the evidence so far does not suggest that computers have a substantial impact on the economic and educational outcome of students. . . . Despite numerous claims by politicians and software vendors to the contrary, the evidence so far suggest that computer use in schools does not seem to contribute substantially to students' learning of basic skills such as math or reading. (4)

At the same time, Rouse and Krueger (2004) performed a detailed study of a popular reading software package, Fast ForWord. They stated the following: "Although schools across [the U.S.] are investing heavily in computers in the classroom, there is surprisingly little evidence that they actually improve student achievement." The authors conclude that, although the software may improve some aspects of language skill, "it does *not* appear that these gains translate into a broader measure of language acquisition, or into actual reading skills." Their study demonstrates, among other things, the exaggerated claims made by the software developer, Scientific Learning Corporation (SLC): "1.5 to 2 years of progress in reading skills after just 6–8 weeks of training." Instead, the authors found "no statistically detectable effect" of the software when doing a controlled study.

Perhaps most troubling, SLC themselves admit that their software has the ability to literally alter the brain physiology of students using it. Pointing to a 2003 MRI study on dyslexic children, SLC notes that "activation of the children's brains fundamentally changed" after using Fast ForWord for 100 minutes per day for eight weeks. Interestingly, they consider this brain change, which became "more like good readers," to be a benefit. But one can easily adopt another view: it suggests that powerful physiological effects are at work on children with developing nervous systems.

In the meantime, some school districts were becoming disillusioned. The Liverpool, New York, school district decided to phase out student laptops in 2007 after having them in place since 2000. "After seven years," said the school board president, "there was literally no evidence it had any impact on student achievement—none."[28] In fact, he called it "a distraction to the educational process." One exasperated parent exclaimed, "I felt like I was ripped off."

Then came a devastating report by the U.S. Department of Education: "Effectiveness of Reading and Mathematics Software Products" (Dynarski et al. 2007). The authors examined 16 leading ET packages, ranging from first to ninth grades, covering 132 schools and 9,400 students in 33 districts—one of the largest studies ever performed. The results were concise and emphatic: "Test scores were not significantly higher in classrooms using selected reading and mathematics software products." This was striking, given that the ET corporations put forth their most successful products, those with the best reputation for improved performance. The suppliers even nominated the school districts that they believed would be most amenable to improvement. Even with all this advantage, the results were the same: "no statistical significance."

At the first-grade level, the researchers looked at five effect sizes: four were slightly positive, one ("sentence reading") was negative, overall no significant difference. For fourth-grade reading, they examined three effect sizes, two small positive and one zero; overall "not statistically significant." Two effect sizes ("procedures" and "problem solving") were evaluated for sixth-grade math scores, and again no gain. Results of the algebra products were particularly surprising: two of three effect sizes were slightly *negative* ("concepts" and "procedure"), and overall, as before, not statistically significant. Note: "not statistically significant" is *very* significant in reality; it tells us that the highest-rated software products did nothing for the students, potentially even reducing their performance.

As expected, the report drew criticism from the ET industry. Dynarski and colleagues were charged with "lumping together the outcomes from many different products" and criticized for testing effectiveness after only one year of use. This issue was resolved in 2009, when a follow-up study was released (Campuzano et al. 2009), looking at individual products and two-year progress. The results of the first study were largely confirmed: *no statistically significant gain.* For both first- and fourth-grade reading products, "the measured product effect in the second year is not statistically significantly larger than the effect in the first year" (with first grade shifting from slightly positive to slightly negative). For sixth-grade math, a negative first-year outcome became *more negative* in the second year. The algebra score shifted from slightly negative to slightly positive. Finally, the researchers looked specifically at the effect of 10 leading products on increasing test scores: "One product had a positive and statistically significant effect. Nine did not have statistically significant effects on test scores. *Five of the insignificant effects were negative,* and four were positive" (xxiv; italics added). Given the gloomy outcome, Dynarski tried to remain optimistic. "There might be a lot of positive things going on in the classroom," he said, "But it's not leading to the kind of results that people want, which are better test scores."

The technophiles of the ET world are sorely disappointed. Two decades of intense rollout of the latest devices and software, and nothing to show for it. The *New York Times* ran an insightful story in late 2011 titled "In Classroom of Future, Stagnant Scores."[29] "In a nutshell," they write, "schools are spending billions on technology, even as they cut budgets and lay off teachers, with little proof that this approach is improving basic learning." The article quotes a leading ET investor, Tom Vander Ark: "The data are pretty weak. It's very difficult when we're pressed to come up with convincing data." But despite all this, school districts across the country cannot help themselves; more technology is always demanded and always forthcoming.

It has become so bad that academic and governmental leaders are bending over backward to turn lemons into lemonade. University of Michigan hypertechnophile Elliot Soloway blithely ignores the data, proclaiming the present era as the "Age of Mobilism." "Smartphones: that's what the future is," he sagely declares.[30] Karen Cator, director of ET at the U.S. Department of Education, offers this remarkable statement: "In places where we've had a large implementing of technology and scores are flat, I see that as great."[31] Great? "Look at all the other things students are doing"—like Internet research, organizing their work, learning to use writing tools, learning to collaborate; surely these justify the billions of dollars spent, she suggests. But she forgot to add a few other associated activities: plagiarizing, surfing porn sites, cyberbullying each other to death, "sexting," game playing, and

Facebooking to a state of addiction. . . . Yes—the irrationality of the technophiles comes full circle. Technology must progress, no matter the data, no matter the cost, and no matter the harm. We are a technological society, after all. *We have no choice.*

Meanwhile, performance continues to stagnate, if not decline, and the costs continue to mount. In September 2012, it was reported that SAT reading scores hit a 40-year low; math scores were a minuscule 1% higher over the same period, and at the same level as in 2000. Virtual (online) schools have been popping up across the country, to big promises. Typical, though, is the situation in Tennessee, where state lawmakers recently blasted officials of the Tennessee Virtual Academy for its "abysmal test results." "It seems like we're using best practices," said one representative, "but we're still getting poor results."

And out west, an undaunted Los Angeles Unified school district recently awarded Apple a $30 million contract for $700 iPads for all 662,000 students. In June 2013, the school board voted unanimously to authorize the two-year plan. Citizens in LA were evidently unconcerned that school Superintendent John Deasy had previously appeared in Apple promotional videos. Sadly, things did not go as planned. In September it was reported that the school was recalling the first wave of 2,100 tablets. The reason? "Students skirted security measures that were intended to block free browsing of the Internet." In other words, it was Internet porn and video games 24/7 in LA. As of the end of the month, only about two-thirds of the devices had actually been returned—hard to put that genie back in the bottle.

But Americans are nothing if not unfazed by adversity. Guildford County in Greensboro, North Carolina, launched a rollout in June 2013 of 15,450 tablet computers for 18 middle schools. The supplier was a company called Amplify, which advocates the idea that American education is "ripe for disruption." Reporting on the situation there, Carlo Rotella notes that schools are seeking to add to the "seven-plus hours per day that the average middle schooler spends gazing at a screen outside of school."[32] As always, "companies with vested interests are pitching themselves as the solution to the country's educational problems"; but he adds, quoting one source, "we don't have the research proving it's true." But not to worry, say the good folks in Guildford County, who will be "looking for higher test scores in two or three years"—as will we all.

Thus we see that educators all across the country are bound and determined to promote educational technology despite the costs and despite the perennial lack of supporting data. It is a fine example of technological determinism at work: no one is forced to do such things, and yet we do them all the same, in spite of compelling reasons to the contrary. When we add the specter of health risks to children—addressed next—the entire situation becomes truly bizarre. Every rational argument supports a dramatic reduction in ET, and yet we relentlessly plow ahead. The same futile reasons continue to rain down: competitiveness, college readiness, "personalized learning," "it's a technological world out there"—endless variations on the same theme. Once again, this situation demands a metaphysical explanation; unlike other contemporary philosophies of technology, the thesis of the Pantechnikon provides this.

CASE THREE: RISKS TO HUMAN HEALTH

Humans are physical beings, and mental beings, and moral beings. In all three dimensions we may flourish or suffer. Technology has a definite impact on all three. Given that it has progressed substantially in recent years and that it allegedly is under our control and works for our best interest, we ought reasonably to see substantial improvements in human well-being on all three fronts. In fact, this could be regarded as a test of technological determinism: a technology that we control will clearly work toward our betterment and, conversely, a technological system out of control will work to our detriment.

When we examine the data, we find, for the most part, significant declines in all three of these areas. Furthermore, we see new problems appearing that did not and could not exist in previous decades. This is consistent with the determinist claim that technology works toward its own ends. And it is consistent with the pantechnical thesis that such determinism is in the process of transitioning from an initial and relatively benign phase to a much more dangerous second phase, wherein human well-being becomes virtually irrelevant to the advancement of the system.

Let me begin with a quick overview of the long-term American health situation. First the good news; it will be brief. Adult smoking rates are down, from around 42% in 1965 to about 18% today. Life expectancy is up to an all-time high of 78.7 years, from 70 years in 1960. Infant mortality is at a record low of 6.14 per thousand live births (0.614%), down from 9.2 in 1990. The five-year survival rates for certain cancers are up. And the 50-year mortality rate trend is significantly lower for two major causes of death: stroke (down 76%) and heart disease (down 66%).[33] Beyond these few items, little is encouraging.

But even this good news comes with qualifications. Keeping people alive is one thing; maintaining a high quality of life is another. Vitality statistics tell us nothing about the quality of people's lives. If people are suffering, long life is not a virtue. And in fact, much of what follows suggests that life quality is declining. Second, such numbers partly explain why the American population is soaring, from around 315 million at present to an estimated 450 million by 2050—a more than 40% increase. With greater numbers of high-consuming Americans comes inevitable competition for resources, increased wastes and pollution, loss of open space, and overcrowded urban conglomerations. In a technological society, good news is rarely unconditional.

And then there is the bad news—which, unlike the good, could fill a small volume in itself. With an aging population, first of all, mortality rates for age-related afflictions have naturally increased: Parkinson's disease by 340% since 1960 and Alzheimer's disease by an astounding 8,330%, just since the late 1970s.[34] Other diseases show a stubborn persistence over the past few decades. Diabetes mortality rates are virtually unchanged since 1950; rates for liver disease are only slightly lower over the same period, while those due to chronic respiratory diseases are up substantially. From a technological perspective, this is problematic: we have endlessly improving diagnostic and treatment options for all these diseases, and yet they persist or increase. If technology really was working for our benefit, we should expect to see improvements in all areas of health. The fact that we see no gain in several areas means, in reality, a net loss: we incur the cost and risk of technology for no benefit.

Of particular interest is cancer. First of all, we tend to forget that cancer is a modern, *technological* disease. It is not natural for humans or other animals to get cancer at all. Capasso (2005: 3) notes that "wild living mammals' neoplasms are rare." The same is true of premodern human populations. Of thousands of Neanderthal bones on record, only one indicates a possible cancerous lesion. This near cancer-free situation held true for much of early human civilization. As Capasso observes, "The prevalence of primary and metastatic cancers was very low everywhere throughout all of pre-Medieval times" (8). He cites the work of Strouhal, who argued that "malignant neoplasms in ancient European populations . . . are very rare." "Only after the Middle Ages was there a significant increase in the number of malignancies observed." Certainly an older population accounts for some of the increase, but it is less a factor than one might assume. Capasso states that "the aging of the human population is not sufficient to explain the tremendous increase in prevalence of cancer in human populations over the past century" (10). He concludes that our modern lifestyle has degraded our condition, noting that high cancer rates in domestic animals that share "our almost completely artificial environments . . . seems to confirm this picture." Overall he is led

> to the conclusion that so-called "civilization" is associated with the increase in the prevalence of cancer over the past century. In addition, paleopathological evidence seems to confirm that a greater prevalence of malignancies is associated with highly civilized populations, high social aggregation levels, and a prevalence of indoor habits. (11)

In other words, with technological society.

A more recent study by David and Zimmerman (2010) reiterates this view. They note that "there are few, often uncertain, references to cancer in Egyptian and Classical literature" (728), though it was identified and studied to some degree. As to the reason for this situation, they write as follows:

> Carcinogenic environmental factors have been linked to up to 75% of human cancers, and the rarity of cancer in antiquity suggests that such factors are limited to societies that are affected by modern lifestyle issues such as tobacco use and pollution resulting from industrialization. (731)

At present, and despite an intense focus on the disease and a suite of new treatment options, we are making little progress. In 2010, it was reported that 41% of all Americans will eventually get cancer, and 21% will die from it.[35] Surprisingly, overall cancer mortality rates have been virtually unchanged for the past 50 years.[36] This suggests a strange situation: that modern technological society is *producing* cancer at the same rate that it is curing it. We make no gains because progress in treatment depends upon advanced technologies, which, in turn, and in other indirect ways, produce yet more cancer. We are running faster on the treadmill but gaining no ground. Technology benefits—as always—but we do not.

Then consider another disease of civilization: obesity.[37] This is an important indicator because it is associated with a significantly higher risk of several other major diseases and maladies, including diabetes, heart disease, stroke, hypertension, cancer, kidney disease, and arthritis. As with cancer, obesity was virtually nonexistent in primitive and premodern cultures. The word 'obesity' did not even exist in the English language until the 1600s and was not recognized as a health problem until the 1700s—aligning perfectly with the rise of industrialism.[38] By 1962, around 13% of American adults were obese. As of 2004, the figure had risen to 30.6%. According to the latest data, it is now at 35.7%, and if current trends hold, it will hit 50% by 2030.[39] If we consider the broader category of "overweight or obese" (BMI > 25), this presently encompasses more than two thirds (69%) of all adults. Looking just at children and teens aged 10 to 17, we find that well over one third are overweight or obese.

Diabetes is especially worrisome due to its chronic nature and long-term detrimental effects; it arises not only from obesity but also from excess sugar consumption. There were about 5.8 million Americans diagnosed with diabetes in 1980. In 2006, the number was 18.2 million and in 2010, 26 million. Current estimates are projecting as many as 44 million cases by 2034. And more than twice these numbers are considered to be prediabetic—people who face elevated risks of other complications and could easily slip into a full diabetic condition.

There are other troubling signs. New diseases have emerged in recent years, including HIV/AIDS, SARS, bird flu, and "mad cow" disease. The effectiveness of antibiotics is diminishing, resulting in such chronic problems as methicillin-resistant staph (MRSA) and "extensively drug resistant" tuberculosis (XDR-TB). Hospital-acquired infections now kill more than 100,000 people annually. And Americans generally seem to be sicker than ever; we receive some 3.7 billion prescriptions per year, or an average of about 12 for every man, woman, and child—nearly double the figure of 1992.

All these issues could be classified as physical disorders. In addition, there is another whole class of health problems arising with the mind. By all accounts, mental disorders are widespread and growing in American society. Journalists have taken to describing the situation as an "epidemic of mental illness."[40] For example, Kupfer and Regier (2010: 1974) report that, among people in a primary healthcare setting, "approximately 30% to 50% of patients have prominent mental health symptoms or identifiable mental disorders." Similarly, a random survey of adults found 46% to have met criteria for at least one mental illness during their lives.[41]

Depression is a major problem; currently 11% of Americans over age 12 are taking antidepressants. Among the white population, the figure is 14%, and for women ages 40 to 60, it reaches 23%. Overall the rate of antidepressant use has increased nearly 400% over the past 20 years.[42] Once again we are dealing with a relatively new phenomenon. Harris (2001: 16) notes that, for the ancient Greeks and Romans, "depression was largely absent." They certainly experienced anger, rage, frustration, faintheartedness, and related emotions, but apparently nothing like the persistent depressed state that afflicts so many in the modern world. One telling sign, he says, is that Seneca's discussion in *De tranquillitate animi* "is accompanied by an express admission that there is no name for this condition" (17). "In antiquity, there being no depression, or not much," mental frustrations were directed elsewhere.

One consequence of depression is suicide. Fromm (1955: 150) recalls statistics showing dramatic increases throughout the latter stages of the Industrial Revolution. From the mid-1830s to the period 1890 through 1910, England, Sweden, Prussia, and France experienced rate increases of 77%, 127%, 140%, and 355% respectively. Suicide rates in the U.S. have held relatively steady over the past 60 years, though the period between 2000 and 2010 saw an increase of 20% for the overall population and of nearly 30% for middle-aged men;[43] this period, notably, corresponds with the recent rapid growth of information technology, as I explain below.

Autism is another area of great concern, one that researchers are also calling an epidemic. In 2001, the official estimate was that about 0.55% of children were afflicted with autism spectrum disorder (ASD). The following year it rose to 0.66%. In 2009, the figure increased to 1.1%—double that of just eight years earlier.[44] In 2013, the CDC stated that the parent-reported rate was up to 2.0% of all children, and the rate for boys alone was 3.23%.[45] This is not inconsistent with international data; for example, it was estimated in 2011 that the South Korean rate was 2.64%.[46]

And there is a variety of other adverse trends. Medical office visits resulting in antipsychotic prescriptions increased dramatically over a recent 10-year period, rising from 0.24% to 1.83% for children, 0.78% to 3.76% for adolescents, and 3.25% to 6.18% for adults.[47] Bipolar disease has skyrocketed in children and adolescents, rising from 25 to 1,003 diagnoses per 100,000 between 1994 and 2003—an astounding 4,000% increase in less than one decade.[48] And attention deficit hyperactivity disorder

(ADHD) is on the increase, at an accelerating rate. From the mid-1990s through mid-2000s, diagnoses grew at a 3% annual rate; more recent data point to a 5.5% yearly growth rate. Today 11% of all school-age children have been diagnosed with it, including almost 20% of high school boys. Revealingly, sales of ADHD drugs more than doubled between the years 2007 and 2012 alone.[49]

THE TECHNOLOGICAL CONNECTION: INFORMATION TECHNOLOGY AND HEALTH

American health is in a perilous condition, and we inevitably look to new technology to cure us. New medicines, new diagnostic machines, new drugs, new forms of surgery—technology is promoted on all fronts as our medical salvation. And yet we scarcely understand the causes of these many ailments. Often, all we can do is treat the symptoms. In a rational society, the medical profession would work diligently to determine root causes and then do everything possible to remove or mitigate these. Everyone realizes that prevention is superior to cure, but we do almost nothing to follow this dictate.

There are at least two reasons for this. The first is money. Doctors, hospitals, pharmaceutical companies, and health care deliverers all profit from illness, not health. There is little or no financial incentive to prevent disease. The second reason is technology—in two ways. The quest for new treatments, new drugs, and new diagnostic equipment is a major driver of technological advance. The health care industry is a $2.7 trillion-per-year enterprise, and it is capable of funding a large array of technical developments. Sickness is good news for technology.

But more than this is the fact that technology itself lies at the root of many of our problems. From toxic chemicals to high-speed automobiles, from unhealthy artificial living conditions to workplace hazards, from sedentary lifestyles to industrially produced junk foods, technology is a leading cause of human suffering. Given this, the appropriate fix—the true remedy—is to reduce or eliminate the technology itself. But this is unacceptable in a technological society. So we pretend that the cause is something else, or a function of "human choice," or an "unavoidable circumstance"; and we then look to technology to save us—from itself.

There are endless ways in which technology directly or indirectly causes health problems, but here I want to focus on the effects of information technology (IT): computers, email, cell phones, and the Internet. In theory, nothing should be less harmful than IT. It involves no dangerous moving parts, no ingesting of toxic chemicals, and no direct physical effects at all. It is purely "white collar": office environment, clean, shiny, and new. It is, after all, *just information.* And information should pose no danger at all.

Unfortunately this is not the case. It is instructive to look at implications of IT for human well-being in all three areas previously mentioned: physical, mental, and moral. Where available, I will include studies on the most susceptible group of people, our children.

Of the three, physical risks are perhaps the least pernicious. Not counting office computer time, adults spend something like 5 hours 15 minutes per day on digital media and another 4 1/2 hours a day watching television.[50] Children under 18 spend between 8 1/2 and 13 hours per day on such devices.[51] With such lengthy exposure times, even small risks become magnified. There are obvious problems with sedentary behavior, carpal-tunnel syndrome, and eyestrain related to keyboard, keypad, and screen exposure. Lumeng and colleagues (2006) reported that young children exposed to two extra hours of television per day were nearly three times as likely to be overweight. A toxic dust of polybrominated diphenyls (PBDEs) has been found on keyboards and computer screens, which have been linked to reproductive and neurological disorders and thyroid dysfunction.

In a landmark study, Nakazawa and colleagues (2002) examined more than 25,000 workers over a three-year period. He found an array of physical, mental, and sleep-related problems correlating with daily time spent in front of a computer terminal. Physical maladies included headache, eyestrain, arthralgia (joint pain), stiff shoulders, back pain, and general fatigue. Symptoms were directly proportional to time spent on a computer each day, a condition that held true even adjusting for complicating factors. They were consistent for age group, gender, and over the duration of the study. Thus it is clear, as the media described it, that "too much screen time can make you sick." (I discuss the nonphysical problems later.)

One area of physical effect that has drawn much attention in recent years is the question of electromagnetic radiation. By its very nature, it has a complex relationship with biology. Radiation varies according to two quantities: field strength and frequency. It is the combination of these two factors, with perhaps other conditions, that determines degree of health risk. Electromagnetic frequency covers a vast range, running from essentially zero (a static field) to the 60 hertz (cycles per second) fields given off by power lines and home appliances, to 100 million hertz radio and TV waves, to 10 billion hertz microwaves, to 10^{13} hertz infrared (heat) waves, to 10^{15} hertz waves of visible light, to 10^{18} hertz x-rays and interstellar gamma rays. Heat and light radiation normally cause few problems but at high intensities can result in surface burns, sunburns, and, over the long term, skin cancer. They are relatively harmless because we have evolved to live with solar light and heat, and our bodies have defenses against these forms of radiation. Not so, however, with other frequencies. We have little evolutionary experience with 60-hertz waves, or strong microwaves or radio waves; and x-rays, as we know, can be cancerous and deadly. But thanks to the electrification of society over the past century, we and all the planet are now continuously exposed to electromagnetic radiation of various frequencies and strengths. The effects are poorly understood, but evidence suggests a clear potential for harm.

High-power radio broadcasting began in the 1920s and expanded over the following decades to include AM and FM radio, television, and wireless communication. Beginning in the early 1970s, there were reports of leukemia associated with living near high-power transmission wires. With the advent of home and office computers in the 1980s and wireless devices in the 1990s, radiation levels have risen continuously—as have signs of problems. In 2010, one Canadian school installed a wireless wifi network, and numerous children reported headaches, dizziness, insomnia, and other symptoms. A study by Li and colleagues (2011) demonstrated an elevated risk of childhood asthma for mothers who were exposed to higher magnetic fields during pregnancy; the mothers with the highest exposure gave birth to children with a 3.5-fold increased risk.

But perhaps the most problematic radiation concern is that pertaining to cell phones. Cell phones nominally radiate at radio frequencies (800 Mhz), but digital transmission necessarily includes components of much higher frequencies. Compounding the problem is the fact that we use them for hours each day pressed directly against our heads. The potential for damage is clearly significant, particularly for children.

In 2000, *Scientific American* reported on a computer simulation model showing that cell phone radiation penetrates far more deeply into the brain of a child than of an adult, heating up the corresponding brain tissue.[52] In 2004, a European study ("Reflex") conducted by 12 research groups in seven countries found that cell phone radiation damaged DNA in a laboratory setting. In 2008, a report by Divan and colleagues showed that pregnant women using cell phones gave birth to children at a 54% higher risk for hyperactivity and other behavioral problems. Later that same year, the head of a prominent cancer research center issued a warning to his faculty and staff to limit their personal cell phone use due to increased cancer risk.[53]

Shortly thereafter came "the most comprehensive study to date" on cell phone risks, published by Hardell and colleagues (2007). They analyzed the results of 11 long-term studies of people who had used cell phones for more than a decade, finding "a consistent pattern of increased risk for acoustic neuroma and glioma," two forms of brain tumors. Additionally, they could not rule out increased risk for other types of tumors.[54] In a revealing comment, one of the researchers remarked, "I find it quite strange to see so many official presentations saying that there is no risk. There are strong indications that something happens after 10 years."[55] Not so strange, actually; restriction on cell phone use would have a notable impact on the further progression of technological society, and this is unacceptable.

And yet critical data continued to emerge. Hardell reported to a London symposium in 2008 that children using a cell phone for just one year experienced a 5.2-fold increased risk of malignant glioma.[56] Volkow and colleagues (2011) found that cell phone radiation accelerates neural glucose metabolism, suggesting one mechanism by which brain cells might be damaged. Then in 2011 came an official recognition: the World Health Organization's

International Agency for Research on Cancer (IARC) declared that cell phones posed a "carcinogenic hazard," on par with industrial toxins. This was reconfirmed in its 2013 report, which stated that "positive correlations have been observed between exposure to radiofrequency radiation from wireless phones and glioma, and acoustic neuroma."[57] Finally, something of a legal precedent was established in 2012 by the Italian Supreme Court, when it ruled that an executive's brain cancer was caused by his cell phone use over a 12-year period.

Despite these warnings and the growing body of evidence, there are more active cell phones than ever. America alone has roughly 330 million wireless devices (phones plus tablet computers) in use, more than one per person. This, oddly enough, is not unusual; at last count roughly half the nations of the world had more active wireless devices than people. Globally there are about 6.8 billion such devices in use at present; soon there will be more active wireless devices on Earth than human beings. About 78% of American teens have cell phones, with nearly half of those being smart phones; and 23% have wireless tablets. Collectively we are happy to ignore the warnings and carry on as usual. And should we be one of the unfortunate victims of technology-induced brain cancer, not to worry; we can always look to the latest technology for a cure.

PSYCHOLOGICAL HARM

In 1998, six researchers at Carnegie Mellon University published a paper titled "Internet Paradox: A Social Technology That Reduces Social Involvement and Psychological Well-Being?" (Kraut et al. 1998). They studied 169 people during their first two years of online experience, concluding that "greater use of the Internet was associated with declines in participants' communication with family members in the household, declines in the size of their social circle, and increases in their depression and loneliness" (1998: 1017). The results were labeled "unexpected," with Kraut adding that he was "surprised to find . . . such anti-social consequences."[58] This was among the first major studies to suggest that information technology could have adverse psychological effects.

Then in 2002, the milestone Nakazawa study, cited earlier, found a variety of mental symptoms that correlated with daily screen time at work. These included lethargy, anxiety, depression, and a number of more complex subjective problems, such as "difficulty enjoying life," "difficulty getting along with coworkers," and "reluctance to go to work." These problems were most strongly correlated with people spending five hours or more on a computer—a figure that today encompasses much of the white-collar workforce.

Since that time, numerous other studies have emerged, pointing to a wide variety of concerns. Several problem areas are summarized below.

Behavioral Issues

This is a broad heading that encompasses various issues. In 2002, a major British marriage counseling institution reported that roughly 10% of couples seeking help cited the Internet as a major source of concern. According to one executive, "excessive use of the Internet is now seen by experts as a key manifestation of a relationship with problems." Counselors report that "increasingly, the Internet is a relationship breaker."[59] Two years later, the *Independent* reported that British divorces had hit a seven-year high; Internet usage was cited as a causal factor.[60] And the problems are not simply the result of incidental or accidental online meetings; organizations now exist specifically dedicated to infidelity. The site Ashley Madison (www.ashleymadison.com), which actively promotes sexual affairs among married people, claims to have more than 25 million members.[61]

A 2010 study by Case Western Reserve found that teenage hypertexting and hypernetworking were linked with a range of risky behaviors, including smoking, drinking, drug use, and sexual activity. Hypertexting, defined as sending more than 120 messages a day, applied to 20% of all teens. Risk increases ranged from 40% more likely to smoke and use drugs, to 350% more likely to have had sex. Hypernetworking, defined as more than three hours per day on sites like Facebook, applied to 12% of all teens and was associated with a 60% greater likelihood to have had multiple sex partners, 84% more likely to use drugs, and 94% more likely to have been in a physical fight.

Earlier it was shown that cell phone usage can cause brain cancer. We should not be surprised, therefore, to find psychological effects as well. A 2012 study of 28,000 children in Denmark found that cell phone use by pregnant women increases the likelihood of behavioral problems in children at age 7. Divan and colleagues (2012) followed up on an earlier study that found similar results in a smaller cohort of 13,000; those children showed an increased propensity for hyperactivity, attention deficit, and social interaction problems. The problems recurred in the larger group, with the greatest effects shown by children with both pre- and postnatal exposure.

Another behavioral issue is technology-induced violence in both children and adults. Huesmann (2007: S10) performed a review of multiple studies, concluding that "experiments unambiguously show that viewing violent videos, films, cartoons, or TV dramas or playing violent video games 'cause' the risk to go up . . . This is true of pre-schoolers, elementary school children, high school children, college students, and adults." Specifically on the question of video games, Huesmann's result was confirmed by a 2010 meta-analysis (Anderson et al.) that showed that "exposure to violent video games is a causal risk factor for increased aggressive behavior, aggressive cognition, and aggressive affect and for decreased empathy."

Pea and colleagues (2012) found that social media use, Internet use, and multitasking adversely affect young girls' social skills. This study, of nearly

3,500 girls aged 8 to 12, correlated time spent on interactive media with social problems. Conversely, girls who spent the most time in face-to-face interaction with friends and relatives had the least social difficulties. The authors speculate that girls pick up considerable clues from facial expressions and bodily movements to develop their social skills—things that are lacking in media interactions.

Attention Deficit

Attention deficit/hyperactivity disorder (ADHD) affects up to 12% of American children. For young children, their primary IT device is the television. Christakis and colleagues (2004) analyzed data from nearly 2,600 children between the ages of 1 and 7, finding that "early television exposure is associated with attentional problems at age 7." The result, they added, was "robust and stable over time."

For teens and adults, the problem is exacerbated by email, cell phones, and the Internet. Similar effects were noted by journalist Nick Carr in a widely discussed essay of 2008, "Is Google Making Us Stupid?" "Over the past few years," he says, "I've had an uncomfortable sense that someone, or something, has been tinkering with my brain, remapping the neural circuitry, reprogramming the memory. . . . I'm not thinking the way I used to think." The effects are most notable when reading; he loses concentration, his mind drifts, he gets fidgety. Carr speculates that "[the Internet is] chipping away my capacity for concentration and contemplation."[62]

A 2010 study (Swing et al.) examined 1,300 elementary school children and 210 college students, looking for attentional effects due to both television and video games. Children who spent more than two hours per day on video games were found to have a 67% higher likelihood of having attention disorders. A similar effect was found for watching television. For the smaller college student cohort, the correlation was even stronger: those spending two or more hours on games or TV were roughly twice as likely to have attention deficit problems.

Communication Skills

In 2007, the State Education Commission of Ireland issued a report noting the threat caused by texting to language skills among schoolchildren. Reviewing exam performance of 15-year-olds, the commission stated that texting "would appear to have impacted on standards of writing." "Text messaging . . . seems to pose a threat to traditional conventions in writing." This concern was confirmed by Cingel and Sundar (2012), who found "broad support for a general negative relationship between the use of techspeak and scores on a grammar assessment."

Media use—even "educational media"—can likewise have a detrimental effect on infants' language skills. Zimmerman and colleagues (2007) found that the viewing of educational baby DVDs and videos, for children under age 2, not only did not improve language development, it actually hindered it. Each hour of viewing time was associated with a 17-point decline in the Communicative Development Inventory (CDI), equivalent to a six- to eight-word deficit (out of 90) compared with infants who watched nothing.

Stress, Anxiety, and Depression

In 2006, the *Los Angeles Times* reported on a Korean study that found that the top one third of high school cell phone users—those making more than 90 calls per day—"scored significantly higher on tests measuring depression and anxiety" compared with less frequent users.[63] In 2012, the same paper reported on a UC Irvine experiment that examined stress and heart rates in workers while checking email.[64] A selected group of workers had their heart rates measured during "normal" frequent emailing at work, and they were found to have a high, steady, "on alert" heart rate—indicative of a mild stress condition. The same group was then banned from email for five work days. Researchers found a more natural, variable heart rate, comparable to a low-stress condition.

Another link between Internet use and depression was examined by Selfhout and colleagues (2009), who found that, among 307 Dutch teens, Internet use for noncommunication purposes (e.g., gaming, Web surfing, videos) "predicted more depression and more social anxiety." A year later, Morrison and Gore (2010) studied the connection between excessive Internet use—bordering on addiction—and depression. They determined that, "across the whole data sample, there was a close relationship between Internet addiction tendencies and depression." Social media were found to be strongly implicated in excessive use, and thus emerged a new phenomenon: "Facebook depression."

Chou and Edge (2012) examined this matter, finding that the more-frequent Facebook users had lower self-esteem than the less-frequent users; frequent users believed that others were happier and that others generally had better lives than they did. Most recently, Kross and colleagues (2013) confirmed this conclusion: "On the surface, Facebook provides an invaluable resource for fulfilling the basic human need for social connection. Rather than enhancing well-being, however, these findings suggest that Facebook may undermine it." Specifically, the researchers determined that both affective and cognitive well-being declined with increasing usage. The authors called for further study, noting, however, that Facebook's "addictive properties" may make it difficult to conduct controlled interventions.

Sleep Disorders

Nakazawa's team (2002) studied three aspects of sleep disorder: difficulty in falling asleep, sleep interruption, and early waking. All three showed a marked increase in prevalence for those spending five or more hours per day on their office computers. Confirming data came from a 2011 poll by the National Sleep Foundation, which found a correlation between late-night technology use and poor sleeping habits. One of the researchers put forth the hypothesis that "more active technologies are worse for sleep because of the psychological effect of being stimulated at night." They speculated that light emissions from screens adversely affected melatonin levels.

Addiction

Perhaps most troubling of the mental disorders is the potential for compulsive or even addictive use of the Internet, or technology generally. Even television has long been known to promote compulsive viewing. Kubey and Csikszentmihalyi (2002) argued, in *Scientific American*, that "television addiction is no mere metaphor," citing a number of parallel behaviors with drug abuse. A few years later, Aboujaoude and colleagues (2006) conducted the first large-scale random-sample survey of adult Internet usage, finding that roughly 12% exhibited at least one sign of problematic use. Warning signs included "relationships suffered" (5.9%), "concealed use from others" (8.7%), "used longer than intended" (12.4), and "hard to stay away" (13.7%). By 2007, some in the business community were worried that employees addicted to smart phones supplied by their employer could have grounds for lawsuits—see Kakabadse and colleagues (2007).

Meanwhile, recognition has been growing globally that Internet addiction (IA) is a major social problem. A 2009 study of 2,293 Taiwanese middle schoolers found that nearly 11% had developed an addiction and that this was linked to greater occurrence of ADHD, hostility, depression, and social phobia.[65] The following year it was reported that roughly 30% of Korean adults were addicted to Internet gaming. Studies of adolescents in Shanghai reported IA rates of 8.8%, and 14% for urban Chinese youth. In response, many Asian countries have created national treatment centers. The U.S. was somewhat slower to respond; our first Internet addiction center opened in 2009 near Redmond, Washington, and the first inpatient treatment center opened in 2013 in Bradford, Pennsylvania.

Other scientists sought physiological effects of IA. Yuan and colleagues (2011) and Zhou and colleagues (2011) found decreased grey matter and microstructure brain abnormalities in addicted adolescents. Other researchers have dedicated entire chapters (Aboujaoude 2011: 214–234) and even complete anthologies (Young 2011) to IA. The phenomenon continues to spread—one British doctor warned that children as young as age 4 were being treated for IA.[66] Overall prevalence in the U.S. and Europe is estimated to range between 1.5% and 8.2%, with certain subpopulations possibly as high as 38%.[67]

Recent research specifically on video games is particularly disturbing. Gentile and colleagues (2011) report that between 8% and 12% of all gamers are addicted. These "pathological players" exhibit a range of problem symptoms, including depression, anxiety, social phobia, and decreased school performance. Gentile argues that it is the gaming itself that is the causal factor: "pathological gaming is not simply a symptom. . . . [The] problems seem to increase as children become more addicted. . . . [W]hen children stop being addicted, depression, anxiety, and social phobias decreased as well." As if these problems are not bad enough, new research shows that compulsive gamers can actually hallucinate, sometimes for hours afterward. Ortiz de Gortari and Griffiths (2014) examined 656 self-reported hallucinatory experiences, categorizing them into three groups: digitally induced images, perceptual distortions, and visual misperceptions. The first was the most common and involves the visualization of game elements—shapes, figures, colors—in the real world. Other problems relate to motion sickness, temporal distortion, and confusing game features with real life. These effects can persist for up to 24 hours, with unknown long-term consequences.

Occasionally the popular media take notice of these broad trends toward ill health. Earlier I cited Carr's prominent 2008 essay "Is Google Making Us Stupid?"—one of the first to examine the topic. *Newsweek* ran a cover story in mid-2012 titled "iCrazy," suggesting, rather explicitly, that the Web is driving us mad:

> [T]he proof is starting to pile up. The first good, peer-reviewed research is emerging, and the picture is much gloomier than the trumpet blasts of Web utopians have allowed. The current incarnation of the Internet—portable, social, accelerated, and all-pervasive—may be making us not just dumber or lonelier but more depressed and anxious, prone to obsessive-compulsive and attention-deficit disorders, even outright psychotic. Our digitized minds can scan like those of drug addicts, and normal people are breaking down in sad and seemingly new ways. (Dokoupil 2012: 26)

Books such as Carr's *The Shallows* (2010), Powers's *Hamlet's Blackberry* (2010), Turkle's *Alone Together* (2011), Aboujaoude's *Virtually You* (2011), and Rosen's *iDisorder* (2012) all explore the same theme, in an equally populist vein.

For all this discussion, the proposed solutions are remarkably uniform and tame: We need to "moderate" our usage, supervise our children, and watch for the "warning signs." The problem is never the technology itself, only our misuse of it—a classic neutralist argument. There is never any indication that there might be deeper problems here, problems that are unaddressable in the normal context of a technological society. Carr's article, for example, ends nonchalantly: "Maybe I'm just a worrywart." He offers nothing, though, by way of a concrete response. The *Newsweek* cover story concludes by asking what we should do, noting that "Some would say nothing," and adding that our "days of complacency should end." Yes—and then what?

Why, for that matter, should information technology pose any mental issues at all? It is allegedly only a tool for communication, information, and entertainment. Books and newspapers pose no such problems; nor do public lectures, debating societies, or town hall forums. Email, text messages, and Web surfing are only more of the same. Why should they present us with these dire psychological issues? Conventional metaphysics has no solution to this riddle. When confronted with the literal growing insanity of technological society, orthodox thinking can only point to vague generalities like stress, overwork, social pressures, and the like. It can only propose the mildest of reform actions.

A pantechnical metaphysics, on the other hand, spells out a different picture. Technology is logos, and all logos is something mind-like. Modern information technology in particular is a deep manifestation of the universal Logos; it is both a result of and embodiment of cosmic reason. As a concentrated and intense form of mind, information technology cannot but have psychological effects on us—who are, ourselves, also particularly concentrated and intense forms of mind. In our construction of the Internet, we act as the agent of Logos, helping to manifest a yet higher articulation of rationality. The Internet is a self-construction of Logos; it is mind-produced mind. As the technological system grows in power, it will increasingly affect us in varying ways—to our detriment, physically and mentally.

MORAL CONFLICTS

Physical and mental pressures from information technology pose serious enough problems in themselves. But technology is unrelenting. As a new order of being, the technological system manifests new values—values distinct from and increasingly at odds with traditional human values. As such, we should expect to see growing value conflicts in our interactions with technology. More specifically, we should expect pressure to compromise and undermine traditional human values. As the forefront of the technological system, information technology should therefore be expected to cause a variety of moral dysfunctions.

Many such cases are widely covered in the news. Hacking, cybercrime, cyberwarfare, government spying scandals—such things are standard fare in current headlines. But these are cases of individual criminals or institutional malfeasance. More subtle and more troubling are the ethical pressures imposed upon ordinary people in their daily lives. Some are obvious. Already by the mid-1990s, it was clear that students could use technology to cheat in school; for example, the *San Francisco Chronicle* carried a cover story, "Students Look to Internet for New Ways to Cheat."[68] In 2000, a Rutgers study of 21 colleges found that about 10% of students had picked up passages for papers from the Internet and that another 5% had taken large blocks of text or even full papers. Plagiarism is a constant threat for any college writing class, and instructors have to go to extreme measures to circumvent such efforts. Adding to these problems is the increasing use of iPods and earbuds to cheat on exams.

Equally obvious are the sexual temptations. Internet pornography is a leading category of all Web traffic and poses a range of ethical problems, particularly for youth. In 2001, it was reported that 19% of 1,500 children aged 10 to 17 had been solicited for sex via the Internet.[69] It was quickly realized that such solicitations, along with Web porn sites, are extremely difficult to block. The U.S. National Research Council reported in 2002 that there is "no one fix" for children and pornography, and that the whole concept of technological supervision is overrated. By its very nature, the Internet allows multiple, diverse, and constantly changing access routes to Web sites; the result is that nearly every blocking strategy can be defeated, often by clever teens. Even accidentally, teens are exposed to online porn; Finkelhor and colleagues (2007) reported that 42% of children aged 10 to 17 had encountered Internet porn in the past 12 months, with a majority stating that it was unexpected and unwanted. And beyond concerns with viewing, there is some evidence that access to video-posting sites like YouTube encourages and promotes sexual assaults, particularly by teens who seek peer-group approval and Internet fame by filming themselves committing such crimes.

Other problems have recently emerged. Researchers at USC reported in 2009 that rapid-fire media tools, such as Twitter, could damage the brain's "moral compass" by circumventing needed time for moral reflection. This in turn could lead to a sense of indifference to human suffering.[70] In a similar vein, two Canadian researchers found that frequent texters appeared to be "shallower" than less-frequent ones. Those who sent more than 100 texts per day were 30% less likely to value leading an ethical, principled life.[71] And one UCLA researcher, Gary Small, asked if the Internet is not "killing empathy" by turning us all into detached voyeurs.

That American college students are less empathetic in recent years was demonstrated by a 2011 study (Konrath et al.). The researchers found a marked decrease in empathic concern, particularly since the year 2000—precisely the time when social media became widespread. They speculate that "one likely contributor to declining empathy is the rising prominence of personal technology and media use in everyday life" (9). Online relationships are clearly different than those of real life, with the result that "interpersonal dynamics such as empathy might certainly be altered." Media role models tend to be increasingly narcissistic, and growing media violence desensitizes people to pain and suffering. By focusing on the content rather than the technology per se, however, such analysis lends credence to the neutralist perspective. If, on the other hand, depersonalization is intrinsic to the technology, as I claim, then no modification of content can improve the situation.

This, then, presents something of the case for a link between technology and numerous risks to human well-being. I emphasize that this is only a sampling of the relevant research; a full assessment would require a book-length treatment in itself. Also, my focus here is on information technology, which represents the cutting edge of technological advancement. There are obviously many other aspects of technology that cause direct and indirect harm, including weaponry, toxic chemicals, pollution, high-speed and high-energy machinery, artificial living conditions, industrial food—the list goes on. The recent boom in autism, for example, has been strongly linked to toxic chemical exposure during pregnancy—see Engel and colleagues (2010) and Landrigan (2010), for example—even as others have connected it with compulsive Internet use[72] and watching television.[73] The technological web is so complex that there may be multiple interacting factors for any given problem area. And though we may not always be able to find the precise causal mechanism, we can be sure that it is ultimately a technological cause.

TECHNOPHILE REPLIES

Faced with the cited facts, defenders of modern society have a handful of ready replies. These tend to fall along three main lines of argumentation. First: "Medical problems are not actually getting worse; it is simply that our methods of diagnosis are getting better, and thus we only seem to have more cases than ever." This is a popular response to the prevalence of depression, ADHD, autism, bipolar disorder, and selected other cases of mental illness. Certainly some portion of the growth of these afflictions could conceivably be attributed to better diagnosis, but when depression increases 400% over two decades, or autism rates double and bipolar disorders jump by 4,000% in less than one decade, we surely cannot attribute more than a small fraction of this increase to improved medical detection. There must be real, underlying growth in occurrence rates. Other concerns, such as those with physical injury, cancer, diabetes, and newly emergent infectious diseases, are by and large not subject to the "better diagnosis" response.

Second: "In all these cases, researchers have only shown correlation, not causation. We have no reason to believe that technology is the causal factor." This is a valid concern, at least for the mental and ethical disorders. In fact, some scientists have speculated that, rather than technology causing, say, depression, it is simply the case that the more-depressed individuals turn to technology for relief. The same could be said for attention-deficit children, or the less ethical among society. But once again, this likely applies, at most, to only a minority of cases. There are simply too many instances, too many independent studies, and too many distinct maladies for there to be no causal link from the technology.

We need to keep in mind that nearly all human harm is a combination of two factors: genetic susceptibility and environmental triggers. The genetic side is fixed—at least, until we start reprogramming our own genetic code. Our genes have evolved to protect us from a wide variety of natural environmental threats, but they have little or no ability to respond to new, technological factors. The fact remains that the well-being of humanity and nature is declining in several ways and by numerous distinct measures. This decline furthermore appears to be accelerating over the past few decades, and dramatically so, in some cases, over just the past 10 years. With a fixed genetic susceptibility and rising technological stresses, we can see that technology must ultimately be the primary causal factor. If chemical exposure *in utero* causes autism, and then the autistic children spend more time on the Internet, it does no good to exclaim that "the Internet was not the causal of their autism." The cause is still technology; we simply debate which aspect of technology it is.

Finally, once these two are out of the way, the technophiles may retreat to a third and final reply: "Yes, it appears that technology causes some hardship to people and to nature. But this hardship must be balanced against the benefits of advancing technology. No pain, no gain." To which I reply: What precisely are the gains, and how do they contribute to human well-being? After all, what is the point of advancing technology, if not human and environmental well-being? The "no pain, no gain" response holds only if it can be shown that, overall, humans and nature are benefitting from technological progress—which they are not. If the pain outweighs the gain, this argument fails. In reality it is more complex, of course. Often it is a case of short-term gain at the expense of long-term pain. By the calculus of short-term thinking, technology always looks like the best solution. But few concern themselves with the long-term effects, which, in any case, often cannot be evaluated due to the profoundly complex nature of technological-environmental-human interaction.

Once these three are dispatched, the only remaining stance is that technology *must* progress, that it is an end in itself, and that its importance exceeds that of any merely human concerns. This, at least, is honest. Most often, this view is couched in subtle political or adversarial terms. But the underlying message is clear: "If we do not promote technology, others will": the Japanese, or the Germans, or the Chinese, or the Russians, or the "terrorists"—and thus we will lose our "competitive edge" in economics, military power, or culture. But it is never shown that our "competitive edge" is in fact improving our quality of life. Rather the opposite is the case: the quest for an ever-greater technological edge is destroying our well-being, along with that of nature.

In early 2013, the BBC published a story titled "How Are Humans Going to Become Extinct?"[74] It refers to a recent Oxford paper, "Existential Risk as a Global Priority," which looks at a variety of risk factors ranging from pandemics to catastrophic natural disasters. Because we, and the planet, have evolutionary experience with such things, all of them are survivable. Technology, however, presents us with new and unprecedented risks. Citing program director Nick Bostrom, the article states that "the advance of technology has overtaken our capacity to control the possible consequences."

Much like Bill Joy argued, Bostrom marks out synthetic biology, nanotechnology, and ultra-intelligent computers as unique risk factors. Such technologies could interact and create chain-reaction feedback effects that would be profoundly dangerous. "We're at the level of infants in moral responsibility," says Bostrom, "but with the technological capability of adults." In such a situation, how can we possibly, rationally, evaluate the risks to our own well-being? Surely we cannot. Like children, we chronically underestimate the risks and overestimate the gains.

Given that most of the Oxford staff had extensive prior experience with technology and thus were predisposed to see its benefits, I was curious how the researchers themselves responded to such dire conclusions. In response to this story, I contacted one of the staff philosophers, Toby Ord. Ord told me that "many of us are from technology-related backgrounds and were previously very optimistic about it. Now, however, I think we mainly dread it, due to its downside risks."[75] A rational conclusion—but then, how to respond? Ord told me that they "mainly talk about moving forward as safely as possible" while recognizing that the benefits and the "edge" go to those who proceed rapidly and recklessly. From a larger perspective, it is a no-win situation—and the future of humanity is at stake.

As a moral and psychological phenomenon, modern technology presents us with a profound challenge. The technological system is, like us, a mental being; it exerts its will and thereby affects our will. It is also a moral being; it operates with certain values, and these impinge upon our distinct and often contrary human values. As a concrete manifestation of a teleological universe, it functions with certain ends in mind, and these are only incidentally related to human ends. Technology mirrors us in many ways—not because "we made it" but because we share common ontological underpinnings. We are both, alike, manifestations of a universal Technê-Logos. But because it is a more recent evolutionary development and represents a higher and more complex realization of Logos, it out-powers us on all fronts.

Psychologically and morally, then, technology necessarily acts to undermine our native sense of health, well-being, and wholeness. It must progress, and it must do so at our expense. It must subvert our normal states of mind and persuade us to work on its behalf. We would not do so if it were obviously contrary to our own interests, and thus technology must convince us that its own advancement is, indeed, to our benefit. Technology must present its own development as an advantageous situation for all, one that no reasonable person could oppose. And every small advance, taken individually, does indeed seem to be a gain. Yet the collective effect is to progressively bury humanity and nature in ever-deepening layers of technological dross.

In undermining our ability to think clearly about our own long-term interests, technology functions as the equivalent to AIDS. As we know, the HIV virus is especially pernicious because it attacks the human immune system—that very process by which we would normally fight off infection.

The virus cleverly targets that which would be the source of its demise. In an analogous manner, technology attacks the moral and psychological basis of our humanity, and persuades us—despite manifest evidence to the contrary—that it is not only not a threat but rather the key to our salvation. From the standpoint of technology, this is a necessary condition; otherwise we might begin to act counter to its best interests and thus impinge upon its further advancement.

Technology, then, functions as a kind of 'mental AIDS.' It erodes our critical thinking processes, native intelligence, capacity for sustained and focused thought, psychological independence, and moral autonomy—the very things that might be used against it. And in a complementary move, it promotes protechnological thinking, protechnological values, and protechnological actions wherever possible. Like the HIV virus, it is amazingly clever at manipulating the human organism to achieve its ends. And like the virus, it risks dying along with its host.

Obviously, this erosion of critical thinking is not absolute. Obviously, there are individual exceptions. But *en masse*, and via the power structures of society, it is astonishingly effective. The level of intelligence of a technological mass society is so low and the corrupting benefits so great that the relative few critical thinkers have no hope of stemming the tide. Our so-called leaders praise the glories of technology, the masses are enthralled with the latest toys and amusements, and technology continues its relentless advance. It acts out of neither maliciousness nor contempt nor hatred but simply as all things do—out of a sense of self-preservation, self-fulfillment, and self-realization. Technology is not evil. It is powerful, and it is indifferent, and that is why it is dangerous.

NOTES

1. Recall Kurzweil's prediction of 2045 as the singularity date. Others give even earlier estimates.
2. See Diamond (1987)—"The Worst Mistake in the History of the Human Race."
3. Ellul (1964, 1980, and 1990).
4. The extermination of large mammals by early humans did not occur until relatively recently—the past 200,000 years or so.
5. The domestication of wheat began around 8500 BC, and selective or controlled planting may have occurred well before that. Of course, Plato had no firm evidence of this timing; he must have been recounting a traditional view, albeit one that was remarkably accurate.
6. Vitousek and colleagues (1986). Other estimates place this figure in the range of 20% to 32%.
7. Sanderson and colleagues (2002).
8. Food and Agriculture Organization (2006).
9. Goncalves and colleagues (2012).
10. Hoffmann and colleagues (2010).
11. Cicerone (2005).

12. Pearce (2005).
13. "The Earth Is About to Catch a Morbid Fever," *Independent* (July 21, 2006).
14. "Global CO_2 Emissions Rising Faster than Worst-Case Scenarios," *Washington Post* (November 4, 2011).
15. The Millennium Project (2009), *State of the Future.*
16. Sometimes abbreviated as 'edtech.' Also, I take this to be synonymous with 'instructional technology.'
17. State of the Union address, 1996.
18. Remarks by the president at Mooresville Middle School, June 6, 2013.
19. "L.A. Unified Takes Back iPads as $1 Billion Plan Hits Hurdles," *Los Angeles Times* (September 30, 2013).
20. State of Michigan, "Freedom to Learn Evaluation Report," January 14, 2004 (p. 3).
21. *State of Michigan Educational Technology Plan, 2006–2010* (p. iii).
22. Generally speaking, effect sizes under 0.35 are considered small, 0.35 to 0.65 are medium, and more than 0.65 are large.
23. "Computers Don't Make Grade," *Detroit News* (March 15, 1995).
24. "Computers in School: Do Students Improve?" *San Jose Mercury News* (January 14, 1996).
25. "Technology Is Still a Promise, Not a Panacea," *Los Angeles Times* (June 8, 1997).
26. "A Laptop for Every Kid," *Time* (May 1, 2000).
27. August 24, 2000.
28. "Seeing No Progress, Some Schools Drop Laptops," *New York Times* (May 4, 2007).
29. September 3, 2011.
30. *University Record* (February 4, 2013).
31. *New York Times,* op. cit.
32. "No Child Left Untableted," *New York Times Magazine* (September 12, 2013).
33. National Center for Health Statistics (2011): "Health, United States, 2010," Table 24.
34. U.S. CDC (2011): "National Vital Statistics Report" (vol. 60, no. 3), Figure 6.
35. National Cancer Institute (2010): "Reducing Environmental Cancer Risk."
36. Op. cit. (note 29).
37. Defined as a body mass index (BMI) of 30 or greater.
38. See Eknoyan (2006).
39. Trust for America's Health (2013): "F as in Fat" (p. 23).
40. See, for example, "The Epidemic of Mental Illness: Why?" *New York Review of Books* (June 23, 2011).
41. Ibid.
42. CDC Data Brief #76 (October 2011).
43. "Suicide Rates Rise Sharply in U.S." *New York Times* (May 2, 2013). Data are from the CDC *Morbidity and Mortality Weekly Report,* 62: 321–325.
44. CDC *Morbidity and Mortality Weekly Report* (2012); 61(3).
45. CDC *National Health Statistics Report,* 65 (March 20, 2013).
46. Kim and colleagues (2011).
47. Olfson and colleagues (2012).
48. Moreno and colleagues (2007).
49. "ADHD Seen in 11% of US Children as Diagnoses Rise," *New York Times* (March 31, 2013).
50. eMarketer survey, "Digital Set to Surpass TV in Time Spent with US Media" (August 1, 2013).

51. Northwestern University, Center on Media and Human Development: "Children, Media, and Race" (2011).
52. "Worrying about Wireless," *Scientific American* (September 2000: 20).
53. "Cancer Expert Warns Employees on Cell Phones," Associated Press (July 24, 2007).
54. Khurana and colleagues (2009) confirmed this finding, showing that a decade of phone use "approximately doubles the risk of being diagnosed with a brain tumor."
55. "The Hidden Menace of Mobile Phones," *Independent* (October 7, 2007).
56. Cited in Sage and Carpenter (2009: 234).
57. "Non-Ionizing Radiation" (vol. 102): 419.
58. The same group published a 'revisit' of the paradox in 2002, negating some of their earlier findings. In particular, they found that, over time, the more extroverted participants reported positive outcomes. But by this time, more studies were appearing that confirmed many of the initial conclusions.
59. *The Scotsman* (April 16, 2002).
60. *Independent* (September 1, 2004).
61. As of April 2014.
62. *The Atlantic* (July/August 2008). Carr elaborates in his book *The Shallows* (2010).
63. May 24, 2006.
64. "Work Emails Are Bad for Your Health, Study Finds," (May 3, 2012).
65. Ko and colleagues (2009).
66. "Girl Aged Four Is Britain's Youngest Known iPad Addict," *Daily Mirror* (April 21, 2013).
67. Cash and colleagues (2012: 293).
68. December 16, 1997.
69. Mitchell et al. (2001).
70. Immordino-Yang and colleagues (2009).
71. University of Winnipeg *NewsCentre* (April 11, 2013).
72. Finkenauer (2012).
73. Waldmann et al. (2006, 2012).
74. April 24, 2013.
75. Personal communication (April 25, 2013).

12 Technology and Human Destiny

Humanity once held a privileged position on this Earth. From the late Neolithic through the early agricultural period and into the era of the first true civilizations, we deployed benign technology: simple, natural, nontoxic tools and various organizational techniques. Access to just three or four metals and a handful of other minerals allowed us to build pyramids and the Parthenon, the Hanging Gardens, and the Colosseum. A global population of perhaps 200 million was widely dispersed, and even the largest cities numbered less than 1 million. Local environments may have been depleted, but the vast majority of the planet was a vibrant, diverse, clear-running wilderness. At its best, human civilization supplied everything needed to live a good life. It was *our* time, the Era of Humanity.

Certainly there were low points—wars, disease, cruelty, injustice. These things have always been with us, and they are with us still, even in our advanced technological age. Some have suffered terribly in the past—as they suffer today. If such ills are worse now than ever, this can only be due to our increased technical power. And if we take into account the exploding population in recent centuries, it is unquestionable: *total* human suffering is vastly greater today than in the past.

There is no reason to believe that the average premodern person, no matter where on Earth they lived, had a less satisfying existence than their modern-day equivalent. Quite the contrary: we have good reason to think that human life was, in many ways, more satisfying in the past. The life of the forager, farmer, or small-scale urbanite was closely aligned with 2 million years of human evolutionary history. We were well suited for such modes of living, and simple tools and social techniques took the hard edge off daily existence. For those in the cities, culture generally thrived; arts, philosophy, theater, literature, poetry all flourished, in some cases achieving heights rarely attained since. Looking back, we can see how little it would have taken to make things better still. And this is the point: A rational, human-scale society in charge of its own destiny can achieve greatness, with even the simplest of tools.

Progressively, as humanity moved into the Middle Ages, the third wave of technological determinism began to take hold. Modern energetic devices emerged and spread. This set us on the road to cancerous growth, loss of control, reduced freedom, ecological destruction, and possibly self-annihilation. Modern technology, loosely defined as that which emerged during this time, allowed the creation of a level of social and technological existence that far outstripped human evolutionary capabilities, carrying us far beyond the era when human beings—*individual* human beings—largely determined their own living conditions.

Technological society is a new order of being, its *own* order of being, and it exists by and for its own sake. Modern technology advances of its own initiative; human concerns are incidental or irrelevant. This situation demands a metaphysical explanation. On my view, it is explicable only in a universe in which all creation is a manifestation of Technê-Logos, a natural process operating throughout the cosmos.

In a universe conceived as a Pantechnikon, then, modern technology works contrary to human interests. Having far exceeded the Era of Humanity, we are now well into the Technological Era, and every step forward in technical progress leaves us, and nature, farther behind. Every gain for the system is a loss for the planet.

If we understood this fact, we would not support the advancement of technology. We would not rush to buy its latest products or lavish such praise upon it. In fact, we might actively oppose it. But doing so would bring progress to a grinding halt. Therefore technology must present itself to us as precisely *the opposite of what it is*. It must be fundamentally deceptive. It must portray itself, everywhere, as promoting human interests, as improving health, as increasing happiness, wealth, and well-being. We willingly take the bait, convinced of the benefits to ourselves or our fellow man or nature, and then are puzzled when things turn out for the worse.

Examples are manifold. Many technologies promise to do things for us faster, and yet we have less time than ever. Technologies promise to make life more exciting, and yet we are constantly bored. They promise us health, and yet we are sicker than ever. They promise to humanize us, and yet we are as brutal and vice ridden as ever. They promise us wealth, and yet, for the vast majority, we are poorer than ever. They promise *a good life*, and yet we are sadder, lonelier, and more exhausted than ever before. All this is explainable, as I said, in a pantechnical universe, one in which the surge of evolutionary development has passed us by, and we are now left in its wake—decaying, declining, and distraught.

A MODEL OF REALITY

It is fitting that the Great Pyramids are the decisive ancient monument to human ingenuity. Surely this is no accident. In their ancient structures, the Egyptians created a model of reality. This is my claim: Nature itself is constructed rather like a pyramid. A base material layer is laid down, and in the presence of abundant energy, successive layers of complexity and organization are built upon it. A natural hierarchy is created, out of the Technê-Logos, and it continues to develop as long and wherever conditions allow.

In the case of our solar system, the base layer consisted of clouds of hydrogen and helium and countless tons of heavy-matter dust that were discharged into space when older stars exploded. Through gravitational attraction and perhaps other forces that we do not yet understand, the gas cloud condensed, and the dust particles agglomerated. At some point around 5 billion years ago, the atomic nuclei in the gas cloud suddenly fused together, initiating the nuclear reaction that became our sun. The previously 'free' atomic particles were now bound into an ordered system.

As it continues to burn, the sun produces increasing amounts of heavier elements—notably including the key components of life: carbon, oxygen, nitrogen, silicon. Layers of these heavier elements build up in the core, adding to the sun's order and complexity. In the process of fusion, simpler elements are destroyed, energy is radiated away into space, and atomic particles are reconstructed as higher-order, heavy elements. A pyramidal-like hierarchy is being constructed in the sun.

An analogous process is occurring on the Earth. When the sun ignited, the force of the blast stripped away any lingering dust particles, leaving a solid material core. After about 1 billion years, liquid water began to collect on the surface, a likely product of both outgassing from the Earth's core and from a steady rain of water-bearing meteorites. The solid matter, liquid water, and thin preoxygenated atmosphere thus form the base of our terrestrial pyramid.

With these basic conditions in place, and with the steady energy flow from the sun, the production of further order was inevitable. The first single-celled life forms fed off the ambient heat, light, and abundant ocean minerals; they constituted the second layer of the pyramid. In time they grew in complexity, becoming multicellular organisms. At the same time, other organisms—animals—evolved to eat the simpler plants; animal life was a third layer. Then a fourth layer appeared, one of animal-eating animals, or carnivores.

We see what is happening: Each new layer of the pyramid imposes constraints on the lower orders, lives off them, and in some cases destroys ("eats") them. At a minimum, the higher layers exact a toll—freedom and autonomy are reduced, order is imposed, pressure increases. The lower orders must bear the increasing load of those above them. Just as a pyramid under construction places a progressively heavier weight on the lower levels as each new layer is added, so too in the cosmos: over time, the lower orders come to exist primarily to support the upper. They sustain the growing structure and carry an ever-greater weight. But they do not benefit thereby. At each point in time, it is the uppermost layer that basks in the full sun, enjoying its time as the peak of existence. The upper layer is the king of creation; it sees itself as god-like, almost divine, as the endpoint of universal evolution. And for awhile, it is. Soon enough, though, the great builder, the Demiurge, comes around again. Technê-Logos adds yet one more layer to the ontological pyramid, and a new king is crowned.

During the Human Era, as I have called it, we ourselves were that uppermost layer of existence. From roughly 10,000 BC to 1200 AD, humanity basked in the glory of the evolutionary sun. This was our Eden. It was not perfect, it was not trouble free—no existence ever is—but it was our time at the top. We were constructed in the image of God/Logos himself, constantly in touch with the divine. God/Logos gave everything into our hand. The world was ours to do with as we liked.

Since then, progressive layers of social and technical complexity have been laid upon us—by our own efforts. Each layer is translucent; each one blocks a bit more of the Big Light. With each succeeding development we live in a bit more darkness, even as our burden grows. Our daily lives thus become ever dimmer as we increasingly function simply as support structures, weight bearers, for the emerging layers of complexity above.

This is no one's fault. There is no one to blame. *Technology is not evil.* It acts not out of malevolence but out of necessity. It is an inevitable consequence of life on a planet of superabundant energy. Technology is the most recent layer on the terrestrial pyramid, and it is now enjoying its time in the sun—to the detriment of the rest of us.

If events continue to unfold as they have in the past, technology's day will also come to an end. It too will be usurped, perhaps by 'technology-made technology,' perhaps by some technological-biological fusion, perhaps by something as yet inconceivable. The process will continue to advance until there is no more free energy available, in which case growth will come to a halt.

There is an alternate future. Instabilities in the pyramidal structure may cause it to collapse of its own weight. Based on our study of nature and of the growing stresses on humanity, this seems an increasingly likely outcome. It will be a catastrophic event, particularly for those creatures—ourselves—at the top. Being at the highest level, we have the farthest to fall. If this happens soon, a fair remnant of humanity will likely survive. If it happens some decades hence, it may be our end. In any case, if it occurs, nature will dust herself off, clear away the rubble, and begin again.

I emphasize, too, that this pyramid building is a naturalistic process of ontological continuity. The higher layers are marked by a reordering of mass and energy, and an increased dynamism vis-à-vis the lower orders—all in an ontologically continuous manner. There is no break in the metaphysical order, no brute emergence of radically new features of reality. The intrinsic qualities of all levels are *the same*—the universe is a monism, after all. Fundamental qualities that we find at one level are of necessity present in all.

Thus, for example, intrinsic qualities in matter—spin, charge, mass, quantum state—exist at all levels. This is why material laws of nature operate with consistency throughout the universe. More strikingly, intrinsic qualities that we find *in ourselves* are likewise present at all levels. The chief of these is consciousness, which I take to be a relatively sophisticated manifestation of something more basic, namely, subjectivity or experientiality. Every layer of existence possesses increasingly sophisticated modes of subjectivity; it is lacking in none. Hence panpsychism obtains, as I argued in Chapter 3. This is no accident; it is the inevitable consequence of an intelligent Logos realizing itself at each phase of universal evolution. Order, complexity, and transcendence are built into the structure of the cosmos. These constitute the *being* of intelligence, of mind. As such, they reside in all layers of existence.

We may press further and note that, in addition to the experiential, our mental lives include a volitional or intentional component. (To the best of our present knowledge, these two aspects—the experiential and the intentional—seem to comprise the two irreducible components of mentality.) Thus we can infer, as Spinoza, Leibniz, Schopenhauer, Nietzsche, and others have done, that all levels of material reality also possess an intentional aspect—again, one that ultimately derives from the Logos, which can be its only source.

This helps to explain why technology acts with such drive and force: there is an intrinsic intentionality to it. It imposes itself ever more powerfully; it seeks to expand and evolve; and it presses forward relentlessly. This is no mere anthropomorphic metaphor. When we speak of technology's wants and desires, we express a metaphysical truth. Technological society is not alive; it is a supra-life, a new higher order of being that embodies the intrinsic qualities of life without the limitations of life. It eats but energy. It reproduces by expansion. It evolves by the material realization of progressively more sophisticated technical devices. It never sleeps, ever.

THE DECLINE AND FALL OF HUMANITY

A pyramid is a monument to creativity, but it is also a tomb. We are becoming entombed by technology. Each new layer that is laid upon us becomes like one more coat of golden varnish on a fine oak coffin. Inside, still alive, we gasp for air. And not only us—all of nature is progressively being buried alive. At best we can hope to bear the growing weight with a modicum of dignity; at worst, we and all the planet become consumed by this expanding autonomous power.

It takes only the barest of insight to realize that signs of decay are all around. Declining health, environmental destruction, exploding populations . . . and quality of life always on the wane. Consider human culture—the highest expression of our humanity. During the 10,000-year Era of Humanity, in different ways and at different times, it attained a high degree of realization. Notable peaks occurred in Egypt, Mesopotamia, Athens, Rome, India, and China. In Europe, a kind of cultural momentum carried over into the Renaissance, yielding the brilliant accomplishments of Leonardo da Vinci and Michelangelo in art, Shakespeare in literature, Tallis and Praetorius in music. But even then, the decline had set in. By the 13th century, Western civilization had peaked; it then entered a period of gradually accelerating decay. Certainly by the onset of the Industrial Revolution—say, 1750—human culture definitively began a marked and steady decline, worldwide, in all areas of endeavor.

Obviously, this is not to say that there were no cultural achievements since that time. Decay, like ascension, is never monotonic; individual peaks continue to break through. Even into the industrial age, humanity was still able to produce the likes of Voltaire and Rousseau, Kant and Hegel, Bach, Mozart, and Beethoven. But these, of course, were the brilliant exceptions; the overall trend, for hundreds if not thousands of years, has been downward.

Through time, as each field became increasingly technologized, decay accelerated. The advent of motion pictures and television caused theater to coarsen and falter. Radio, record players, tape machines, CDs, and iPods in turn had a comparable effect on music. In literature, the move from handwriting to typewriters to word processors led to a similar outcome. Even painting and sculpture, minimally affected by advancing technology, have long been in decline; such is the culturally corrosive effect of modern technological society.

Culture today has become the glory of *technology*, not of humanity. Social media are technological media; architecture has become a paean to technical brilliance rather than artistic or organic vision; motion pictures rely to an ever-greater degree, and often completely, on computer-generated effects, while subtleties of plot or character development languish. And consider what has become of modern music. It is fully electrified and digitized, complete with Auto-Tune's "mechanical modulations," and synthesized versions of every instrument. Rap music is little more than crude and mechanically repetitive chanting; it competes with such genres as "techno," "heavy metal," "industrial," and "electronica" for the lucrative youth market. Even musicians themselves are coming to realize the extent of the problem; in late 2013, alternative music icon David Byrne stated that "the Internet will suck the creative content out of the whole world."

Protests to the contrary are fruitless. "What about the cruelties of ancient Rome?" some may say. "Certainly gladiator fights and throwing people to the lions were brutal forms of culture, were they not?" "Ritual sacrifice was a cultural fixture in several ancient civilizations; surely we have progressed since then?" But it is not at all clear that watching a few actual people die is more harmful to one's individual or social well-being than watching the thousands of fake deaths and dismemberments that occur on television, the cinema, video games, and the Internet. A real death is at least reality, and it impresses upon the observer the vicissitudes of organic existence. Fake death is depersonalized and abstract death. It desensitizes the viewer to the pain and suffering of others. It supports such virtuous activities as the killing of 'enemies' in far-away nations via high-tech military drones.

What about women and minority rights? Don't these show signs of progress over the centuries? Yes and no. It is obvious that previously disenfranchised groups now have a variety of formal rights, and this removes them from certain past abuses. But this process has a downside: it allows more people to become drawn into the technological system and thus subject to new indignities. Women and minorities are now free to work 60-hour weeks, to submit to the humiliations of the corporate world, to become 'competitive,' and to abandon their children to endless daycare. "Only if they want it," comes the reply. Yes, but to an ever-greater degree, it is no longer a truly free choice; two full-time jobs are often needed just to make ends meet, healthcare and education costs are rising, and social pressure is increasing to be a success. The alleged progress of women and minorities can best be accounted for, I claim, as evidence that the technological system is increasingly able to pull in all available talent. A century ago, the system eliminated child labor and sent them to school; now the bill comes due.

The situation is comparable to the American military accepting homosexuals and (soon) women into combat ranks. They do this not because they suddenly became enlightened or as a sign of moral progress, but rather for strictly technical reasons: the military needs all available bodies. Gays and women may feel 'empowered' by this, but it does them little good. Yes, they now have equal rights—to kill, and to die. Hardly a sign of progress.

And even if we grant that, in a handful of isolated ways, culture has become less cruel, this does not invalidate my general conclusion. By nearly any measure, and according to nearly any objective criterion, culture has been digressing for several centuries. And this fact, I claim, can only be explained by the advance of technology and its broadly dehumanizing effects.

Even the most superficial analysis shows that, culturally and sociologically, things are getting worse. In talking to my students, we like to contrast the quality of their lives today with that of earlier generations—such as my own. On the surface, little has changed. They have friends, and we had friends. They keep up on the news, we kept up on the news. They go to parties, we went to parties. They enjoy sports, we enjoyed sports. They have personal dramas, we had personal dramas. They struggle with coming of age, we struggled with coming of age. As the discussion plays out, we realize that *in no sense* is the quality of their lives any higher or any better than that of my generation or earlier ones.

But there is one large difference: their technological burden has greatly increased. To do the same things that previous generations did, they are now 'required' to carry a cell phone day and night, to create and maintain Facebook pages, to service Twitter accounts, to send and comment on countless pointless photos, to send and reply to dozens or hundreds of text messages daily. The average undergrad spends something like two and a half hours per day on mobile technology devices alone. The time burden is huge. And the financial burden is not inconsiderable: a typical smart phone costs $200 or $300, and the user incurs monthly fees of $100 or more—all to achieve the same ends that earlier generations did with minimal time commitment and no cost. On top of this, the psychological burden increases as well, as social pressure mounts to respond immediately, stay current, and be online.

What are young people getting for their large and growing investment of time, money, and stress? Are their friends better now than before? Are their lives richer? Are they morally stronger? Are they deeper thinkers? No. Even if we accept that quality of life is roughly comparable, that today's rap music is no different than yesterday's rock, that friends are always friends, that the young have always tended toward shallow thinking, superficial pleasures, and short-term rewards—even then, *they have lost ground.* They have taken on an undeniably greater burden for no gain in benefits. People today pay a higher price for, at best, the same quality of goods—which yields an absolute decline. Burdens are increasing on all fronts, and this could be justified only if the gains were also increasing on all fronts; but they are not. *If things are not unquestionably getting better, they are undeniably getting worse.*

One thing, however, does get better. One thing increases, day by day: technology—and the wealth and power of those who serve it.

CREATIVE RECONSTRUCTION

We presently live under a condition of technological determinism, but one in its first phase of existence—that of *dependence without control.* As explained previously, it is obvious that human technê, at the moment, is fully dependent on humanity for its existence and operation. Were we all to die tomorrow, the technological system would disintegrate soon thereafter. Our products, residue, and wastes would persist for thousands of years, but for all practical purposes, if we die, our technology dies with us. But despite this obvious fact, humanity cannot meaningfully control technology's progress or future development. It is autonomous but functionally dependent.

This condition will not last long. Soon—perhaps in a few decades—we will enter a second phase of determinism. Here, technology will become self-making and self-evolving. It will not need us, and in all likelihood it will not want us. It will achieve a radical autonomy at both the operational and existential levels. Needless to say, this situation, should we allow it to occur, will be one of extreme danger. The biosphere has no evolutionary experience with such ultra-technology and will lie defenseless before it. Science fiction could become a horrific reality, with the worst outcomes being realized. Any sane person, and any sane society, would do everything possible to avoid such a future scenario.

Earlier I explained that the terrestrial pyramid could evolve down two possible paths: growing order and complexity until all free energy is exhausted, or collapse due to intrinsic instabilities. There is, however, a third possibility. Though we cannot control its progress, we may control its regress. This involves humanity taking advantage of its innate creativity and will to preservation that derives from the Logos itself, redirecting the process of hierarchical evolution. This act I call *creative reconstruction*. Its goal would be to recreate, deliberately, the conditions of the Human Era. Not out of selfishness or some anthropocentric mania to rule the world, but simply out of a desire for self-preservation, for sanity, and to avoid a self-inflicted enslavement—and perhaps destruction. Call it a true nostalgia: a longing to return home (*nostos*, coming home).

Let me be specific. Creative reconstruction will involve, at a minimum, three things: a partial dismantling of the technological layer, a substantial global population reduction, and the restoration of a majority of the Earth's land area to true wilderness. These three would be brought back in line with conditions that existed during the Human Era. In parallel with these primary goals, other related objectives would follow, including reducing energy consumption and taking all necessary actions to forestall the second phase of technological determinism. A detailed discussion is beyond the scope of this work, but I offer here a few thoughts on the process.

The call to deconstruct "the machine" is an old one, reaching back at least to Butler's striking essay of 1863. In the 1960s, Lewis Mumford insisted upon "a deliberate large-scale dismantling of the Megamachine." Similar pleas have been issued by Marcuse, Illich, Sale, Kaczynski, and a host of environmentalists, eco-activists, and neo-Luddites. As expected, the meaning of the concept, and the means by which this is to occur, vary widely. In all cases, details are sorely lacking. Even so, I think we can provide a few specifics.

For one, deconstruction clearly does *not* mean the attempted elimination of all technology—which, as I have emphasized throughout, is not only undesirable but impossible. But it does mean a dramatic retrenchment of the contemporary technosphere. This brings us back to the old debate about so-called modern technology as a kind of turning point. The pantechnical analysis I presented earlier suggests that our peak time in history ended around 1200 AD, and that therefore this ought to be the standard by which a manageable technology could be maintained. A case could be made for Renaissance-era technology, but certainly it would call for preindustrial techniques.

Second, the focus is simply on *the technology* and not the many various and often irrational social practices of the past. This point should be obvious, but based on experience, it is not. An endorsement of, say, medieval technology does not imply, in the least, an endorsement of witch burning, bloodletting, Inquisitions, dogmatic theocracies, and so on. The inability of people, even educated ones, to separate such things is disturbing. Clearly it is possible to achieve an enlightened, rational, civilized, even *advanced* society using only the most basic of tools and techniques. The Athenians did far better with much less, and we could do better still. This would be our only aim.

Third, deconstruction can, in principle, be as gradual and carefully managed as we wish. Nothing entails a sudden, chaotic, or violent process. If we have any pretense to being a rational animal, we have many options at our disposal. We have rationally built up the technological system over hundreds of years, and we may, if desired, rationally deconstruct it over hundreds of years.

Then we have the question of overpopulation. This issue has been perennially cited as a problem without solution. This is misleading. The solutions are known; it is the means of implementation that are the challenges. The global figure is currently 7.1 billion, growing by some 200,000 daily. It will almost certainly reach 9 or 10 billion by mid-century. Stringent, coordinated, global action would be required simply to flatten out the growth curve. Toward this end, there are many rational, compassionate, and non-coercive options available. Some measures are obvious: better education on family planning, freer discussion of birth control options, and government funding of contraceptives, for a start. Taxation laws could be changed to penalize those with more than two children rather than encouraging them, as at present. Beyond these basic steps, governments could provide free sterilization for childless adults—or even more, offer financial incentives for such a procedure. Alternatively, a system of birth quotas or licenses could be implemented—tradable and negotiable rights to have children.[1] Collectively, such options would undoubtedly have an impact.

The vital point, again, is that none of this has to be implemented rapidly or harshly. It needs to be *initiated* soon, but the process itself can be reasonably gradual. As with a technology drawdown, if the rate of population decrease matched the past rate of increase, it should be acceptable to society. We took four centuries to grow from 500 million to present levels, and we might well need the same to return to that state. All well and good—if we have four centuries to spare.

On the wilderness issue, the challenges are equally severe. Many developed nations have virtually none left. Others have extensive undeveloped land, but this may only superficially count as wilderness. For example, the U.S. has, by all appearances, vast untouched land in the west. Much of this, though, is 'managed' land with extensive lumbering, mineral extraction, hunting, and cattle grazing. Of the roughly 2 billion acres of land in the continental U.S., less than 1% is true wilderness—defined as large areas with unpaved roads, free-burning fires, no permanent human settlements, no hunting, no direct pollution, and no resource extraction. Two centuries ago, more than 75% of the U.S. was in such a state. Today, we have the means to recover it. Large areas of land are under governmental control, and these could, over time, be restored to true wilderness. As before, given a century or two, we could surely convert much of the land to a near-original condition.

I add here that I make no claim about the actual feasibility of these actions. In fact, by any conventional accounting, they are utterly impossible. I include them simply as part of an ideal and rational approach that would follow from taking the analysis seriously. And furthermore, in light of the current situation, we have few alternatives. All the so-called reasonable and practical solutions that have been proposed fail to match the unprecedented severity of the problem. Not only will they not work, in many cases they are actually counterproductive; they ameliorate some local condition while allowing the underlying technological causes to grow worse. Creative reconstruction follows as a logical consequence of a pantechnical reading of the universe. Protests about its unreasonableness are irrelevant. I make no claim that it is reasonable—only that it is necessary.

At present I see no acceptable alternative. A radically smaller, detechnified human species is the planet's future, if we are to *have* a future. We can get there through a slow, careful, and rational process. If we do not, nature will likely drive us there herself—and she will be neither slow, nor careful, nor rational. When nature acts against errant life forms, she is ruthless.

I am not optimistic that we will act as we must. The collective being is too crude and too short sighted to rationally execute its own reduction, and to restrict the technological system that is its sole source of existence. Movement toward reconstruction will happen only through the actions of individuals or small groups. Whether this will be sufficient to have an impact remains to be seen.

WHENCE THE FUTURE?

Metaphysics has consequences. It helps to explain why things happened in the past, why they are happening at present, and how they are likely to proceed in the future. The pantechnical thesis offers an explanation of past evolution, including the appearance of life on Earth, of complex life, of human beings, and of the technical sphere. As such, it functions as an immutable law of nature. But this does not mean that it cannot be circumvented, at least in certain aspects and for limited spans of time. Gravity is also an immutable law, ubiquitous and eternal. And yet things defy gravity all the time. Every animal that climbs or flies defies it. Every time we take an airplane trip, go up in an elevator, or merely stand up, we defy it. But we know the old saying: What goes up must come down. Our defiance can only last for a limited time.

In an analogous fashion, we can likewise defy pantechnical evolution—at least for a time, and in limited ways. Creative reconstruction is one proposal to defer, for an extended time, the more pernicious effects of technological advance on this planet. For how long and to what degree of success, we cannot be sure. But it seems to be a viable option, and I doubt that we have better alternatives. To continue on the present path is suicidal. It means to yield totally and completely to the evolutionary forces that portend our demise. It would be like a child who wishes to be old; we laugh at such immaturity, knowing full well that age will come soon enough and that youth is a blessing. But we as a species are in the same condition. We wish for technological maturity, which brings certain powers but also carries us that much closer to old age and death.

To recall an old metaphor: If time is a river leading to a tumultuous waterfall, and we are midstream in a boat, what ought we to do, rationally? The river carries us forward to the inevitable end—but should we also paddle forward, hastening that day? At present we are paddling forward, furiously, like madmen. Is this rational? Is it sane? Shouldn't we rather, at least, stop paddling? Or better: paddle in reverse? We cannot stop the river, but we might at least maximize our time on it.

A truly normative metaphysics has an obligation to offer some commentary on human prospects for the future. As I have said, there is very little chance that large-scale societies—nation-states or the global community—will act in time. Such social beings are too new, evolutionarily speaking, and have no ability to anticipate subtle dangers, to act morally (in a human sense), or to exercise preemptive self-restraint. They act only under the crudest of self-survival motives or via a brute will to power. But the dangers posed by technology, including large-scale environmental crises and overpopulation, will only be acted upon when the danger is unmistakable—at which time, of course, it will be far too late.

Those who might claim that large-scale organizations *do* have such anticipatory capabilities must of necessity believe that the dangers are low or nonexistent. This would be the only explanation for the fact that we are doing almost nothing about them. Society is constantly monitoring world events for such perilous threats and has evidently decided that serious technological dangers are limited to one or two categories: military threat or terrorist attack. And in the face of such dangers, society has concluded that it must *relentlessly advance its technical capabilities*, to track and keep ahead of the enemy. By such logic, technology is not the enemy; it is our protector and savior. Technology must advance. We have no choice.

A similar thought process relates to the dangers of environmental destruction and overpopulation. Yes, they say, global climate change is a problem, but this can be solved only by newer and better technologies: wind farms, fuel cells, solar panels, nuclear fusion, and the like. Yes, the world is becoming overcrowded; therefore we need newer and better food production technologies, new energy technologies, new communication and organizational technologies. *Technology is not the problem, it is the solution*—or so we are told. Again, an utterly opposing conclusion to that based on a pantechnical metaphysics. The competing views are clear.

Here, then, is a prediction. Advancing technology will expose humanity and the entire planet to increasing peril. Because the large-scale organizations of the world are incapable of understanding this, little or no action will be taken to preempt the technological system. In fact, precisely the opposite: as the global situation degrades, the rush to develop technology will only accelerate, which will in turn exacerbate the problem. A deadly feedback loop will be established, one that will be nearly impossible to forestall. We will soon enter a downward spiral of rapidly advancing technology combined with rapidly declining quality of life. Technology will accelerate toward true autonomy—phase two determinism, and perhaps a singularity—as humanity and nature plunge toward a breaking point.

Thus we will find ourselves in a race. The question then will be this: Will technology reach the point of autonomy *before* humanity and the global ecosystem collapse? If it does, we can scarcely imagine the outcome; but nearly any conceivable future will be disastrous. If it does not, if humanity collapses and takes down technology with it, what then? Some of us will survive, surely, but perhaps only a few percent of current numbers. The survivors will be forced to live on a depleted and chemically altered planet, and with none of the previous technological aids. They will be like hunter-gatherers in a desert, scrapping for food under the most difficult of conditions. But at least humanity will survive, and the planet will begin to heal. This, sadly, is the best alternative that we have—short of creative reconstruction.

But the Pollyanna technophiles will have none of this. Technology will stay under our control, they say, and will progressively solve all problems. Any difficulties, including climate change and overpopulation, are only of a temporary nature; they will rapidly be overcome as new technologies emerge. Under this scenario, life for all people, as well as the health of the global ecosystem, should *very* quickly get better—*much* better. We should very soon see rapid and undeniable increases in human physical and mental well-being. Climate change should very soon be solved by some combination of carbon-capture devices and by new, emission-free energy and transportation technologies. Extinct species will be restored by advanced biotechnologies and depleted oceans restocked with their native fish. This is what we *must* expect—given that technology is exponentially increasing in power and scope, and assuming that its purpose is to improve the human condition.

This is not an idle dispute. And it is not merely theoretical. We need wait only a few years, perhaps two decades at most, and things will become quite clear. We will know which road we are heading down. And then the value or disvalue of pantechnical metaphysics will be obvious to all.

* * * * *

We live in an age of deceptive normality. So much of life continues as it has for generations: children are born, people get married, we go to school, we pursue our careers, we attend birthdays and funerals. Seasons come and go, the sun rises and sets each day, the sky is the same eternal blue that buoyed the spirits of our distant ancestors. The world falls asleep during the long winter months and reawakens, anew, each spring. Nature carries on, seemingly unperturbed.

But at a deeper level, we know that all is not well. The planet is straining under the growing mass of humanity. Our toxic byproducts pervade every corner of the globe. Man-made radiation saturates the biosphere. The fish are disappearing from the seas. Virgin land is vanishing before our eyes. The chemistry of the atmosphere is being radically altered. Species are dying out every day, never to be seen again. The human psyche is at the breaking point. We are hurried, always hurried, running ever faster, day by day, to win an ever smaller prize.

And technology proceeds apace.

* *

NOTE

1. Also called 'birth credits.' This system has long been promoted by Kenneth Boulding and Herman Daly.

Epilogue

It was on October 15, 2009, that I finally caught up with Bill Joy. I had spent a couple years trying to reach him by letter and email as part of some early research on the present work. I was curious about two things. First, I wanted to know how his thinking had changed since the publication of his famous essay of 2000 and, second, why he had written no follow-up essays,[1] and indeed why he had publicly said nothing further at all since that time.

Having had no success reaching him, I was pleased to learn that he would be coming to my home institution, the University of Michigan—his alma mater—to accept a lifetime achievement award from the College of Engineering. As part of the ceremony, Joy would give a short speech on his latest activities and answer a few questions from the audience.

As his talk progressed, we learned that he departed from Sun Microsystems in 2003 and was now, since 2005, a partner at the venture capital firm Kleiner Perkins. He spoke about recent developments at Kleiner, with special emphasis on his work on green technologies: high-efficiency batteries, wind turbines, and advanced solar cells. In effect, Joy was helping Kleiner to profit from the new, 'good' technologies that will save us from the dangers wrought by the old, 'bad' technologies. Notably, there is no profit in relinquishment.

It was an unexceptional talk and lacked the slightest reference to his past controversial writing. Upon completion, I rose and asked my two questions: why he had written nothing since, and what his latest thinking was on the looming GNR (genetics, nanotechnology, robotics) dangers. He gave a short and circumspect reply, one that amounted to the following: "I tried to raise the concern, and no one wanted to listen, so I dropped it." I was surprised—could it really be the case, with the world-destroying dangers posed by ubiquitous GNR technologies, that someone as prominent and knowledgeable as he could simply "drop it?" One attempt; failure; that's it?

At one level, of course, his piece was a great success; it was one of the most widely cited and discussed essays on technology, ever. As I stated in Chapter 8, it could have been the ideal launching point for a new, serious discussion on the limits of modern technology. But needless to say, this did not happen. Even though Joy earned the admiration of many technology skeptics, his critics evidently held the upper hand. In the rarefied tech circles in which he travels, Joy apparently found little or no support. Industry had no interest in relinquishment, technology insiders had no interest, government had no interest—of course. The agents of the system are in no position—theoretically, psychologically, or financially—to oppose the very system that they themselves have constructed and profit from. That Joy himself could do such a thing once was something close to miraculous. Evidently he learned that, in a quintessentially rational system, miracles do not exist.

In the end we are torn over the 'Bill Joy affair.' Should we praise him as one who had the courage to speak out on the widely ignored but potentially catastrophic dangers of technology? Or should we condemn him for folding so quickly? Perhaps a bit of both. Of course he could reengage with the subject at any time—though this appears unlikely. In a recent and very brief interview in *Wired*, the interviewer recalls Joy's famous article of 2000 and then asks him about his present worries.[2] Declining to take up the issues of bioengineered pathogens or nanoreplicators, Joy's leading worry these days is "self-replication on the Internet, where botnets can take down the whole network." Now, we need to be clear on this: His main concern at present is not the well-being—let alone the survival—of humanity or nature, but rather *the integrity of the Internet*. The health of the technological system itself is now at the top of his list. The interview closes with Joy explaining the wonders of the custom, high-tech yacht that he is building for himself. By all indications, he will not be rejoining the antitech crusade any time soon.

The original fears, however, have not gone away. In his article of 2000, Joy elaborated on the many dangers of superintelligent computers. These would arrive, he warned, by the year 2030. At the time, that was three decades away. We are now halfway there. The window of opportunity closes rapidly.

****

I will close this brief epilogue and this book with a handful of final observations.

- Throughout the present work, I have spoken of the technological system as a quasimental or semiconscious being, endowed with a will and manifesting certain desires. This is not a metaphor, and it is not careless anthropomorphism. Though I am unable to elaborate here, there are good philosophical reasons and a strong historical precedent for viewing the techno-social system as a thing in itself, embodying mind and will.

  At first glance this seems unlikely. And yet, is it any less probable that a collection of some 100 billion neurons, each of them a relatively simple cellular mechanism, can manifest a singular human intelligence, complete with conscious awareness, desire, and will? This latter fact is undeniable. Why, then, is the notion of a comparably unified, experiential entity, formed by the collective activity of some 7 billion complex human beings in conjunction with their billions of complex machines, so far fetched? I submit that it is not. It is the outcome of entirely parallel forces and processes. This idea is not mere idle speculation; it goes a long way toward providing an explanation of how and why the system functions as it does.

- The determinism and universal scope of the Pantechnikon, as I have outlined it, might suggest to some a form of fatalism: that there is nothing we can do, that the system will swamp us, and that the planet is doomed. In one sense, of course, this is true. We all are 'doomed,' given that we all will die someday. The planet too is doomed, in that it will eventually be swallowed up by our expanding sun at the end of its life. Perhaps the universe is doomed as well—it may either evolve to some static end-state or be recycled as part of some new cosmic order. These facts are simply physical realities for any concrete object or system.

  More to the point, though, it is true that man-made technology may well overwhelm us long before any of these grand cosmic trends materialize. But this has not happened yet, and notwithstanding the exponential growth of technology, we still have time for action. Bear in mind: It is *never* the case that there is *nothing* we can do. As long as living, breathing human beings exist in anything like our present form, we can take action to affect the course of future events. Even if we are destined to lose in the long run, it may be a *very* long run, and any rational species worthy of the name would seek to sustain itself and its surroundings in the most diverse, stable, and just manner possible. Pantechnical metaphysics is no basis for either quietude or despair. On the contrary; it is a call to action.
- Further to the foregoing, it may appear to some that I have adopted an overly pessimistic tone, in that I address the harsh downsides of modern technological life. It strikes me how unusual this approach is today. Apart from the occasional sci-fi action film, modern society is relentlessly optimistic about technology. Schools promote it as a key to social progress and enlightenment. Corporations promote it as the means to economic prosperity. Governments promote it as central to national strength and security. Tech corporations, like Google and Apple, promote their products as essential to the good life; lately they have taken to running television commercials parading a sequence of happy, dancing young people, and even children, gazing lovingly into the face of their newest iProduct. This, of course, is how it must be: In a technological society, *everyone* must be upbeat about technology. Naysayers are not allowed. In fact, what we witness in this process is nothing less than the self-promotion of technology. The system is very good at propagandizing on its own behalf.
- Some may say: What about the good side of technology? Hasn't this been left this out? Where is the whole picture? In a sense, I have addressed this. For one thing, the history of philosophy of technology is almost exclusively a story of critical remarks. It is not as if I picked out only the negative commentaries; positive ones do not exist. It seems

that any thinking person, regardless of era, who examined the question of technology found it problematic.

Second, the metaphysics of technology necessarily points to detrimental outcomes for humanity. This is not my choice; it is simply a consequence of an unbiased reading of the workings of nature. Despite what we may wish, and despite our religious anthropocentrism, the universe does not exist for our sake. Some will call this harsh; I call it reality.

On the pragmatic side, this much is clear: Technology is obviously useful, fun, and practical. Such things are essential to its coming into being. But every new device has an inevitable counterpoint, namely, that it marks one more small step in the emergence of a dehumanizing, suffocating, and perhaps ultimately life-destroying system. Even if we grasp this point, we still resist taking action: the step is so small, and the benefit comparably so much greater, that we think it harmless. What's the problem, we say, with upgrading a cell phone? Or buying one more CD? Or sending a few text messages? Or keeping up on Facebook once in awhile? And yet—billions of people, making billions of small, incremental decisions, relentlessly grow the system.

The Internet was originally designed for people to communicate and access information. Over time, more and more devices, sensors, smart phones, and tablet computers became connected to it—to the point where, today, there are *more devices on the Internet than people.* This phenomenon has a name: the 'Internet of Things.' At present there are roughly 25 billion devices on the Internet, a figure that will double by 2020. Compare this to the 2 billion people who use the Internet, and we can see that humanity is outnumbered by an order of magnitude. And the gap will only grow. The Internet now exists primarily for machines, not for men. This is one more sign of an autonomous system coming into being.

- My synthetic metaphysical approach is, as I have emphasized, deeply normative. An important aspect of this is the passing of judgments—evaluative, pragmatic, and moral. This too is something that is evidently quite difficult for contemporary philosophers. Decades of objectivist, analytic thinking have seemingly eroded our collective capacity and stomach for making such judgments.

  When we are judgment averse, we avoid making assessments. We become unable to discriminate—and discrimination is certainly a key part of any healthy intellectual outlook. Philosophers of technology are no more immune to this problem than others. Consider this fact: Hardly a one today will declare technology neutral. This is an almost universal philosophical consensus: *technology is not neutral.* And yet, *none dare criticize it*—at more than a superficial level. As I explained throughout, even the so-called critics of technology can barely muster the mildest words of condemnation. To judge by the writings of major figures like Borgmann, Ihde, and Feenberg, and lesser names like

Tabachnick and Agar, not to mention faux-philosophers like Turkle, Morozov, Powers, and Carr, a fundamental critique is all but taboo.

Again, this serves a purpose: In a judgment-averse, if not to say judgment-incapable society, even the grossest indignities are perpetuated, and even the gravest injustices go unpunished. In the short term, this attitude benefits the technological system, which advances unhindered. In the long run, it is suicidal. It prevents us from making the critical assessments needed to confront the technological power structures of modern society and from making the necessary corrections to our present course of action. Let us hope that normativity can be restored to political and moral philosophy, at a minimum.

The prospects, though, are not good. As a society, decades of inclusionism, relativism, and political correctness have severely undermined our general taste for judgmental talk. And the dominant, passivist strain of Christianity is certainly a factor as well. Needless to say, I disagree. "Judge, and be judged"—that's *my* sermon on the mount.

- The addictive and hypnotic power of technology is certainly impressive. Orwell compared it to a drug: useful, dangerous, and habit forming. Kaczynski drew an analogy to a barrel of wine in front of an alcoholic.[3] The data I presented in Chapter 11 strongly suggest an increasingly potent addictive force. I called it a 'mental AIDS,' in that it has the power to destroy the means of thinking about solutions. And it is all the more dangerous given that it functions as a force of nature, operating with the imperatives of evolution.

  As I have repeatedly noted, contemporary thinkers do not see things this way. "No technology is, has been, or will be a natural force," says the historian David Nye, confidently.[4] Philosophers of technology undoubtedly agree with him. Who is right? Sooner than we like, we shall see.
- At several points in the present work, I have argued, explicitly or implicitly, that technology produces an effect on the mind of the user. Metaphysically, this is so because technology is embodied logos. All creation, in fact, is a reification of mind. This cannot but produce a psychological effect.

  And yet the common view is that technology is tools, and tools are simply there for us to use as we see fit. Any psychological effect can only come from the maker or the user—that is, from human beings. In a remarkably archaic bit of philosophizing, Pitt (2000) flat out endorses this crudely instrumentalist view. "It is not the machine that is frightening," he says reassuringly, "but what some men will do with the machine" (99). He finds a ring of truth in the old canard: 'Guns don't kill, people do.' Technological autonomy "is no problem." "The tools by themselves do nothing."

  I think I have shown the inadequacy of this stance. The example of handguns is instructive. If it is admitted that we in America have a problem with gun violence—something like 30,000 people die each year by firearms—we then ought, rationally, to take action. The problem, though, is not simply one of hardware; it is also a psychological problem. Like every piece of technology, a gun is a psychological device. It exists primarily for its psychological effects. It provides a sense of security to the owner, presenting the threat of deterrence to one's adversary. A gun makes one *feel powerful* and thus act on the basis of that feeling. Guns create a 'gun as solution' mentality. It is very difficult to alter this psychological component, which is deeply rooted in our genetic structure. Therefore we are compelled to address the technology itself if we hope to alleviate the problem. The solution to gun violence is not *better* guns or *safer* guns, and certainly not *more* guns, but rather—*fewer guns*.

It is the same with technology in general. Technology poses severe problems, creating complex and deeply rooted psychological effects. It makes one feel secure and powerful. It creates a 'technology as solution' mentality. On a large scale, these psychological effects are almost impossible to address. Thus the only true solution to the problem of technology is not better technology or safer technology, and certainly not more technology, but rather—less technology. This is the general sentiment underlying my proposal for creative reconstruction.

- On January 24, 2014, Pope Francis declared that "the Internet . . . is something truly good, a gift from God." There is a reason philosophers do not consult theologians.

I am repeatedly struck by passing comments from notable thinkers on the problem of technology. Consider Henry Adams, grandson and great-grandson of presidents and prominent historian in his own right. Writing to his brother Charles in 1862, he describes science as a horse upon which we ride, but which no longer heeds our command:

> Man has mounted science, and is now run away with. I firmly believe that before many centuries more, science will be the master of man. The engines he will have invented will be beyond his strength to control. ([1862] 1982: 290)

Adams rightly anticipates that loss of control over science and technology entails disaster: "Some day science may have the existence of mankind in its power, and the human race commit suicide, by blowing up the world." This is a remarkable early observation, one that appeared a year before even Butler's initial prophetic essay.

Nearly a century later, Wittgenstein offered one of the more oblique aphoristic criticisms in history:

> It isn't absurd to believe that the age of science and technology is the beginning of the end for humanity; that the idea of great progress is

> a delusion (*Verblendung*), along with the idea that the truth will ultimately be known; that there is nothing good or desirable about scientific knowledge, and that mankind, in seeking it, is falling into a trap. It is by no means obvious that this is not how things are. ([1947] 1980: 56e)

A backhanded critique, to be sure, and unfortunately without further commentary.

But a much blunter and more direct statement came from the pen of Albert Einstein. Writing to his friend Heinrich Zangger in 1917, he expresses disgust at the state of European society, which was in the latter stages of the protracted World War I. "How is it possible that this culture-loving time is so terribly amoral?" he laments. Perhaps it would be best, Einstein says, for Europe to completely collapse. And then a striking line: "Our entire much-praised technological progress, and civilization generally, could be compared to an axe in the hand of a pathological criminal".[5] An appropriate epitaph for the technological age.

* * * * *

Heidegger, drawing on Dessauer, said that technology is a window on reality. They both were right. Heidegger, drawing on Juenger, said that technology is the supreme danger. Again, they were right. Ellul said that the technological system is an autonomous force in the world, beyond our control. He was right. These are the three most important metaphysical insights of the past century. And yet they do not explain anything. A much deeper metaphysical infrastructure is demanded, and my thesis of the Pantechnikon is one such proposal.

Where we are not ascending, we are declining. This is a law of evolution. By failing to rein in the pernicious forces at work in the world, we hasten our downward spiral. The costs of our technological civilization mount, and we gain precious little in return. Technology is a net loss to human well-being. As it advances, we decline.

Our time of ascendancy ended 10,000 years ago. For several thousand years thereafter, we basked in the glory, resting on our laurels. All the while, the usurper built up its power. A thousand years ago it came to the fore—in the humble form of the mechanical clock—declaring itself the new world-force. And we obeyed. Thus on the whole, and despite periodic exceptions, humanity has been on a downward trajectory for ten millennia. We have the power to change this. Our time is not yet at an end. The decline can be forestalled.

Plato held that there were four prime virtues: wisdom, courage, temperance, and justice. Wisdom—the ability to see deeply into reality. Courage—to envision an uplifting and noble future for humanity. Temperance—to know moderation. And justice—that respects the dignity of all beings. These were once seen as the keys to our salvation. We call upon them again.

END

NOTES

1. Though this is not quite true. One short and inconsequential piece did appear the following summer; see Joy (2000b).
2. "Bill Joy on venture capital, clean tech, and big boats," *Wired* (April 16, 2013).
3. Kaczynski (2010: 103).
4. Nye (2006. 19).
5. "*Unser ganzer gepriesene Fortschritt der Technik . . . ist der Axt in der Hand des pathologischen Verbrechers vergleichbar*" ([1917] 1998: 562). I had long written off this phrase as Internet legend until I was able to track down the original text.

References

Aboujaoude, E. (2011), *Virtually You,* New York: W. W. Norton.
Aboujaoude, E. et al. (2006), "Potential markers for problematic Internet use," *CNS Spectrums,* 11: 750–755.
Adams, H. (1982), *The Letters of Henry Adams* (vol. 1), Cambridge, MA: Harvard University Press.
Agar, N. (2010), *Humanity's End,* Cambridge: MIT Press.
Agrippa, H. ([1530] 1676), *Of the Uncertainty and Vanity of the Worldly Arts and Sciences,* London: S. Speed.
Alexander, S. (1920), *Space, Time, and Deity,* New York: Macmillan.
Anderson, C. et al. (2010), "Violent video game effects on aggression, empathy, and prosocial behavior in Eastern and Western countries: A meta-analytic review," *Psychological Bulletin,* 136: 151–173.
Angrist, J. and Lavy, V. (2002), "New evidence on classroom computers and pupil learning," *The Economic Journal,* 112: 735–765.
Aristotle. (1984), *The Complete Works of Aristotle* (2 vols.), Princeton, NJ: Princeton University Press.
Arthur, B. (2009), *The Nature of Technology,* New York: Free Press.
Ayala, F. (1970), "Teleological explanations in evolutionary biology," *Philosophy of Science,* 37: 1–15.
Bacon, F. ([1620] 2000), *The New Organon,* Cambridge: Cambridge University Press.
Baines, L. (1997), "Future schlock: Using fabricated data and politically correct platitudes in the name of educational reform," *Phi Delta Kappan,* 78: 492–498.
Bak, P. (1996), *How Nature Works,* New York: Copernicus.
Barnes, J. (1987), *Early Greek Philosophy,* New York: Penguin.
Baudrillard, J. ([2001] 2012), *Telemorphosis,* Minneapolis: University of Minnesota Press.
Bergson, H. ([1896] 1911), *Matter and Memory,* London: Allen and Unwin.
Bergson, H. ([1907] 1911), *Creative Evolution,* New York: Holt.
Berman, M. et al. (2008), "The cognitive benefits of interacting with nature," *Psychological Science,* 19: 1207.
Berman, M. et al. (2012), "Interacting with nature improves cognition and affect for individuals with depression," *Journal of Affective Disorders,* 140: 300–305.
Bird, C. and Emery, N. (2009), "Insightful problem solving and creative tool modification by captive nontool-using rooks," *Proceedings of the National Academy of Sciences,* 106: 10370–10375.
Blackburn, S. (2000), "Enquivering," *The New Republic,* 223: 43–48.
Blackburn, S. (2005), *Truth,* Oxford: Oxford University Press.
Blamauer, M. (ed.) (2011), *The Mental as Fundamental,* Frankfurt: Ontos Verlag.

Blauner, B. (1964), *Alienation and Freedom*, Chicago: University of Chicago Press.
Bohm, D. (1986), "A new theory of the relationship between mind and matter," *Journal of the American Society of Psychical Research*, 80: 113–135.
Bohm, D. (1990), "A new theory of the relationship between mind and matter," *Philosophical Psychology*, 3: 271–286.
Borgmann, A. (1984), *Technology and the Character of Contemporary Life*, Chicago: University of Chicago Press.
Borgmann, A. (1992), *Crossing the Postmodern Divide*, Chicago: University of Chicago Press.
Borgmann, A. (1999), *Holding on to Reality*, Chicago: University of Chicago Press.
Borgmann, A. (2000), "Reply to my critics," in E. Higgs et al. (eds.), *Technology and the Good Life?* Chicago: University of Chicago Press.
Borgmann, A. (2006), *Real American Ethics*, Chicago: University of Chicago Press.
Borgmann, A. (2007), "Albert Borgmann," in J. Olsen and E. Selinger (eds.), *Philosophy of Technology: Five Questions*, Automatic Press/VIP.
Borgmann, A. (2008), "Does the crisis of global warming require a spiritual solution?" *Truth and Grace*, 25: 38–43.
Borgmann, A. (2012), "Science, ethics, and technology and the challenge of global warming," in D. Scott and B. Francis (eds.), *Debating Science*, Amherst, NY: Humanity Books.
Bos, A. (2003), *The Soul and Its Instrumental Body*, Boston: Brill.
Bos, A. (2008), *Aristotle, On the Life-bearing Spirit*, Boston: Brill.
Braun, E. (1995), *Futile Progress*, London: Earthscan.
Brown, L. (2009), "Could food shortages bring down civilization?" *Scientific American*, 300(5): 50–57.
Bruno, G. ([1584] 1998), *Cause, Principle, and Unity*, Cambridge: Cambridge University Press.
Butler, S. ([1863] 1968), "Darwin among the machines," in S. Butler, *The Shrewsbury Edition of the Works of Samuel Butler* (vol. 1), New York: AMS Press.
Butler, S. ([1865] 1968), "Mechanical creation," in S. Butler, *The Shrewsbury Edition of the Works of Samuel Butler* (vol. 1), New York: AMS Press.
Campuzano, L. et al. (2009), *Effectiveness of Reading and Mathematics Software Products*, Washington, DC: U.S. Department of Education.
Capasso, L. (2005), "Antiquity of cancer," *International Journal of Cancer*, 113: 2–13.
Caputo, J. (1988), "Demythologizing Heidegger," *Review of Metaphysics*, 41: 519–546.
Carlyle, T. ([1829] 2007), "The signs of the times," in G. Himmelfarb (ed.), *The Spirit of the Age*, New Haven, CT: Yale University Press.
Carnap, R. (1935), *Philosophy and Logical Syntax*, London: Kegan Paul, Trench, Trubner and Co.
Carr, N. (2008), "Is Google making us stupid?" *The Atlantic*, 302: 56–63.
Carr, N. (2010), *The Shallows*, New York: W. W. Norton.
Cash, H. et al. (2012), "Internet addiction: A brief summary of research and practice," *Current Psychiatry Reviews*, 8: 292–298.
Cecchetto, D. (2013), *Humanesis*, Minneapolis: University of Minnesota Press.
Cerbone, D. (2008), *Heidegger: A Guide for the Perplexed*, New York: Continuum.
Chaisson, E. (1997), "The rise of information in an evolutionary universe," *World Futures*, 50: 447–455.
Chaisson, E. (2001), *Cosmic Evolution*, Cambridge, MA: Harvard University Press.
Chaisson, E. (2006), *Epic of Evolution*, New York: Columbia University Press.
Chaisson, E. (2011a), "Energy rate density as a complexity metric and evolutionary driver," *Complexity*, 16: 27–40.
Chaisson, E. (2011b), "Energy rate density, II: Probing further a new complexity metric," *Complexity*, 17: 44–63.

Chaisson, E. (2012), "A singular universe of many singularities," in A. Eden et al. (eds.), *Singularity Hypotheses,* Berlin: Springer-Verlag.
Chaisson, E. (2013), "Using complexity science to search for unity in the natural sciences," in C. Lineweaver et al. (eds.), *Complexity and the Arrow of Time,* Cambridge: Cambridge University Press.
Christakis, D. et al. (2004), "Early television exposure and subsequent attentional problems in children," *Pediatrics,* 113: 708–713.
Chou, H. and Edge, N. (2012), "'They are happier and having better lives than I am,'" *Cyberpsychology, Behavior, and Social Networking,* 15: 117–121.
Chroust, A.-H. (1960), "The origin of 'metaphysics,'" *Review of Metaphysics,* 14: 601–610.
Chuang-Tzu. (1999), "Heaven and Earth," in S. Hamill and J. Seaton (eds.), *The Essential Chuang-Tzu,* Boston: Shambhala.
Cicerone, R. (2005), "Current state of climate science," retrieved from http://www7.nationalacademies.org/ocga/testimony/Climate_Change_Science_and_Economics.asp
Cingel, D. and Sundar, S. (2012), "Texting, techspeak, and tweens," *New Media and Society,* 14: 1304–1320.
Cleve, F. (1969), *The Giants of Pre-Sophistic Greek Philosophy,* The Hague: Martinus Nijhoff.
Conway, J. and Kochen, S. (2006), "The free will theorem," *Foundations of Physics,* 36: 1441–1473.
Conway, J. and Kochen, S. (2009), "The strong free will theorem," *Notices of the AMS,* 56: 226–232.
Cooper, D. (1995), "Technology: Liberation or enslavement?" in R. Fellows (ed.), *Philosophy and Technology,* Cambridge: Cambridge University Press.
Cordes, C. and Miller, E. (eds.) (2000), *Fool's Gold: A Critical Look at Computers in Childhood,* College Park, MD: Alliance for Childhood.
Cuban, L. (1996), "Techno-reformers and classroom teachers," *Education Week,* 16.
Cullen, B. (1979), *Hegel's Social and Political Thought,* New York: St. Martin's Press.
Dalton, T. (2010), "Nietzsche and the origins of Christianity," retrieved from www.theoccidentalobserver.net/authors/Dalton-Nietzsche2.html
David, R. and Zimmerman, M. (2010), "Cancer: An old disease, a new disease or something in between?" *Nature Reviews Cancer,* 10: 728–733.
Descartes, R. ([1637] 1985), "Discourse on the method," in *The Philosophical Writings of Descartes* (vol. 1), Cambridge: Cambridge University Press.
Descartes, R. ([1641] 1985), "Meditations," in *The Philosophical Writings of Descartes* (vol. 2), Cambridge: Cambridge University Press.
Dessauer, F. ([1927] 1972), "Technology in its proper sphere," in C. Mitcham and R. Mackey (eds.), *Philosophy and Technology,* New York: Free Press. Partial translation of *Philosophie der Technik.*
deVries, S. et al. (2003), "Natural environments—healthy environments?" *Environment and Planning A,* 35: 1717–1731.
Diamond, J. (1987), "The worst mistake in the history of the human race," *Discover,* 8(5): 64–66.
Divan, H. et al. (2008), "Prenatal and postnatal exposure to cell phone use and behavioral problems in children," *Epidemiology,* 19: 523–529.
Divan, H. et al. (2012), "Cell phone use and behavioural problems in young children," *Journal of Epidemiology and Community Health,* 66: 524–529.
Dokoupil, T. (2012), "Tweets. Texts. Email. Posts. Is the onslaught making us crazy?" *Newsweek,* 160(3): 24–30.
Dreyfus, H. (1984), "Between techne and technology," *Tulane Studies in Philosophy,* 32: 23–35.

Dreyfus, H. and Spinosa, C. (2003), "Further reflections on Heidegger, technology, and the everyday," *Bulletin of Science, Technology, and Society,* 23: 339–349.

Dubos, R. (1968), *So Human an Animal,* New York: Scribner.

Dummett, M. (1978), *Truth and Other Enigmas,* Cambridge, MA: Harvard University Press.

Dusek, V. (2006), *Philosophy of Technology,* Oxford: Blackwell.

Dynarski, M. et al. (2007), *Effectiveness of Reading and Mathematics Software Products,* Washington, DC: U.S. Department of Education.

Dyson, F. (1979), *Disturbing the Universe,* New York: Harper and Row.

Eaton, W. (1974), "Residence, social class, and schizophrenia," *Journal of Health and Social Behavior,* 15: 289–299.

Edwards, P. (2004), *Heidegger's Confusions,* Amherst, NY: Prometheus.

Einstein, A. (1998), *The Collected Papers of Albert Einstein* (vol. 8a), Princeton, NJ: Princeton University Press.

Eknoyan, G. (2006), "A history of obesity, or how what was good became ugly and then bad," *Advances in Chronic Kidney Disease,* 13: 421–427.

Ellul, J. ([1954] 1964), *The Technological Society,* New York: Random House.

Ellul, J. ([1977] 1980), *The Technological System,* New York: Continuum.

Ellul, J. ([1983] 1989), "The search for ethics in a technicist society," *Research in Philosophy and Technology,* 9: 23–36.

Ellul, J. ([1988] 1990), *The Technological Bluff,* Grand Rapids, MI: Eerdmans.

Engel, S. et al. (2010), "Prenatal phthalate exposure is associated with childhood behavior and executive function," *Environmental Health Perspectives,* 118: 565–571.

Farris, R. and Dunham, H. (1939), *Mental Disorders in Urban Areas,* Chicago: University of Chicago Press.

Feenberg, A. (1991), *Critical Theory of Technology,* Oxford: Oxford University Press.

Feenberg, A. (1999), *Questioning Technology,* London: Routledge.

Feenberg, A. (2002), *Transforming Technology,* Oxford: Oxford University Press.

Feenberg, A. (2010), *Between Reason and Experience,* Cambridge: MIT Press.

Fellows, R. (ed.) (1995), *Philosophy and Technology,* Cambridge: Cambridge University Press.

Ferre, F. (1988), *Philosophy of Technology,* Englewood Cliffs, NJ: Simon and Schuster.

Finkelhor, D. et al. (2007), "Unwanted and wanted exposure to online pornography in a national sample of youth Internet users," *Pediatrics,* 119: 247–253.

Finkenauer, C. (2012), "Brief report: Examining the link between autistic traits and compulsive Internet use in a non-clinical sample," *Journal of Autism and Developmental Disorders,* 42: 2252–2256.

Finn, J. et al. (2009), "Defensive tool use in a coconut-carrying octopus," *Current Biology,* 19: R1069–R1070.

Finnegan, W. (1998), "Defending the Unabomber," *The New Yorker,* 74(4): 52–63.

Florman, S. (1975), "In praise of technology," *Harper's,* 251(1506): 53–72.

Florman, S. (1976), *The Existential Pleasures of Engineering,* New York: St. Martin's Press.

Friedman, T. (2007), *The World Is Flat,* New York: Farrar, Straus, and Giroux.

Food and Agriculture Organization, U.N. (2006), "Livestock a major threat to environment," retrieved from www.fao.org/newsroom/en/news/2006/1000448/index.html

Fox, N. (2002), *Against the Machine,* Washington, DC: Island Press.

Freeman, K. (1948), *Ancilla to the Pre-Socratic Philosophers,* London: Blackwell.

Fromm, E. (1955), *The Sane Society,* New York: Holt, Rinehart, and Winston.

Fuchs, T. and Woessmann, L. (2004), "Computers and student learning," CESifo Working Paper no. 1321, retrieved from www.SSRN.com.

Fuller, R. et al. (2007), "Psychological benefits of greenspace increase with biodiversity," *Biology Letters,* 3: 390–394.

Gentile, D. et al. (2011), "Pathological video game use among youths," *Pediatrics,* 127: e319–e329.

Goebbels, J. (1939), "*Ueber die Bedeutung der Technik,*" *Deutsche Technik,* March 1939. English translation in J. Herf, *Reactionary Modernism* (1984: 196), Cambridge University Press.

Goncalves, M. et al. (2012), *Justice for Forests,* Washington, DC: The World Bank.

Goodwin, B. (1994), *How the Leopard Changed Its Spots,* New York: Simon and Schuster.

Graham, G. (1999), *The Internet: A Philosophical Inquiry,* London: Routledge.

Greulach, V. and Adams, J. (1967), *Plants,* New York: Wiley.

Guthrie, W. (1962–1981), *History of Greek Philosophy,* Cambridge: Cambridge University Press.

Habermas, J. ([1968] 1970), *Toward a Rational Society,* Boston: Beacon Press.

Hameroff, S. (1998), "Funda-mentality: Is the conscious mind subtly linked to a basic level of the universe?" *Trends in Cognitive Science,* 2: 119–127.

Hameroff, S. (2009), "The conscious connection," in D. Skrbina (ed.), *Mind That Abides,* Amsterdam: Benjamins.

Hanks, C. (ed.) (2010), *Technology and Values,* Oxford: Wiley-Blackwell.

Hardell, L. et al. (2007), "Long-term use of cellular phones and brain tumors," *Occupational and Environmental Medicine,* 64: 626–632.

Hare, E. (1956a), "Family setting and the urban distribution of schizophrenia," *Journal of Mental Science,* 102: 753–760.

Hare, E. (1956b), "Mental illness and social conditions in Bristol," *Journal of Mental Science,* 102: 349–357.

Harman, G. (2002), *Tool-Being,* Chicago: Open Court.

Harman, G. (2009), "The McLuhans and metaphysics," in J. Olsen et al. (eds.), *New Waves in Philosophy of Technology,* New York: Palgrave Macmillan.

Harman, G. (2010), "Technology, objects and things in Heidegger," *Cambridge Journal of Economics,* 34: 17–25.

Harris, W. (2001), *Restraining Rage,* Cambridge, MA: Harvard University Press.

Heidegger, M. ([1954] 1977), *The Question Concerning Technology and Other Essays,* New York: Harper and Row.

Heidegger, M. ([1976] 1981), *Heidegger: The Man and the Thinker,* Chicago: Precedent.

Heidegger, M. (1993), "Building dwelling thinking," in M. Heidegger, *Basic Writings,* New York: HarperCollins.

Heidegger, M. (1999), *Contributions to Philosophy: From Enowning,* Bloomington: Indiana University Press.

Heilbronner, R. (1967), "Do machines make history?" *Technology and Culture,* 8: 335–345.

Higgs, E. et al. (eds.) (2000), *Technology and the Good Life?* Chicago: University of Chicago Press.

Hoffmann, M. et al. (2010), "The impact of conservation on the status of the world's vertebrates," *Science,* 330: 1503–1509.

Hood, W. (1972), "The Aristotelian versus the Heideggerian approach to the problem of technology," in C. Mitcham and R. Mackey (eds.), *Philosophy and Technology,* New York: Free Press.

Hoyle, F. (1964), *Of Men and Galaxies,* Seattle: University of Washington Press.

Huesmann, L. (2007), "The impact of electronic media violence," *Journal of Adolescent Health,* 41: S6–S13.

Huesemann, M. and Huesemann, J. (2011), *Techno-Fix: Why Technology Won't Save Us or the Environment,* Gabriola Island, BC: New Society Publishers.

Hume, D. ([1779] 2007), *Dialogues Concerning Natural Religion*, Cambridge: Cambridge University Press.
Ihde, D. (1979), *Technics and Praxis*, Boston: D. Reidel.
Ihde, D. (1983), *Existential Technics*, Albany: SUNY Press.
Ihde, D. (1990), *Technology and the Lifeworld*, Bloomington: Indiana University Press.
Ihde, D. (2002), *Bodies in Technology*, Minneapolis: University of Minnesota Press.
Ihde, D. (2010), *Heidegger's Technologies*, New York: Fordham University Press.
Ihde, D. and Selinger, E. (eds.) (2003), *Chasing Technoscience*, Bloomington: Indiana University Press.
Illich, I. (1973), *Tools for Conviviality*, New York: Harper and Row.
Illich, I. (1974), *Energy and Equity*, New York: Harper and Row.
Immordino-Yang, H. et al. (2009), "Neural correlates of admiration and compassion," *Proceedings of the National Academy of Sciences*, 106: 8021–8026.
International Agency for Research on Cancer (2013), "Non-Ionizing Radiation, Part 2" (vol. 102). Conference paper from the IARC Working Group on the evaluation of carcinogenic risks to Humans, Lyon, France, May 24–31, 2011.
Jantsch, E. (1980), *The Self-Organizing Universe*, New York: Pergamon.
Jaspers, K. ([1931] 1959), *Man in the Modern Age*, London: Routledge.
Jaspers, K. ([1949] 1953), *The Origin and Goal of History*, New Haven, CT: Yale University Press.
Jensen, D. (2006), *Endgame*, New York: Seven Stories Press.
Jensen, D. and Draffan, G. (2004), *Welcome to the Machine*, White River Junction, VT: Chelsea Green Publishing.
Johnson, D. (2006), "Computer systems: Moral entities but not moral agents," *Ethics and Information Technology*, 8: 195–204.
Johnson, D. and Wetmore, J. (eds.) (2009), *Technology and Society*, Cambridge: MIT Press.
Jonas, H. (1979), "Toward a philosophy of technology," *Hastings Center Report*, 9: 34–43.
Jonas, H. (1984), *The Imperative of Responsibility*, Chicago: University of Chicago Press.
Joy, B. (2000a), "Why the future doesn't need us," *Wired*, 8.04.
Joy, B. (2000b), "Act now to keep new technologies out of destructive hands," *New Perspectives Quarterly*, 17(3): 12–14.
Juenger, F. ([1946] 1956), *The Failure of Technology*, Chicago: Regnery.
Kaczynski, T. (2010), *Technological Slavery*, Port Townsend, WA: Feral House.
Kakabadse, N. et al. (2007), "Addicted to technology," *Business Strategy Review*, 18: 81–85.
Kant, I. ([1785] 1996), "The groundwork of the metaphysics of morals," in I. Kant, *Practical Philosophy*, Cambridge: Cambridge University Press.
Kant, I. (2012), *Natural Science*, Cambridge: Cambridge University Press.
Kaplan, D. (ed.) (2009), *Readings in the Philosophy of Technology*, Lanham, MD: Rowman and Littlefield.
Katz, E. et al. (eds.) (2003), *Controlling Technology*, Amherst, NY: Prometheus.
Kauffman, S. (1993), *The Origins of Order*, New York: Oxford University Press.
Kauffman, S. (1995), *At Home in the Universe*, New York: Oxford University Press.
Kelly, K. (2010), *What Technology Wants*, New York: Viking.
Khurana, V. et al. (2009), "Cell phones and brain tumors," *Surgical Neurology*, 72: 205–214.
Kim, Y. et al. (2011), "Prevalence of autism spectrum disorders in a total population sample," *American Journal of Psychiatry*, 168: 904–912.
Kirk, G. et al. (1983), *The Presocratic Philosophers*, Cambridge: Cambridge University Press.

Kizza, J. (2013), *Ethical and Social Issues in the Information Age,* London: Springer.

Klimov, P. and O'Connor, B. (2013), "Is permanent parasitism reversible?" *Systematic Biology,* 62: 411–423.

Ko, C.-H. et al. (2009), "Predictive values of psychiatric symptoms for Internet addiction in adolescents," *Archives of Pediatrics and Adolescent Medicine,* 163: 937–943.

Koblitz, N. (1996), "The case against computers in K–13 math education," *The Mathematical Intelligencer,* 18: 9–16.

Koch, C. (2009), "A theory of consciousness," *Scientific American,* 20(4): 16–19.

Koch, C. (2012), *Consciousness,* Cambridge: MIT Press.

Konrath, S. et al. (2011), "Changes in dispositional empathy in American college students over time: A meta-analysis," *Personality and Social Psychology Review,* 15: 180–198.

Krabbendam, L. and van Os, J. (2005), "Schizophrenia and urbanicity," *Schizophrenia Bulletin,* 31: 795–799.

Kraut, R. et al. (1998), "Internet paradox: A social technology that reduces social involvement and psychological well-being?" *American Psychologist,* 53: 1017–1031.

Kross, E. et al. (2013), "Facebook use predicts decline in subjective well-being in young adults," *PLosOne,* 8(8).

Kubey, R. and Csikszentmihalyi, M. (2002), "Television addiction is no mere metaphor," *Scientific American,* 286(2): 74–76.

Kulik, J. (1994), "Meta-analytic studies of findings on computer-based instruction," in E. Baker and H. O'Neil (eds.), *Technology Assessment in Education and Training,* Hillsdale, NJ: L. Erlbaum Associates.

Kuo, F. (2001), "Coping with poverty," *Environmental Behavior,* 33: 5–34.

Kupfer, D. and Regier, D. (2010), "Why all of medicine should care about DSM-5," *Journal of the American Medical Association,* 303: 1974–1975.

Kurzweil, R. (1999), *The Age of Spiritual Machines,* New York: Viking.

Kurzweil, R. (2003), "Promise and peril," in A. Lightman et al. (eds.), *Living with the Genie,* Washington, DC: Island Press.

Kurzweil, R. (2005), *The Singularity Is Near,* New York: Viking.

Kurzweil, R. (2012), *How to Create a Mind,* New York: Viking.

Landrigan, P. (2010), "What causes autism?" *Current Opinion in Pediatrics,* 22: 219–255.

Lao Tzu. (1963), *Tao Te Ching,* New York: Penguin.

Lederbogen, F. et al. (2011), "City living and urban upbringing affect neural social stress processing in humans," *Nature,* 474: 498–501.

Leibniz, G. (1989a), "On body and force," in R. Ariew and D. Garber (eds.), *Philosophical Essays,* Indianapolis: Hackett.

Leibniz, G. (1989b), "New system of nature," in R. Ariew and D. Garber (eds.), *Philosophical Essays,* Indianapolis: Hackett.

Leibniz, G. (1989c), "Against barbaric physics," in R. Ariew and D. Garber (eds.), *Philosophical Essays,* Indianapolis: Hackett.

Leiter, B. and Weisberg, M. (2012), "Do you only have a brain? On Thomas Nagel," *Nation,* 295(17): 27–31.

Lewis, G. et al. (1992), "Schizophrenia and city life," *The Lancet,* 340(8812): 137–140.

Li, D. et al. (2011), "Maternal exposure to magnetic fields during pregnancy in relation to the risk of asthma in offspring," *Archives of Pediatrics and Adolescent Medicine,* 165: 945–950.

Long, A. (1974), *Hellenistic Philosophy,* New York: Charles Scribner's Sons.

Long, A. (1996), *Stoic Studies,* Cambridge: Cambridge University Press.

Long, A. and Sedley, D. (1987), *The Hellenistic Philosophers,* Cambridge: Cambridge University Press.

Loux, M. (2006), *Metaphysics*, London: Routledge.
Lumeng, J. et al. (2006), "Television exposure and overweight risk in preschoolers," *JAMA Pediatrics,* 160: 417–422.
Malpas, J. (2006), *Heidegger's Topology*, Cambridge: MIT Press
Mander, J. (1978), *Four Arguments for the Elimination of Television,* New York: Quill.
Mann, J. et al. (2008), "Why do dolphins carry sponges"? *PLoSOne,* 3(12).
Marcuse, H. (1964), *One Dimensional Man*, Boston: Beacon Press.
Marcuse, H. (1966), "Ethics and revolution," in R. DeGeorge (ed.), *Ethics and Society,* Garden City, NY: Anchor Books.
Marx, K. ([1844] 1961), *Economic and Philosophic Manuscripts of 1844,* Moscow, Russia: Foreign Languages Publishing.
Marx, K. ([1857] 1971), *The Grundrisse,* New York: Harper and Row.
Marx, K. ([1867] 1976), *Capital* (vol. 1), London: Penguin.
Marx, K. (1975–2004), *Collected Works,* New York: International Publishers.
Matthew, K. (1997), "A comparison of the influence of interactive CD-ROM storybooks and traditional print storybooks on reading comprehension," *Journal of Research on Computing in Education,* 29: 263–275.
McGinn, R. (1980), "Nietzsche on technology," *Journal of the History of Ideas,* 41: 679–691.
McKenzie, J. (2000), "Beyond technology," *Electronic School,* 3.
McKibben, W. (1989), *The End of Nature,* New York: Random House.
McKibben, W. (2004), *Enough,* New York: Holt.
Meyer-Lindenberg, A. (2013), "Big city blues," *Scientific American Mind,* 24(1): 58–61.
Mitcham, C. (1978), "Types of technology," *Research in Philosophy and Technology,* 1: 229–294.
Mitcham, C. (1994), *Thinking through Technology,* Chicago: University of Chicago Press.
Mitcham, C. and Mackey, R. (eds.) (1972), *Philosophy and Technology,* New York: Free Press.
Mitchell, K., Finkelhor, D. and Wolak, J. (2001), "Risk factors for and impact of online sexual solicitation of youth," *Journal of American Medical Association,* 285(23): 3011–3014.
Morat, D. (2007), *Von der Tat zur Gelassenheit,* Goettingen: Wallstein Verlag.
Moreno, C. et al. (2007), "National trends in the outpatient diagnosis and treatment of bipolar disorder in youth," *JAMA Psychiatry,* 64: 1032–1039.
Morozov, E. (2011), *The Net Delusion,* New York: PublicAffairs.
Morozov, E. (2013), *To Save Everything, Click Here,* New York: PublicAffairs.
Morrison, C. and Gore, H. (2010), "The relationship between excessive Internet use and depression," *Psychopathology,* 43: 121–126.
Mumford, L. ([1934] 1963), *Technics and Civilization,* New York: Harcourt, Brace.
Mumford, L. (1966), "Technics and the nature of man," *Technology and Culture,* 7: 303–317.
Mumford, L. (1970), *The Myth of the Machine* (vol. 2), New York: Harcourt, Brace.
Naess, A. ([1976] 1989), *Ecology, Community, and Lifestyle,* Cambridge: Cambridge University Press.
Nagel, T. (2012), *Mind and Cosmos,* Oxford: Oxford University Press.
Nakazawa, T. et al. (2002), "Association between duration of daily VDT use and subjective symptoms," *American Journal of Industrial Medicine,* 42: 421–426.
Nietzsche, F. ([1878–80] 1996), *Human, All Too Human,* Cambridge: Cambridge University Press.
Nietzsche, F. ([1881] 1997), *Daybreak,* Cambridge: Cambridge University Press.
Nietzsche, F. ([1886] 1973), *Beyond Good and Evil,* London: Penguin.

Nietzsche, F. ([1887] 1967), *On the Genealogy of Morals,* New York: Random House.
Nietzsche, F. ([1888] 1954), *Antichrist,* in W. Kaufmann (ed.), *The Portable Nietzsche,* New York: Penguin.
Nietzsche, F. (1967), *The Will to Power,* New York: Random House.
Northwestern University (2011), "Children, media, and race." Center on Media and Human Development, School of Communication.
Nusselder, A. (2009), *Interface Fantasy,* Cambridge: MIT Press.
Nye, D. (2006), *Technology Matters,* Cambridge: MIT Press.
Olfson, M. et al. (2012), "National trends in the office-based treatment of children, adolescents, and adults with antipsychotics," *Archives of General Psychiatry,* 69: 1247–1256.
Olsen, J.-K. et al. (eds.) (2009), *New Waves in Philosophy of Technology,* New York: Palgrave Macmillan.
Olsen, J.-K. and Selinger, E. (eds.) (2007), *Philosophy of Technology: Five Questions,* Automatic Press/VIP.
Oppenheimer, T. (1997), "The computer delusion," *Atlantic Monthly,* 280(1): 45–62.
Orr, H. (2013), "Awaiting a new Darwin," *New York Review of Books,* 60(2).
Ortega y Gasset, J. ([1939] 1941), *Toward a Philosophy of History,* New York: W.W. Norton.
Ortiz de Gortari, A. and Griffiths, M. (2014), "Altered visual perception in game transfer phenomena," *International Journal of Human-Computer Interaction,* 30: 95–105.
Orwell, G. ([1937] 1958), *The Road to Wigan Pier,* New York: Harcourt, Brace.
Pacey, A. (1999), *Meaning in Technology,* Cambridge: MIT Press.
Pea, R. et al. (2012), "Media use, face-to-face communication, media multitasking, and social well-being among 8- to 12-year-old girls," *Developmental Psychology,* 48: 327–336.
Pearce, F. (2005), "Climate warming as Siberia melts," *New Scientist,* 187(2512): 12.
Peirce, C. (1891), "The architecture of theories," *Monist,* 1: 161–176.
Pepi, D. and Scheurman, G. (1996), "The emperor's new computer," *Journal of Teacher Education,* 47: 229–236.
Plato. (1997), *Complete Works,* Indianapolis: Hackett.
Pitt, J. (2000), *Thinking about Technology,* New York: Seven Bridges Press.
Popper, K. (1962), *Conjectures and Refutations,* New York: Basic Books.
Powers, W. (2010), *Hamlet's Blackberry,* New York: Harper.
Preen, J. et al. (2010), "The current status of urban-rural differences in psychiatric disorders," *Acta Psychiatrica Scandinavica,* 121: 84–93.
Prigogine, I. (1977), *Self-Organization in Nonequilibrium Systems,* New York: Wiley.
Prigogine, I. (1980), *From Being to Becoming,* San Francisco: W.H. Freeman.
Prigogine, I. and Stengers, I. (1984), *Order out of Chaos,* New York: Bantam.
Raunig, G. (2010), *A Thousand Machines,* Cambridge: MIT Press.
Reeves, T. (1998), "'Future schlock,' 'the computer delusion,' and 'the end of education': Responding to critics of educational technology," *Educational Technology,* 38(5): 49–53.
Reich, C. (1970), *The Greening of America,* New York: Random House.
Rhodes, R. (ed.) (1999), *Visions of Technology,* New York: Simon and Schuster.
Rist, J. (1979), *The Mind of Aristotle,* Toronto: University of Toronto Press.
Rockmore, T. (1992), *On Heidegger's Nazism and Philosophy,* Berkeley: University of California Press.
Rodman, J. (1977), "The liberation of nature?" *Inquiry,* 20: 83–145.
Rosen, L. (2012), *iDisorder,* New York: Palgrave Macmillan.

Rouse, C. and Krueger, A. (2004), "Putting computerized instruction to the test," National Bureau of Economic Research, Working Paper no. 10315, retrieved from http://nber.org/papers/w10315
Rousseau, J.-J. ([1750] 1993), "A discourse on the arts and sciences," in J.-J. Rousseau, *The Social Contract and the Discourses,* New York: Knopf.
Roszak, T. (1969), *The Making of a Counter Culture,* Garden City, NY: Doubleday.
Roszak, T. (1972), *Where the Wasteland Ends,* Garden City, NY: Doubleday.
Roszak, T. (1978), *Person/Planet*, Garden City, NY: Anchor.
Roszak, T. (1986), *The Cult of Information*, New York: Pantheon.
Rushkoff, D. (2013), *Present Shock,* New York: Current.
Sagan, C. (1994), *Pale Blue Dot,* New York: Random House.
Sage, C. and Carpenter, D. (2009), "Public health implications of wireless technologies," *Pathophysiology,* 16: 233–246.
Sandbach, F. (1975), *The Stoics,* New York: W. W. Norton.
Sanderson, E. et al. (2002), "The human footprint and the last of the wild," *BioScience,* 52: 891–904.
Sava, S. (1997), "Electronic genie," *Vital Speeches of the Day,* 64(2): 55–57.
Scharff, R. and Dusek, V. (eds.) (2003), *Philosophy of Technology: The Technological Condition,* Oxford: Blackwell.
Scheler, M. ([1915] 1994), *Ressentiment,* Milwaukee, WI: Marquette University Press.
Schiller, F. (1891), *Riddles of the Sphinx*, London: S. Sonnenschein.
Schmookler, A. (1995), *The Parable of the Tribes,* Albany: SUNY Press.
Schopenhauer, A. (1992), *On the Will in Nature,* Oxford: Berg.
Schroeder, R. (2007), *Rethinking Science, Technology, and Social Change,* Stanford, CA: Stanford University Press.
Schumacher, E. (1973), *Small Is Beautiful,* New York: Harper and Row.
Sclove, R. (1995), *Democracy and Technology,* New York: Guilford Press.
Segal, H. (1994), *Future Imperfect: The Mixed Blessings of Technology in America,* Amherst: University of Massachusetts Press.
Seidenberg, R. (1950), *Posthistoric Man,* Chapel Hill: University of North Carolina Press.
Selfhout, M. et al. (2009), "Different types of Internet use, depression, and social anxiety," *Journal of Adolescence,* 32: 819–833.
Sennett, R. (ed.) (1969), *Classic Essays on the Culture of Cities,* Englewood Cliffs, NJ: Prentice-Hall.
Shaw, W. (1978), "Marx's technological determinism," in W. Shaw, *Marx's Theory of History,* Stanford, CA: Stanford University Press.
Shaw, W. (1979), "'The handmill gives you the feudal lord': Marx's technological determinism," *History and Theory,* 18: 155–176.
Silva, R. et al. (2013), "Global premature mortality due to anthropogenic outdoor air pollution and the contribution of past climate change," *Environmental Research Letters,* 8: 1–11.
Simmel, G. ([1900] 1990), *The Philosophy of Money,* London: Routledge.
Simondon, G. (1958), *Du mode d'existence des objets techniques*, Aubier: Editions Montaigne.
Skolimowski, H. (1966), "The structure of thinking in technology," *Technology and Culture,* 7: 371–383.
Skolimowski, H. (1967), "The 'monster' of technology," *Center Diary,* 18: 56–59.
Skolimowski, H. (1968), "Technology and philosophy," in J. Kilbansky (ed.), *Contemporary Philosophy,* Florence, Italy: Nuovo Italia.
Skolimowski, H. (1969), "Human space in the technological age," *Architectural Association Quarterly,* 1: 81–83.
Skolimowski, H. (1970), "Technology: The myth behind the reality," *Architectural Association Quarterly,* 2: 21–31.

Skolimowski, H. (1971), "Technology, the seducer," *Building Design,* December 3: 6.

Skolimowski, H. (1972), "Science and the modern predicament," *New Scientist,* 104: 435–437.

Skolimowski, H. (1973), "Technology v. nature," *Ecologist,* 3(2): 50–55.

Skolimowski, H. (1974a), "The myth of progress," *Ecologist,* 4(7): 248–258.

Skolimowski, H. (1974b), "The scientific worldview and the illusions of progress," *Social Research,* 41: 52–82.

Skolimowski, H. (1975), "A way out of the abyss," *Main Currents,* 31: 71–76.

Skolimowski, H. (1978), "Technological alienation," *Indian Philosophical Annual,* 13: 46–56.

Skolimowski, H. (1981), *Eco-Philosophy: Designing New Tactics for Living,* London: Marion Boyars.

Skolimowski, H. (1983), *Technology and Human Destiny,* Madras, India: University of Madras Press.

Skolimowski, H. (1991), "The eco-philosophy approach to technological research," in P. Durbin (ed.), *Critical Perspectives on Nonacademic Science and Engineering,* Bethlehem, PA: Lehigh University Press.

Skolimowski, H. (2005), *Philosophy for a New Civilisation,* New Delhi, India: Gyan.

Skrbina, D. (2003), "Panpsychism as an underlying theme in Western philosophy," *Journal of Consciousness Studies,* 10(3): 4–46.

Skrbina, D. (2005), *Panpsychism in the West,* Cambridge: MIT Press.

Skrbina, D. (2009a), "Panpsychism in history: An overview," in D. Skrbina (ed.), *Mind That Abides,* Amsterdam: John Benjamins.

Skrbina, D. (2009b), "Technological anarchism," *Chromatikon* V: 191–202.

Skrbina, D. (ed.) (2009c), *Mind That Abides,* Amsterdam: John Benjamins.

Skrbina, D. (2010), "Whitehead and the ubiquity of mind," *Chromatikon* VI: 181–189.

Smith, M. (1994), "Recourse of empire," in M. Smith and L. Marx (eds.), *Does Technology Drive History?* Cambridge: MIT Press.

Smith, M. and Marx, L. (eds.) (1994), *Does Technology Drive History?* Cambridge: MIT Press.

Sober, E. (2012), "Ending science as we know it," *Boston Review,* 37(6): 50–55.

Spengler, O. ([1931] 1976), *Man and Technics,* Westport, CT: Greenwood Press.

Spinoza, B. ([1677] 1989), "Ethics," in E. Curley (ed.), *A Spinoza Reader*, Princeton, NJ: Princeton University Press.

Steiner, G. (1979), *Martin Heidegger,* New York: Viking.

Stivers, R. (1999), *Technology as Magic: The Triumph of the Irrational,* New York: Continuum.

Stivers, R. (2004), *Shades of Loneliness: Pathologies of a Technological Society,* Lanham, MD: Rowman and Littlefield.

Stoll, C. (1995), *Silicon Snake Oil,* New York: Doubleday.

Sturken, M. et al. (eds.) (2004), *Technological Visions,* Philadelphia: Temple University Press.

Sullivan, W. et al. (2004), "The fruit of urban nature," *Environmental Behavior,* 36: 678–700.

Swing, E. et al. (2010), "Television and video game exposure and the development of attention problems," *Pediatrics,* 126: 214–221.

Tabachnick, D. (2013), *The Great Reversal: How We Let Technology Take Control of the Planet,* Toronto: University of Toronto Press.

Teilhard, P. ([1952] 1970), *Activation of Energy,* New York: Collins.

Teilhard, P. ([1955] 1959), *The Phenomenon of Man,* New York: Harper and Row.

Tenner, E. (1996), *Why Things Bite Back: Technology and the Revenge of Unintended Consequences,* New York: Knopf.

Thiele, L. (2011), *Indra's Net and the Midas Touch,* Cambridge: MIT Press.

Thoreau, H. ([1854] 1992), *Walden,* New York: Knopf.
Tononi, G. (2008), "Consciousness as integrated information," *Biological Bulletin,* 215: 216–244.
Torrey, B. and Bowler, A. (1991), "Geographical distribution of insanity in America," *Schizophrenia Bulletin,* 16: 591–604.
Trust for America's Health (2013), "F as in Fat." Washington, DC: U.S. Government Printing Office.
Turkle, S. (1995), *Life on the Screen,* New York: Simon and Schuster.
Turkle, S. (2011), *Alone Together,* New York: Basic Books.
Ulrich, R. (1984), "View through a window may influence recovery from surgery," *Science,* 224: 420–421.
University of Winnipeg (2013), "Study supports theory on teen texting and shallow thought," *NewsCentre* (April 11).
U.S. Centers for Disease Control and Prevention (2011), "Antidepressant use in persons aged 12 and over," *NCHS Data Brief,* 76.
U.S. Centers for Disease Control and Prevention (2011), "Deaths: Final data for 2009," *National Vital Statistics Report,* 60(3).
U.S. Centers for Disease Control and Prevention (2011), "Health, United States, 2010." Washington, DC: U.S. Government Printing Office.
U.S. Department of Education (2004), "Toward a New Golden Age in American Education." Washington, DC: U.S. Government Printing Office.
U.S. Department of Education (2010), "Transforming American Education." Washington, DC: U.S. Government Printing Office.
U.S. National Cancer Institute (2010), "Reducing environmental cancer risk." Washington, DC: U.S. Government Printing Office.
van Son, G. et al. (2006), "Urbanisation and the incidence of eating disorders," *British Journal of Psychiatry,* 189: 562–563.
Veblen, T. (1906), "The place of science in modern civilization," *The American Journal of Sociology,* 11: 585–609.
Verbeek, P.-P. (2001), "Don Ihde: The technological lifeworld," in H. Achterhuis (ed.), *American Philosophy of Technology,* Bloomington: Indiana University Press.
Verbeek, P.-P. (2005), *What Things Do,* University Park: Pennsylvania State University Press.
Virilio, P. ([2010] 2012), *The Great Accelerator,* Cambridge: Polity.
Vitousek, P. et al. (1986), "Human appropriation of the products of photosynthesis," *BioScience,* 36: 363–373.
Volkow, N. et al. (2011), "Effects of cell phone radiofrequency signal exposure on brain glucose metabolism," *Journal of the American Medical Association,* 305: 808–813.
Waldman, M. et al. (2006), "Does television cause autism?" National Bureau of Economic Research, Working Paper no. 12632.
Waldman, M. et al. (2012), "Positive and negative mental health consequences of early childhood television watching," National Bureau of Economic Research, Working Paper no. 17786.
Walker, T. ([1832] 1982), "Defence of mechanical philosophy," in M. Folsom and S. Lubar (eds.), *The Philosophy of Manufactures,* Cambridge: MIT Press.
Watts, M. (2011), *The Philosophy of Heidegger,* Montreal: McGill-Queen's University Press.
Waxman, H. et al. (2002), "A quantitative synthesis of recent research on the effects of teaching and learning with technology on student outcomes," North Central Regional Educational Laboratory (NCREL), retrieved from www.ncrel.org
Weber, M. ([1905] 1930), *The Protestant Ethic and the Spirit of Capitalism,* New York: Scribner.
White, L. (1949), *The Science of Culture,* New York: Farrar, Straus.

White, L. (1959), *The Evolution of Culture,* New York: McGraw-Hill.
White, L. Jr. (1967), "The historical roots of our ecologic crisis," *Science,* 155: 1203–1207.
Whitehead, A. ([1925] 1967), *Science and the Modern World,* New York: Free Press.
Whitehead, A. ([1929] 1978), *Process and Reality,* New York: Free Press.
Wiley, D. et al. (2011), "Underwater components of humpback whale bubble-net feeding behavior," *Behaviour,* 148: 575–602.
Williams, R. (2002), *Retooling,* Cambridge: MIT Press.
Winner, L. (1977), *Autonomous Technology,* Cambridge: MIT Press.
Winner, L. (1986), *The Whale and the Reactor,* Chicago: University of Chicago Press.
Wisnewski, J. (2013), *Heidegger: An Introduction,* Lanham, MD: Rowman and Littlefield.
Wittgenstein, L. (1980), *Culture and Value,* Chicago: University of Chicago Press.
Yuan, K. et al. (2011), "Microstructure abnormalities in adolescents with Internet addiction disorder," *PLoSOne,* 6(6).
Young, K. (ed.) (2011), *Internet Addiction,* Hoboken, NJ: John Wiley and Sons.
Zammit, S. et al. (2010), "Individuals, schools, and neighborhood," *Archives of General Psychiatry,* 67: 914–922.
Zerzan, J. (1988), *Elements of Refusal,* Seattle, WA: Left Bank Books.
Zerzan, J. (2002), *Running on Emptiness,* Los Angeles: Feral House.
Zerzan, J. (ed.) (2005), *Against Civilization,* Los Angeles: Feral House.
Zerzan, J. and Carnes, A. (eds.) (1988), *Questioning Technology,* London: Freedom Press.
Zhou, Y. et al. (2011), "Gray matter abnormalities in Internet addiction," *European Journal of Radiology,* 79: 92–95.
Zimmerman, F. et al. (2007), "Associations between media viewing and language development in children under age 2 years," *Pediatrics,* 151(4): 1–5.
Zimmerman, M. (1990), *Heidegger's Confrontation with Modernity,* Bloomington: Indiana University Press.

Index

Routledge Studies in Social and Political Thought

For a full list of titles in this series, please visit www.routledge.com

61 **The Political Philosophy of Michel Foucault**
Mark G. E. Kelly

62 **Democratic Legitimacy**
Fabienne Peter

63 **Edward Said and the Literary, Social, and Political World**
Edited by Ranjan Ghosh

64 **Perspectives on Gramsci**
Politics, Culture and Social Theory
Edited by Joseph Francese

65 **Enlightenment Political Thought and Non-Western Societies**
Sultans and Savages
Frederick G. Whelan

66 **Liberalism, Neoliberalism, Social Democracy**
Thin Communitarian Perspectives on Political Philosophy and Education
Mark Olssen

67 **Oppositional Discourses and Democracies**
Edited by Michael Huspek

68 **The Contemporary Goffman**
Edited by Michael Hviid Jacobsen

69 **Hemingway on Politics and Rebellion**
Edited by Lauretta Conklin Frederking

70 **Social Theory in Contemporary Asia**
Ann Brooks

71 **Governmentality**
Current Issues and Future Challenges
Edited by Ulrich Bröckling, Susanne Krasmann and Thomas Lemke

72 **Gender, Emotions and Labour Markets—Asian and Western Perspectives**
Ann Brooks and Theresa Devasahayam

73 **Alienation and the Carnivalization of Society**
Edited by Jerome Braun and Lauren Langman

74 **The Post-Colonial State in the Era of Capitalist Globalization**
Historical, Political and Theoretical Approaches to State Formation
Tariq Amin-Khan

75 **The Psychology and Politics of the Collective**
Groups, Crowds and Mass Identifications
Edited by Ruth Parkin-Gounelas

76 **Environmental Solidarity**
How Religions Can Sustain Sustainability
Pablo Martínez de Anguita

77 **Comedy and the Public Sphere**
The Rebirth of Theatre as Comedy and the Genealogy of the Modern Public Arena
Arpad Szakolczai

78 **Culture, Class, and Critical Theory**
Between Bourdieu and the Frankfurt School
David Gartman

79 **Environmental Apocalypse in Science and Art**
Designing Nightmares
Sergio Fava

80 **Conspicuous and Inconspicuous Discriminations in Everyday Life**
Victor N. Shaw

81 **Understanding the Tacit**
Stephen P. Turner

82 **The Politics of Expertise**
Stephen P. Turner

83 **Globalized Knowledge Flows and Chinese Social Theory**
Xiaoying Qi

84 **Reconstructing Social Justice**
Lauretta Conklin Frederking

85 **The Dialectics of Inquiry Across the Historical Social Sciences**
David Baronov

86 **How Groups Matter**
Challenges of Toleration in Pluralistic Societies
Edited by Gideon Calder, Magali Bessone and Federico Zuolo

87 **The Politics of Rationality**
Reason through Occidental History
Charles P. Webel

88 **Gramsci, Materialism, and Philosophy**
Esteve Morera

89 **Emotions and Social Change**
Historical and Sociological Perspectives
Edited by David Lemmings and Ann Brooks

90 **John Rawls and the History of Political Thought**
The Rousseauvian and Hegelian Heritage of Justice as Fairness
Jeffrey Bercuson

91 **What Holism Can Do for Social Theory**
Barbara Hanson

92 **The Sociology of Knowledge in a Time of Crisis**
Challenging the Phantom of Liberty
Onofrio Romano

93 **Cosmopolitanism and the Legacies of Dissent**
Tamara Caraus and Camil-Alexandru Parvu

94 **The Metaphysics of Technology**
David Skrbina